INSIDE THE
COLD WAR

Loy Henderson in Baghdad, 1943. Courtesy of Fraser Wilkins.

INSIDE THE COLD WAR

Loy Henderson and the Rise of the American Empire 1918–1961

H. W. BRANDS

New York   Oxford
OXFORD UNIVERSITY PRESS
1991

Oxford University Press

Oxford New York Toronto
Delhi Bombay Calcutta Madras Karachi
Petaling Jaya Singapore Hong Kong Tokyo
Nairobi Dar es Salaam Cape Town
Melbourne Auckland

and associated companies in
Berlin Ibadan

Published by Oxford University Press, Inc.,
200 Madison Avenue, New York, New York 10016

Library of Congress Cataloging-in-Publication Data
Brands, H. W.
Inside the cold war : Loy Henderson and the rise of
the American empire, 1918–1961 / H. W. Brands.
p. cm.
Includes index.
ISBN 0-19-506707-X
1. Henderson, Loy W. (Loy Wesley), 1892–1986. 2. Diplomats—
United States—Biography. 3. United States—Foreign
relations—20th century. 4. Cold War. 5. United States—
Foreign relations—Soviet Union. 6. Soviet Union—Foreign
relations—United States. I. Title.
E748.H412B73 1991
327.73047—dc20   90-39910 CIP

2 4 6 8 9 7 5 3 1

Printed in the United States of America
on acid-free paper

Preface

Dusk on a snowy day at the end of the First World War found Loy Henderson wandering the Lithuanian countryside. Henderson, a captain in the American Red Cross, had set out that morning with Colonel Edward Ryan, his superior officer, and their driver from Mitau, across the border in Latvia, with the goal of reaching Kaunas. Only 130 miles separated the two cities, and allowing for weather and breakdowns the party expected to reach their destination by sunset. Just before noon a heavy, sticky snow had begun to fall. Although automobiles had arrived in this part of the world some years earlier, windshield wipers were unknown, and every several hundred yards Henderson had to get out and clear the glass. Early in the afternoon the three men encountered a greater hindrance, in the form of a long column of German soldiers. The guns on the western front had fallen silent months before, but conditions in the east remained unsettled. These troops, having occupied portions of the emerging Baltic republics, were heading home to Prussia. Uncertainty as to what they would find there counterbalanced a desire to have done with soldiering, and on this day they were in no hurry. The column moved at the speed of the slowest troops and horses, and the pace steadily decreased as men and animals tired and the road grew rutted and soft.

Henderson and the two others had almost managed to make their way past the main body of soldiers when a nail from a horse's shoe caused a flat. With the snow falling harder, repairing the damage took nearly an hour. The job was completed just as a second car, flying a large American flag, pulled alongside. From the rear seat a muffled passenger offered help. When they explained that they had solved their problem, he waved and the car drove on.

A more thorough introduction took place a short while later. Still many miles from Kaunas and with the brief December day rapidly closing, Henderson and Ryan began searching for a farmhouse or other accommodation. Proceeding slowly in the fading light, they saw little that looked promising, until to their surprise they spied the second car ahead, surrounded by three men. Amid the general chaos in the country, banditry seemed a genuine possibility. But the culprit in this case proved to be a snowdrift, not robbers, and the men were the car's driver and passengers, cursing in three languages as they strove to get their vehicle back on the road.

Henderson and Ryan identified themselves to John Gade, an American businessmen recently appointed United States commissioner to the Baltic republics; to John Lehrs, a naturalized American citizen of Russian birth who was acting as Gade's interpreter; and to the hired driver. Having just presented his credentials to authorities in Latvia, Commissioner Gade was heading for Kaunas to do likewise with the Lithuanians. Meantime he was keeping an eye out for commercial opportunities for American firms.

Six backs were no more successful than three in getting Gade's car unstuck, and after some discussion Henderson and Ryan went on to look for help. By now it was full night and snowing harder than ever. Only good fortune guided them to a railroad waystation manned by a lone caretaker. The old peasant would not leave his post, but he roused a rescue party of several workers and horses. The relief crew managed to follow the quickly filling tracks of Henderson's automobile back to Gade's. Hitching ropes to the bumper, the men hauled the commissioner's car to the station.

The combined prestige of a representative of the American government and a colonel in the Red Cross impressed the locals. After assuring his guests that a train would arrive in the morning to transport them to Kaunas, the stationmaster ordered a sleigh to carry the Americans to the nearby palace of a Lithuanian nobleman. This individual, like much of the Baltic barony, had made himself unpopular by monopolizing thousands of acres of the best land in the vicinity and reducing the hundreds of peasants who toiled thereon nearly to the status of serfs. Equally galling in a different way was the fact that he and his class had forsaken the ancient culture and language of Lithuania for those of Poland. The war and its attendant anarchy had driven many of the nobles away; where the grandee of this particular manor was residing at the moment, none could tell. Perhaps he was dead. Perhaps with the Polish army. Perhaps in Paris or London trying, like the émigrés who had fled the guillotines of revolutionary France, to reverse the clock and prevent the rise of an uncontrollable and doubtlessly vindictive republicanism. In any event he was not at home, and the Americans had use of the premises for the night.

The building lacked heat and furniture, but it was dry and out of the wind. A caretaker and his wife, who lived in the kitchen, let them in. While she cooked a hot meal he rounded up some old U.S. Army blankets, delivered to Lithuania courtesy of Herbert Hoover and the American relief administration. All things considered, the arrangements proved quite satisfactory.

As good as his word, the stationmaster at dawn produced an engine, a passenger coach, and a flatcar. After their rescuers loaded the two automobiles on the flatcar, the Americans climbed aboard and the train pulled out. By noon they reached Kaunas, where the pick of the burghers welcomed the American commissioner, and a less distinguished group met the officials of the Red Cross.

Such was Loy Henderson's first contact with the personal face of American foreign policy. More than four decades later he found himself in a situation no less confusing; but this time *he* received the honor accorded the representative of the United States. And as much as Henderson's standing had changed, so had that of his country. The America that paid John Gade's salary was retreating from the arena of world affairs, having found its brief appearance there distasteful and frustrating. The America of the early 1960s, on the other hand, was thoroughly enmeshed in international matters, and it did not hesitate to project its power ever further, to the very ends of the earth. During the summer of 1960 Henderson, now deputy undersecretary of state, headed a delegation of American officials to Africa. With the European empires in a condition of terminal decay, Henderson went to deliver America's respects to the successor regimes.

But more, or perhaps less, than respect motivated American policy. Even as the diplomats congratulated the Africans on their newly won independence, American agents worked by covert methods to ensure that Africa not become *too* independent. Africa possessed much in the way of mineral and human resources, of potentially great use in America's struggle with the Soviet Union. While the Europeans, allied to the United States, had controlled the continent, Africa seemed secure from the communists. But with Europe now on the run, Washington perceived the need to move in before Moscow did. The United States would not construct a formal empire in Africa; the Europeans' experience had demonstrated the futility of direct foreign control in an age of militant nationalism. Besides, Americans had no stomach for colonialism as such. But as they had been doing ever since World War II, American leaders would work through clients and proxies, expanding a network of alliances and a system of influence that amounted, in operational terms, to an informal empire. By 1960 the American sphere included nearly all the western hemisphere (Cuba's status was an open question Washington would soon try to close). It embraced Europe to the Elbe. It ran from the Mediterranean across Asia to the South China Sea. It stretched in the Pacific from New Zealand to the Aleutians. Covering every time zone and all inhabitable (and some uninhabitable) latitudes, it dwarfed anything the world had ever seen.

And in the middle months of 1960, it was reaching into Africa. American concern during that boiling summer focused on the Congo, where Belgium was handing power to an unstable and murderously divided regime. When violence swept the new nation, threatening the lives and property of the thousands of Belgians still residing there, Brussels panicked and dispatched paratroopers to reoccupy the country it had just set free. Belgian copper interests, seeing their opportunity to delay the future beyond a few more annual reports, threw their support to a secessionist movement in Katanga province. The leftist Congolese prime minister, Patrice Lumumba, responded by calling on the United Nations, and then the Soviet Union, for help. The Russians acted first, sending advisers and military equipment. By the time UN troops arrived, Lumumba had changed his mind about them. He told them to go home, denouncing the international peacekeepers as tools of colonialism.

Washington rejoined with efforts to topple Lumumba, whom American leaders considered an ally, witting or otherwise, of the Kremlin. During the next several months the Central Intelligence Agency devised schemes to neutralize him, by means that included assassination. With at least the tacit approval of the White House, the agency put its plans into effect, and hit men in the employ of the American government did their best to eliminate the troublesome radical. In the event, they failed; Lumumba's Congolese rivals did him in before the Americans could. Nevertheless the outcome, in which a pro-American regime replaced Lumumba's, afforded Washington satisfaction.

By 1960, both Loy Henderson and the United States had come far since that snowy evening at the end of the first great war. To a large degree they had traveled the road together, for as well as any individual's, Henderson's career tracked the growth of the American empire. From 1918 to 1961 Henderson witnessed personally or figured professionally in many of the world's major crises. From 1918 to 1921 he worked Central and Eastern Europe for the Red Cross, trying to salvage the human flotsam of the wrecks of Germany, Austria-Hungary, and imperial Russia. After enlisting in the American foreign service in 1922 he apprenticed in Ireland, where Irish republicans waged the anti-British battle they have fought ever since. Further training in Washington and Latvia led to his appointment to the first American diplomatic mission to Moscow in 1934. With grim fascination he watched the unfolding of Stalin's totalitarian designs, including the ravages of collectivization and the terror of the purges. The experience marked Henderson for life—and through Henderson and others of the Moscow mission, it marked American foreign policy for two generations.

From Moscow Henderson also observed the emergence of militarism in Nazi Germany and Japan. He devoted much effort to explaining and predicting the reactions of the Soviets to the fascist menace. By the time World War II began, Henderson had returned to the United States; when Hitler's attack on the Soviet Union in the summer of 1941 and Japan's bombardment of Pearl Harbor in December threw Washington and Moscow into an uneasy embrace, he found himself charged with coordinating assistance to America's improbable ally. Among other activities, he accompanied Averell Harriman and Winston

Churchill to the Soviet Union in 1942 to soften the news that there would be no western front in the near future.

Unfortunately for Henderson's advancement prospects, he failed to hide his distrust of Moscow's intentions. The war's end, he argued, would place the United States no longer in collaboration with the Kremlin, but in competition; and if American leaders desired to secure the goals they had endorsed so hopefully in the Atlantic charter, they must prepare for the day. The Roosevelt White House did not appreciate Henderson's message, and it cast the Cato into exile, to a Baghdad far removed from the center of wartime activity and controversy.

Then Roosevelt died; the spirit of Soviet-American cooperation expired soon after. Meanwhile Henderson was mastering the essentials of the great game in the Middle East. In the spring of 1945 President Harry Truman recalled him to Washington to head the State Department's division of Near Eastern and African affairs. Having warned against the communists for a decade, Henderson was well placed ideologically for the inevitable postwar falling out, and as NEA director when the initial storms of the cold war blew off the mountains of Azerbaijan in the winter of 1946, he was perfectly positioned bureaucratically to press his views regarding an enlargement of American responsibilities. Crises in Greece and Turkey followed quickly upon the Iranian affair; Henderson seized the opportunity to agitate for an immediate and energetic American response. The Truman doctrine, which more than any other document served as the blueprint of America's anticommunist empire, took shape in Henderson's office and under his careful direction.

Henderson's insistence on countering the communists also impelled him to oppose American support for a Jewish state in Palestine. America's future in the Middle East, he contended, lay with the Arabs, not with the Zionists. Matters of morality in the issue canceled out: Arabs and Jews both possessed sufficient grievances to justify their political aspirations. But the Arabs also possessed oil and the strategic positions the United States needed to occupy in its effort to contain the Soviets. Prudence dictated setting American policy by the crescent of Mohammed, not by David's star.

As earlier, Henderson refused to temper his advice to suit the White House, and he became one of the casualties of the 1948 election campaign. This time exile meant India, where he crossed opinions with Prime Minister Jawaharlal Nehru. Henderson spent three years trying to convince Nehru of the illogic, if not the immorality, of India's policy of neutralism. The debate began as something of an academic affair—although Henderson never took such matters lightly—but it gained geopolitical substance following the communist victory in China and the outbreak of the Korean war. As India became a mediator between east and west, Henderson became a mediator between Washington and New Delhi. Henderson felt Nehru's personal magnetism, but he disdained the prime minister's philosophy of world politics, and he cautioned Washington against creeping nonalignment.

By the time Henderson outwore his welcome in New Delhi, a fresh crisis had developed in Iran. Upon assuming the premiership, Mohammed Mosadeq had provoked London by announcing the nationalization of the British-owned Anglo-Iranian Oil Company. To prevent the ensuing conflict from erupting into war, Secretary of State Dean Acheson hustled Henderson to Tehran. Mosadeq had taken power on a popular wave of anti-British passion, and although Henderson believed the British deserved most of what they were getting, he feared that the only victor in the deepening struggle would be the Kremlin. For two years Mosadeq amazed Henderson with his combination of political agility and personal flair. It was with regret that the ambassador concluded in the spring of 1953 that the prime minister had to go. Shortly after a visit to Washington during which Henderson suggested steps for removing Mosadeq, Eisenhower approved a CIA operation that accomplished the recommended result.

The 1953 coup in Iran carried the United States into the Middle East in an unprecedented way; the formation of the American-sponsored Baghdad pact continued the process. Henderson served as the American representative to the Baghdad alliance, assuring its members that although domestic and other constraints precluded direct American participation, Washington backed their anti-Soviet efforts entirely. Henderson proved less persuasive in 1956, following Egyptian President Gamal Abdel Nasser's seizure of the Suez canal. As the American representative on an international mediation mission to Cairo, Henderson failed to convince Nasser that a conflict over the canal would only benefit the communists. War came, leading to the further embroilment of the United States in a region traditionally the cockpit of empires.

Henderson would have retired at sixty-five in 1957 but for special presidential dispensation. Although his last permanent assignment involved primarily administrative work in Washington, he kept in touch with old friends among the conservative regimes in the Middle East. In 1960, when the United States prepared to establish relations with the newly independent countries of Africa, he embarked on his tour of the Congo and neighboring parts of the continent.

By the time he retired at the beginning of 1961, Henderson possessed a reputation unmatched in his profession. In Washington and in American missions overseas he was commonly known as "Mr. Foreign Service." Secretary of State Henry Kissinger, later dedicating the Loy Henderson International Conference Room at the State Department, characterized Henderson as the quintessential "insider," one of the "giants of postwar diplomacy."

Kissinger did not identify Henderson as a principal fabricator of the modern American empire; secretaries of state do not commonly speak in imperial terms—however they, especially those in the Kissinger mold, might think. Besides, Kissinger in the 1970s confronted the decline of American power, at a moment when the empire was showing its age. Henderson already represented another era, a period that saw the United States transformed from a country reluctant to enter the world fray to one whose interests and self-defined responsibilities circled the globe.

Loy Henderson's story, in many respects, is that of the empire he helped build. This double tale is the subject of what follows.

I am indebted to many persons and institutions for assistance during my work on this book. To all I am grateful. I would like especially to acknowledge the help of Robert Divine and Waldo Heinrichs, who read the manuscript with care and offered cogent suggestions for improvement.

College Station and Austin, Texas   H. W. B.
July 1990

Contents

I. TO THE FINLAND STATION

  1. Beyond Arkansas   3
  2. Initiation   17
  3. Recognizing Reality   28

II. THE EASTERN FRONT

  4. Red Tape   49
  5. Moscow Trials   60
  6. Minuet for Dictators   74
  7. Between the Devil and the Deep Blue Sea   88
  8. Comrades in Arms   101

III. EXILE AND RETURN

  9. "Head net, mosquito bar, sun helmet . . ."   115
  10. The Great Game   127
  11. The Battle Joined   147
  12. In the Palestine Labyrinth   165

IV. THE WAGES OF NEUTRALISM

  13. The Most Charming Man He Ever Despised   195
  14. Their Finest Hour   211

V. BUBBLE AND BOIL, TROUBLE AND OIL

  15. Nationalization and Its Discontents   233
  16. The Wily Premier, the Angry Ayatollah, and Hamlet
     on the Peacock Throne   251
  17. What Gentlemen Will Do   269

VI. CONCLUSION

18. The Ends of Empire   295

Notes   315

Index   333

I

TO THE
FINLAND STATION

1

Beyond Arkansas

Henderson entered life during one of the most turbulent decades in American history. The 1890s began with a revolt among American farmers, who responded to falling commodity prices, monopolistic practices by railroads, a perceived loss of social stature, and a catalog of additional woes by abandoning the established political parties in large numbers in favor of their own Populist party. The decade proceeded through a succession of bloody strikes—the bloodiest occurring at Carnegie Steel's Homestead works in Pennsylvania in the summer of 1892, at just the time identical twin boys, Roy Wilmington Henderson and Loy Wesley Henderson, entered the world of the Ozark plateau of northwestern Arkansas—into the deepest depression the United States had ever experienced. The depression produced more strikes and more Populists, as well as other protest movements aiming to bar immigrants, relieve the rich of their wealth, debase the currency, socialize industry, and close saloons.

American foreign affairs reflected the general tumult. Having filled their own continent, or as much of it as they seemed likely to manage to buy or steal, Americans looked abroad for the opportunities they saw diminishing at home—and for distraction from domestic troubles. In 1893 Americans in Hawaii overthrew the monarchy there and asked for annexation to the United States. Grover Cleveland's scruples prevented quick acceptance of the offer, but his successor William McKinley gladly planted the flag halfway across the Pacific. Cleveland didn't disdain to tangle with Britain in 1895 over the boundary between British Guiana and Venezuela. By then the Democratic president was happy for any diversion from the deepening gloom at home. The dispute almost led to war, but since this was more diversion than Cleveland intended he softened at the brink, as did the British, who had plenty of worries of their own. Consequently war had to wait until 1898, when Spanish mistreatment of Cuba provided the pretext American jingoes and other expansionists needed to convert that island into an American protectorate and make outright colonies of Puerto Rico and the Philippines.

By the turn of the new century America had made a fair start on fashioning an empire; and Loy Henderson had made a fair start on life. The assassination of McKinley in 1901 slowed the march of empire not at all, since Theodore Roosevelt, the most ebullient of American expansionists, now became president. A boyhood injury slowed Henderson's progress, though. At nine he broke his arm falling from a fence. Poorly set, the arm never healed properly and remained stiff ever after. The minor disability would have a profound effect on his career.

While the bumptious Teddy swung his big stick around the Caribbean, the Henderson family grew, with two daughters and another son following the twins. Henderson's father, a Methodist minister, had chronic difficulty supporting this crowd, and he repeatedly relocated in search of a more secure pulpit. In 1904, not long after Roosevelt's bludgeon knocked Panama loose from Colombia and a canal zone free from Panama, the Henderson family moved to Ohio. There they remained through the bully days of the Roosevelt corollary and TR's incongruous Nobel peace prize, won for devising an end to the Russo-Japanese war. By 1911, when the Hendersons moved to Kansas, William Howard Taft was experimenting with the economic form of American expansionism—trading dollars for bullets, as the day's phrase went—that would occupy the country for a generation and fill Henderson's waking hours during his first years in the foreign service. Of equal portent for both Henderson and the country, Woodrow Wilson's star had just cleared the trees and gothic towers of Princeton on its passage over Trenton toward the Potomac.

While Wilson was looking south, the Henderson twins were gazing east. The Reverend Mr. Henderson had been a schoolteacher before donning the cloth; his wife also, until the twins arrived. Not surprisingly the parents expected academic excellence from the boys. They obliged. All understood that the children would attend college. Considering where to apply, Roy and Loy initially set their minds on Harvard. But shortly before graduation from high school their mother fell seriously and expensively ill. This ruled out the Ivy League; indeed, for the moment it precluded anything far from home. The twins enrolled at Southwestern College in Winfield, Kansas.

In 1913 Mrs. Henderson died, grieving husband and children but lifting some of the family's financial burden. Loy and Roy transferred to Northwestern University in Illinois. The war in Europe commenced the summer before the twins' senior year. Until 1917 the conflict remained "over there," where most Americans wanted it. But its tentacles slowly drew the United States in. The *Lusitania* sinking outraged Americans. More ominous if less noticed was the sinking of two billion American dollars into British and French war bonds.

On a smaller scale than the allied governments, the Henderson twins also experienced cash-flow problems. College degrees in hand but no solid prospects in sight, they had decided on joint careers in law, and they needed money to continue their education. Postponing their meeting with Blackstone, they took jobs teaching school in Colorado, where their father and siblings now lived. When one year's work in a small mining town did not produce a sufficient grubstake to finance the jump into the legal world, they signed up for a second.

The delay overturned their career plans, for before the second year ended the United States joined the war. No longer too proud to fight, Wilson embraced war as a means of regenerating the earth. He was a compelling rhet-orician, perhaps because he wrote his own speeches, and he carried Congress and the country with him.

Wilson's missionary call to public service touched a chord in the Henderson twins. Roy and Loy were not especially close to their father, who maintained a Victorian aloofness and who in any event had been preoccupied with keeping the family clothed and fed. But they absorbed enough of his ministerial tem-perament to believe that young men like themselves had an obligation to improve the world. They found Wilson's summons to the service of mankind greatly appealing.

The allure of arms also played a role. Grandfather Henderson, a Civil War veteran, had recited the tale of the Battle of Pea Ridge innumerable times, to the boys' endless delight. Fascinated by this latest war's implicit challenge to their manhood—they had turned twenty-two the day the Austrian archduke was assassinated at Sarajevo—they rushed to enlist. Roy, eager and healthy, put on a uniform and entered the army. Loy, equally eager but with a partially disabled arm, flunked the physical exam and went home.

The blow to pride and sense of self stunned Loy; the shock of separation compounded the pain. The twins had rarely been apart and never for long. The double hurt, combined with the conviction of having proved second best in the usually unspoken yet ever-present competition between twins, left Loy devas-tated. Had his physical disability been more obvious, it might have made things easier, for then he would have seemed, to himself as to others, less a shirker. Roy was going overseas, doing his duty, facing death. Loy sat home, a failure.

Several attempts to obtain a medical waiver availed nothing. Assorted efforts to circumvent military regulations succeeded no better. Eventually Loy gave up and went to law school. But after a year at the University of Denver he learned of a recruiting drive by the American Red Cross, specifically targeting individuals medically unfit for military duty. While the ARC was not the AEF, its work did include public service in the war zone, and it involved uniforms, commissions, and other trappings of military life. Henderson seized the opportunity.

By this time, however, the war was nearly over. In fact, in midcontinent on the train ride from Colorado to ARC headquarters in New York, Henderson awoke from a daydream to hear church bells greeting the news of an armistice. At first he feared he had missed his chance, that he would be cheated of the opportunity to serve. But his spirits revived on learning that the reports of a ceasefire were premature. By the time the genuine article flashed across the Atlantic cable, he had reached New York, where Red Cross personnel emphasized that the fighting's end would not halt the ARC's work. Reassured, Henderson embarked for France, intending to do his modest part in reconstructing the torn continent.

II

Henderson just beat Wilson to Paris. He stood among the millions who greeted the victorious American president, now come to complete at the peace conference what he had begun on the battlefield. The sight of the man who had inspired him to public service moved Henderson, as did the enthusiasm of the French populace. Recalling the moment later, Henderson commented that America never enjoyed such popularity in Europe as on that day. "The girls were liberal in the impartial bestowal of embraces and kisses," he added.[1]

Henderson's first assignment took him to the Beau Desert American Hospital Center on the outskirts of Bordeaux. The war was over, but for many of the sick and wounded, especially those with permanent injuries, the armistice brought more uncertainty than cheer. Henderson's task was to visit patients, explain how the American government would care for them, and generally raise morale. At the same time, as a novice in the field of relief he was expected to absorb as much as possible regarding the operation of a medical facility in a war zone.

Henderson found the work gratifying. Despite the difficult conditions in the hospital he felt he succeeded in boosting hopes needing a lift. In his initial report for the ARC he wrote:

> From January 1st until March 10th we worked in this hospital center and saw many thousands of men. I myself personally talked before some 20,000 wounded or diseased soldiers. ... Our work was pleasant because the men were so delighted to hear of the opportunities which were being offered to them, and the surgeons were glad to have us talk. ... They said that we encouraged the men and kept them from becoming morbid and discouraged. ... I felt rewarded for the many days spent walking through the hospitals in the rain and mud by the change of spirits which came over many of them after hearing what the Government was going to do for them. ... The nurses told me that many men who were soured on the Army and on their country and who were hard to handle in the wards would change entirely after they had heard how they were to be cared for.[2]

This training mission soon ended, and Henderson received a new assignment. Dispatched to Berlin as an inspector of German prisoner-of-war camps, he began to gain an understanding of the extraordinary complications the fighting had created in central Europe. Pending the conclusion of a peace treaty, Germany remained technically at war with the allied powers. Although the allies had moved into the southern and southwestern parts of Germany, the occupation had yet to be completed and much of the north was still under the control of the government of the recently established German republic. Henderson, accredited to the Interallied Commission for the Repatriation of Prisoners of War, found himself in the northern zone at a time when the carthaginian shape of the Paris treaty was coming into view. Not surprisingly, the prospect strained relations between German officials and representatives of the allied governments.

A division of labor within the camps did not help matters. The Germans assumed responsibility for maintaining order among the prisoners—not an easy task since the virtual end of the war increased the restiveness of the prisoners even as it diminished their fear of the guards. The repatriation commission had the duty of looking out for the interests of the POWs, especially of ensuring that they received adequate food, clothing, and medical supplies. With most of the German population in various stages of malnutrition, exposure, and sickness—and with the allied blockade of Germany still in effect—such oversight increased the potential for friction.

Henderson encountered a deeper form of unrest on his arrival in Berlin. The abdication of the kaiser in November 1918 had opened the way to the establishment of the republic, but republicanism did not satisfy the many radicals who looked to the Soviet Union for inspiration. The Spartacists, as these revolutionary advocates of alliance with Moscow were called, attracted thousands of converts and took over sections of Berlin. Considering the straitened conditions in Germany at the time and the increasingly apparent vindictiveness of the capitalist allies, it appeared entirely possible that the present moderate regime might prove, as Kerensky's had in Russia, merely a stepping stone to socialism. Meanwhile the uncertainty regarding Germany's future compounded the problems of prisoner relief.

At the end of March, Henderson received instructions to inspect and evaluate nine camps in the vicinity of Berlin. As a representative of the ARC, which had sent packages to American prisoners during the course of the fighting, he would have some control over the distribution of blankets, clothing, and certain other supplies. Although the situation with respect to these goods had improved since the armistice, Henderson's commanding officer suggested that he try to identify prisoners who could sew and encourage them to mend uniforms that could not yet be replaced. To avoid conflict with the Germans, his instructions prohibited him from handing out food or medicine.[3]

Henderson discovered that prisoners from the western countries had fared best in the camps. This had resulted in part from greater efforts by western relief organizations to supplement rations provided by the Germans, and partly as a result of the smaller numbers of men involved. Russian prisoners, by contrast, had had a considerably tougher time. As the eastern front collapsed, Russians had surrendered en masse, sometimes tens of thousands in a day. Under the best of circumstances these hordes would have strained Germany's capacity to feed and house them. Under conditions of impending defeat the Russians not infrequently went hungry and inhabited the most primitive quarters.

The Russian prisoners labored under another disability. With the western front quiet and the allies setting the agenda at the peace conference, nearly all the western POWs were repatriated within a few months of the armistice. In Russia, however, fighting continued. The second stage of the revolution, which had brought the Bolsheviks to power at the end of 1917, had touched off a civil war that raged unabated. Allied intervention exacerbated the confusion. Aiming to reopen the eastern front, which Lenin and Trotsky had closed by the treaty of Brest-Litovsk in March 1918; desiring to protect supplies of war matériel, which sat on the docks of Archangel and other ports of entry; and hoping to aid the anti-Bolshevik forces, which appeared the best, or most readily available, guarantee against the spread of the communist infection, Britain, France, and the United States had invaded Russia from the north. A second American contingent and a Japanese force entered Siberia from the Pacific. Aside from slowing the peace-writing process, the intervention retarded the repatriation of prisoners, since the interventionists did not wish to add to the ranks of Trotsky's Red army.

But the Russian POWS could not stay in Germany forever. In the third week of April 1919 Henderson joined a task group heading east to arrange for shipping the Russians back to their country. On Easter Sunday he and his associates stopped at a camp in western Prussia to pass the word to the Russian prisoners that they would soon see home. The reaction was predictable. "All in good spirits," Henderson jotted in a diary of the trip. Indeed, the young men spent much of the afternoon entertaining the bearers of the welcome news with Russian Easter hymns.[4]

The next morning Henderson and the other allied officials continued east, crossing the German-Russian frontier near midday. If Henderson had found conditions in Germany bleak, the situation in Russia sobered him more. "Unspeakable difference as I cross border," he wrote. The contrast owed principally to the fact that most of the fighting on the eastern front had taken place in Russia. Three and a half years of war had devastated the Russian countryside, which had not appreciably recovered since the shooting, or most of it anyway, stopped. Fields were overgrown, bottomland had reverted to swamp, piles of debris and occasional chimneys marked the sites of villages and farmhouses.[5]

Nor was the end of the country's affliction in sight. As part of an anti-Bolshevik agreement among the negotiators at Paris, the Germans continued to defend a line some hundred miles inside Russian territory. In addition, in the name of self-determination, but more basically in the pursuit of the same anti-communist objective, the allies were encouraging the newly organized army of reviving Poland to root the Bolsheviks out of the country the Poles sought to reclaim. Finally, as if these insults did not sufficiently wound what remained of the pride of imperial Russia, the peoples of the Baltic seaboard were asserting their right to independence and statehood.

Henderson and his party ran smack against this Baltic nationalism when they arrived in Kovno on April 21. For all they knew, Kovno was simply another Russian city under German occupation, and upon entering they presented their credentials to the German commandant. This officer suggested the Metropol hotel as a suitable headquarters for the repatriation commission. Following his advice, Henderson and the others repaired to the hotel, where to their surprise they encountered a guard unit in German uniforms, distinguished from those they had just seen only by the addition of yellow stripes on lapels and caps. Greeting the ARC commission was a pretentious individual who bowed and introduced himself as the "minister of supplies." Puzzled, the leader of Henderson's group, a Major Thornburn, inquired whose minister he was. The Republic of Lithuania's, he replied.[6]

This was news to Thornburn, to Henderson, and to everyone else on the commission, and some time elapsed before they unraveled the details of the arrangement by which the Germans had allowed the Lithuanians to form a provisional government as a buffer against the Bolsheviks to the east. Anxious to please the Americans, the minister of supply dug deep in his storerooms to produce what Henderson described as a "big feast," a rarity for that time and place. During dinner various other Lithuanian officials appeared, including the minister of war, who suggested an interview with his country's president.[7]

Accordingly, Thornburn and Henderson paid a call on Dr. Antanas Smetona. After Smetona expressed his desire that this visit from representatives of the allied nations be simply the first of many, Thornburn inquired about the chances of repatriating Russian prisoners through the German-Russian lines. Smetona replied dubiously, citing the complicated nature of relations among Lithuanians, Germans, Poles, and Russians in the area. But he said he would do what he could to expedite matters.[8]

Following another banquet, at which a band played "Yankee Doodle" and the minister of war and his colleagues toasted America as the "great hope" of their people, the commission set off for the front. On the morning of April 23 they consulted with the German chief of staff, who offered the welcome news that he had communicated with the regional headquarters of the Bolsheviks and been assured that the prisoners would be allowed to cross the lines and return to their homes. To guarantee the smoothness of the operation, Thornburn and Henderson continued the remaining thirty kilometers to the front. The countryside was the most barren Henderson had yet encountered. "Absolute ruins," he wrote. "People like starved animals living mostly in basements of wrecked houses. Deadly quiet."[9]

Fortunately the prisoners' crossing went well. Just on the German side of the line the POWs disembarked from their train and received twelve days' rations. Their officers made speeches about how they ought to conduct themselves as they returned to their homeland. The commanders added that they should remember that they had served their country honorably; they had nothing to fear. Not all the men were convinced. "Many trembling," Henderson noted, "others full of joy." After thanking the commission for its assistance, the officers raised a white flag and the column marched slowly up a hill to the first Bolshevik check point and thence out of sight.[10]

Returning to Berlin with other members of the commission, Henderson reported on the miserable conditions in Lithuania. "Unless relief of some kind comes into this country, next winter will witness terrible suffering," he wrote. Famine would take a human toll; it would also exact a political price. The "Bolshevists," Henderson wrote, were not currently "strong either in numbers or aggressiveness" in Lithuania. Yet the situation might change. "At the present time this strip of Europe is one of the barriers against the Bolshevists, but if the people here are driven to extremities, they are likely to embrace anything which might be offered to them on the ground that it is impossible to make their situation worse." To Henderson's pleasure, this paper elicited a swift response. Within a short while ARC supplies of food and medicine began moving into the devastated region.[11]

After this minor victory, the situation grew more complicated again. Ordered back to the front to arrange the repatriation of more Russian POWs, Henderson arrived at Marienburg simultaneously with accounts of the final terms of the peace settlement. His first inkling of a change in mood among the local inhabitants came when he walked into a cafe, wearing his ARC uniform, and the German-speaking patrons got up and left. He learned the reason when a German officer told him the news of the day. On the street later that afternoon Henderson encountered further hostility, which was not lessened by the fact that a principal aspect of his job involved safeguarding food supplies, intended for Russian prisoners, against a hungry and now embittered populace. Such good feeling toward the Americans as earlier had existed evaporated entirely when Henderson's feeling of responsibility to his job and to the POWS forced him to reject a request from one of the town elders for food for the children. Not for the adults, the man pleaded, just the little ones. But Henderson still said no. Tension reached a point where Henderson's German liaison suggested— "begged," was how Henderson put it in his diary—that he refrain from wearing his uniform in public.[12]

A shift in the military fortunes on the still-active eastern front placed Henderson in another uncomfortable position at the beginning of June. For most of a month he had worked to expedite the repatriation of more Russian prisoners; at the end of May carloads of POWs, most in high spirits, began arriving in Marienburg from the west, awaiting final approval to cross the front. At the last minute, however, an offensive by the Polish army succeeded in filling the gap between the retreating Bolshevik forces and the Germans, with the result that the POWs would have to traverse territory now held by the Poles. The latter, disinclined to hand over to their enemies troops rested and well fed—due to the efforts of Henderson and the ARC—refused to allow the crossing. Henderson assumed the unpleasant task of breaking the bad tidings. Although he feared an outbreak of violence, only stunned silence punctuated by occasional sobs greeted his announcement that the men were not going home after all.

III

The Polish victories put an end to Henderson's repatriation activities, and in August he received orders to join a relief commission for the Baltic states. After some reconnoitering and preparation, he set off with Colonel Edward Ryan and other members for Riga. The ARC caravan comprised an ill-matched collection of persons and equipment. Henderson described the scene:

> The whole affair looked like a circus en-route. Immediately behind the locomotive were two sleeping cars bearing some 25 members of the American Red Cross Commission for Western Russia and the Baltic States; then strung out like the tail of a kite were box cars containing food, gasoline, personal luggage, office equipment, etc.; and finally a long line of flat cars burdened with automobiles of American make, each bearing the inscription in large letters, "The American Red Cross."

As the journey progressed and the first day ended, Henderson grew more pensive.

> The night, though cold, was a wonderful one. On both sides of us were tall slender pine trees, and a full moon reflecting on the road turned it into a shining corridor with high walls of green. Indeed, the spirit of the evening coming upon me, I forgot my troubles for a few moments and found time to ponder over the fact that the inhabitants of such a beautiful country should be afflicted by such fearful misfortunes.[13]

The misfortunes became more apparent as the party moved east. On several occasions battles between Latvians, Germans, and Russians interrupted the journey. When Henderson's group arrived at Riga, Ryan cabled to Berlin that there was "a great deal of shooting going on." Henderson depicted the scene more poetically: "From time to time the whole landscape was made bright by star shells and rockets fired from unseen hands."[14]

By now fighting in the area, at least that part of the fighting involving the Germans, was supposed to be over. It wasn't, though, and renegade German troops from the so-called Iron Division had locked up with the formidable Lettish Rifles. Both sides were seeking control of the vicinity of Riga when Henderson and the commission approached. After a battle of several weeks the Latvians, aided by allied warships ascending the Daugava River, succeeded in fending off the Germans. In the meantime Henderson and his colleagues kept busy driving ambulances and tending to the wounded. Henderson spoke neither Latvian nor Russian, the common tongues of the region, and his command of German, the third language, was limited. With soldiers of various nationalities running around in assorted uniforms, Henderson felt that "by some weird mistake we had suddenly been catapulted into a scene from a Viennese operetta." At one point local troops arrested him for a German, an error he was unable to rectify since he could not speak their languages nor they his. He strongly suspected that his warm coat and boots interested his captors more than his allegiance. Fortunately, just before being stripped and packed off to detention, he spied a Latvian officer he knew, who explained, to the dismay of the enlisted men, that Henderson was one of the good guys.[15]

From Riga, Henderson and Ryan proceeded to Mitau, and after a few weeks of organizing relief efforts they continued to Kaunas. It was on this journey, delayed by snow, Germans, and a flat tire, that Henderson met John Gade. The country through which they traveled was more ravaged than anything they had seen to date. The retreating Germans had carried away crops and livestock and torched most of what they could not steal. Kaunas seemed the nadir of the region's despair. Henderson described conditions there as "very critical." Although everything was in short supply, Henderson, in charge of provisioning local hospitals, cabled most urgently for basics like castor oil, vaseline, ether, and disinfectants.[16]

The Germans had disappeared, but Latvia at the beginning of 1920 remained a country very much up for grabs. Freebooters and adventurers of all stripes roamed the region. About the time he reached Kaunas, Henderson learned of an "American-Lithuanian Legion" reportedly being organized in the United States for defense of the Lithuanian homeland. Shortly afterward an individual who did not look particularly Lithuanian showed up on Henderson's doorstep and introduced himself as Colonel Isabel, chief of staff of the American-Lithuanian Legion. His commander, he said, would arrive shortly, and the rest of the troops would follow.

Colonel Isabel made quite an impression. With martially trimmed mustache and pointed beard he looked, Henderson thought, more like a Spanish conquistador than a Lithuanian liberator, and indeed he claimed to have fought in Latin America before serving with an American tank battalion in France. At a dinner hosted by the chief of the British military mission in the Baltics, Isabel regaled the gathering with his exploits in various theaters of war. When the host raised a glass to Anglo-American cooperation and the success of the American-Lithuanian Legion, here so gallantly represented in the person of Colonel Isabel, the colonel rose to reply. But before speaking he insisted on producing his talisman, which, he said, had delivered him from numerous tight spots. He added, to the amusement of the guests, that he considered speeches before such distinguished gatherings tight spots. Thereupon he reached into his pocket and withdrew a shrunken head. While the guests gazed in astonishment, he explained that he had acquired it in South America, from a tribe of headhunters.

After a toast that seemed, despite its unorthodox introduction, appropriate to the occasion, Isabel emptied his glass, as did the others. Remaining standing, he asked a waiter for a refill and proceeded to toast the charm and graciousness of the hostess. Upon draining this round he announced that it was not his custom, when saluting a beautiful woman, to leave off simply because his glass was empty. He then took a large bite out of the rim, chewed and swallowed, and ate the glass down to the stem. To his host's shocked query whether he was all right, he replied, "Don't be alarmed. I do this frequently. It will do me no harm. As an American soldier I have hardened myself so that I can eat almost everything and can endure all kinds of pain without flinching." To underline his final remark he produced a long hatpin and drove it through one cheek, pushing until it came out the other. Removing it abruptly he stabbed himself once through each arm. Then he calmly sat down.[17]

Disappointingly, after Isabel's heroics, the American-Lithuanian Legion never materialized, and Henderson continued the routine work of stretching insufficient supplies across difficult conditions. In February word arrived from Narva in Estonia of an outbreak of typhus among prisoners of a defeated White Russian army. Of the 25,000 to 30,000 soldiers captured some 10,000 or more were ill. Making a quick trip to the area, Henderson and Ryan toured a hospital in one of the worst-stricken zones. The "hospital," in fact, was nothing but the shell of a textile factory looted by the Bolsheviks. Inside the cold, drafty building Henderson and Ryan found hundreds of prisoners in various stages of typhus, from delirium to death. Filth covered the floor, and the stench, despite the draft, was nearly unbearable. Lice swarmed over the patients, as one of Henderson's associates phrased it, "like ants on an ant-hill."[18]

In his description of the epidemic, Ryan commented that "the one thing that was most noticeable in the whole situation was the lack of discipline and organization." Ryan knew about discipline and organization; Henderson learned much on the subject during this period. The colonel immediately threw the anti-typhus campaign into high gear. Henderson, other ARC personnel and anyone else they could mobilize bathed, shaved, and deloused patients. They mopped, scrubbed, and disinfected the building. They organized activities to educate the populace. They inspected and, where necessary, quarantined refugees.[19]

They succeeded in drastically reducing the mortality rate, but they took a few casualties themselves. In March, Henderson and three others of his group exhibited typhus symptoms. Within days delirium set in. Two men died while Henderson thrashed in fevered semi-consciousness. According to his later recollection, he believed he also was dying, and as he felt himself slipping over the edge he saw his brother Roy and wished him farewell. So vivid was the image that he thought he could feel Roy's hand on his shoulder just before they parted.

He awoke forty-eight hours later. Although relieved, naturally, to find himself still alive, he could not shake the feeling that something beyond the fever had caused his dream. At the beginning of April he received a cable from the United States. Fearful of its contents, he asked Ryan to read it. The latter's face betrayed the message; only the details required filling in. Roy had suffered a kidney ailment while in the army and had died of complications.

Several days after this, a packet of letters from the twins' father arrived. Henderson senior had accepted Loy's decision to join the ARC, although grudgingly, since with the war nearly over it seemed a vain gesture. He thought Loy needed to get started on his life's work. Roy's death increased the urgency. Loy would have to complete his legal education and carry on for both. "You must come home in time for school this fall," Loy's father wrote. "You must be doubly as good a man as you had ever planned to be."[20]

This paternal admonition, combined with the impact of Roy's death—the latter enormously intensified by Loy's dream of his farewell visit—placed a great burden on the younger twin's shoulders. Already he inclined to the conviction that a career should be a calling. Now his father was using Roy's death to make the point again. Moreover, Loy had joined the ARC to prove himself as good as Roy. Roy had died in the service of his country. This set an impossibly high standard for Loy to meet; one can never overtake a ghost. But Loy would spend much of the rest of his years in the attempt.

The eventual result was a near-obsession with duty, a trait that more than any other would characterize Henderson's approach to life and work. In certain respects his devotion to duty would produce positive results. Much of the success Henderson achieved owed to his ability to concentrate on the business at hand, and his insistence on following duty's call inspired an integrity that became almost legendary among his colleagues. Yet it would exact a cost, for his concentration tended to narrow his field of vision, and his belief that he was answering a higher call discouraged the reflection that allows a person to learn from criticism.

IV

Henderson had gone to Europe to catch up with Roy. Now, although Roy was dead, he would return to America for the same reason. But more than a personal feeling of obligation moved him to resign his commission. Even as Roy lay dying and Loy nearly so, the American Senate pondered the fate of the Versailles treaty. For a variety of reasons—the arrogance of the British and French, personal animosity between Wilson and the Senate's Republican leadership, Wilson's inability to compromise, American moral fatigue—the Senate rejected the treaty. To Henderson the rejection seemed to signal an American turning away from Europe. Coming as it did simultaneously with the news of Roy's death, it appeared strong evidence that the proper place for him was back in the United States. With his father adding that he must get home and on with his life, the Red Cross lost its attractiveness. Consequently Henderson laid plans to return to America by the end of the summer of 1920. When Ryan asked him if he would accept a transfer to the ARC office in Berlin, he agreed on condition that he be relieved by the beginning of September.

Things did not happen quite as Henderson planned. After four mildly eventful months operating what amounted to a liaison office between the ARC and the German government, and later running a transportation depot for the Red Cross, Henderson received a request from Ryan to postpone his departure for several months more. Conditions in Germany remained unsettled, with general strikes occurring on a regular basis. The situation in the Baltic, in Poland, and in Russia was more jumbled than ever. Under the circumstances the ARC might soon need someone with Henderson's experience.

The feeling of adventure and significance that had sustained Henderson earlier had vanished. "I am very much bored with Berlin and am very anxious to get home," he wrote in August. "I feel that every minute longer spent in Europe is a wasted one for me." The esprit de corps of earlier days was disappearing as well. "The old Commission is practically disbanded," he remarked. "Practically all the older men are gone and only three or four of the originals are left." Even so, he succumbed to Ryan's appeal and stayed on, commenting, "It seems that it is a good deal harder to get away from Europe than it is to get here."[21]

Part of Henderson's inertia resulted from an inability to decide exactly what he wanted to go home to. At twenty-eight he felt rather old to return to law school, for by the time he finished school and established a practice he would be in his middle thirties. In addition, a principal motive for pursuing a legal career had been the desire to share an office with Roy. That option no longer existed, and its passing left a bitter taste. His inclination to public service and his notion of career as calling made him look down on private enterprise. The past two years had fitted him for an international profession, but with the country apparently rejecting the world, this prospect appeared clouded. Besides, diplomacy—the obvious path for one desiring to serve his country abroad—paid so poorly that individuals without private means could scarcely survive and certainly could not prosper. Members of the less prestigious consular corps, at that time a separate branch of the foreign service, received salaries above subsistence level, but entrance required passing a stiff exam that in turn demanded a course of intense study. Time would slip by while he prepared.

Henderson puzzled over these matters in Berlin, becoming more restless by the day. His work lapsed into tedium. "My conscience has been telling me for a year and a half," he wrote in November 1920, "that in a business way we are gaining nothing in prolonging our stay in Europe." A few months later he added, "I think I have the softest job in the commission, in fact, I think the softest job of the Red Cross in Europe. . . . I cannot afford to waste my time much longer here." Nor did life in Germany's great city afford much diversion. "Berlin is quite dead," he wrote in the summer of 1921.[22]

Occasionally he escaped the monotony. In November 1920 he traveled to the Netherlands to purchase delousing equipment. The trip provided an opportunity to see a part of the continent he had not visited before—and to eat again without feeling guilty. Continued rationing in the east made Henderson appreciate the simple fact of bread on the table. On one occasion a representative of an American commodities trading firm offered him a chance to make use of his connections in the relief field to sell American rice to the Europeans. The compensation offered caused him to consider the proposal, but only for a moment, since it would involve a potential conflict of interest.[23]

Henderson apparently formed at least one romantic attachment during this period—or at least an attachment was formed to him. Henderson's papers include almost nothing of letters of the heart, even to the woman he later married. But the file for Berlin in 1921 contains a photograph of an attractive young woman, with a note: "Dear Mr. Henderson: I send you a little reminiscence of a bad girl. If you have the wish to see me once more, I shall be very glad if you would call on me one evening."[24]

Whether Henderson accepted the invitation is unclear; but his feelings toward the young lady did not lessen his determination to get out of Europe and back to the United States as soon as conveniently possible. The several months Ryan had requested he stay on stretched into the summer of 1921, at which point Henderson made it quite plain he intended to leave. Ryan this time did not try to keep him, suggesting only that he take some vacation before rushing to the States to start a career. In later years Henderson usually skipped vacations, not infrequently driving himself to the point of exhaustion. But now he followed Ryan's advice and detoured south to the German Alps with some acquaintances from Berlin. After two weeks he departed for France, where he caught a steamship for New York.

2

Initiation

Henderson's ship from Cherbourg docked in New York on Labor Day, 1921. In some respects the voyage home had continued Henderson's holiday, for after nearly three years of wearing a uniform and subsisting on ration cards he now got to live like a tourist. The ship included plenty of the latter. For the greater part of a decade, war and revolution had isolated much of Europe from American sightseers and other casual travelers; their pent-up wanderlust burst forth in the early 1920s. With the end of the summer season of 1921, thousands were returning to the United States, crowding Henderson's ship and injecting into the air a frivolity he had not seen for years.

Most of those who knew Henderson in later life agreed that his gifts did not include a lively sense of humor. ("Subtle" was how the more generous among them would describe it.) To some degree Henderson's seriousness reflected the intensity with which he pursued his calling. But it also evinced an unease in crowds, an inability to share popular emotions. The first rumblings of the roaring twenties, which Henderson was witnessing on his return to America, left him cold. "In my opinion," he wrote afterward, "the irresponsibility and grossness in behavior of this era, as well as the lack of consideration for others that its exponents displayed, left an imprint on American life that remained for decades."

In this case the uncertainty hanging over his future added to Henderson's lack of appreciation for the mood of the era. The "irresponsibility" he cited could not have contradicted more completely his notion of duty; at the same time, since he had not found a career or discovered his duty, the irresponsibility he decried in others underlined his own failure to find a meaning in life. The voyage home had afforded a measure of relief. "I had been able," he recalled afterward, "to thrust aside oppressive thoughts about my uncertain future." But upon arrival in the United States these thoughts again crowded his mind. A trip to Colorado to see his father and visit Roy's grave helped some. He spent a month in the west, climbing, fishing, riding horses, and accompanying his father on rounds of churches the minister superintended. By the middle of October,

17

however, the need to set his course in life forced an end to vacationing, and he went back to New York.

At a loss where to start, and discouraged about the feasibility of a career in public service, he visited various business firms engaged in international operations. He spoke to personnel directors at banks, oil companies, shipping lines, and import-export houses. The experience disheartened him further. New York he considered brusque and crass. The city, he wrote, was "more foreign to me than London, Paris, or Berlin had been. The people in the restaurants and on the subway or elevated trains, and those who jostled me in the streets or pushed in front of me in the shops, seemed to have little in common with me." He realized he was not seeking out the worthwhile things New York had to offer, but he was in a hurry. "I had neither the time nor the desire to search for glamor or culture. My first task was to establish myself, to lay the basis for my future life." Interviews did not go well. No one offered him any of the few positions he found relatively appealing, and those who did express an interest did not have what he wanted. "I was determined not to embark on a career to which I could not wholeheartedly devote my life."[1]

Partly from his nagging sense of duty and partly by process of elimination, he arrived at the idea that he ought to try the American foreign service. In Germany he had spoken with Evan Young, recently appointed commissioner to the Baltic states, who had encouraged him to consider a career in the consular corps. At that time contrary advice from other American officials complaining about low pay had deterred him, but as he cast about New York, confronting the prospect of no pay at all, he decided to give Young's suggestion a try. At the beginning of November he wrote to Washington requesting information. Shortly thereafter he received a reply from the director of the consular service, Wilbur Carr, describing openings and informing him that the State Department would next examine candidates in January.[2]

Henderson crammed for two months in German and other subjects he could expect to find on the test. On January 15, 1922, he boarded a train at Penn Station for Washington. The next morning found him at the Civil Service Commission with more than a hundred other hopefuls. For two days the group sat for the written exams, which were followed by interviews. In Henderson's case the interviewers gave him a chance to describe his experiences in Eastern Europe, in addition to asking him more prosaic questions about United States trade with Latin America and the great irrigation projects of the twentieth century.

Several weeks later Henderson received notice that he had scored eighty-six, making him eligible for appointment when an opening appeared. During the next few months the consular service kept him posted regarding his prospects; finally in July he learned of his commission as "vice-consul de carrière, class III," at an annual salary of $2500, and in August he was told to report to Washington in two weeks for training. In a letter accepting his commission Henderson wrote Carr, "I am very grateful for thus having the door to a career in the Consular Service opened to me and am entering the Service determined to do my utmost to prove worthy of it and to uphold its best traditions." He quit a temporary job he had taken, submitting his letter of resignation over the signature "Loy W. Henderson, Vice Consul of the United States."[3]

During a brief orientation at the State Department, Henderson learned the rudiments of a consulate's operation. Asked about his geographical preferences, he requested assignment to a post where he could improve his German or learn French; he also indicated that eventually he would like to return to Eastern Europe. The department sent him to Ireland.[4]

II

A person less devoted to duty might have found Henderson's introduction to consular work dismaying. After a September voyage from New York, during which he struck up friendships with three other freshly minted vice consuls bound for various parts of Europe, Henderson arrived at Cobh in a gray drizzle. From there a coasting packet took him to Dublin.

The natives were friendly enough: one old man, not quite catching his name and rank, greeted him warmly: "It's right glad we are, Mr. Hennessey, to welcome back to Ireland a son of the old sod as the American Consul to Dublin. Won't you come down and have a drop with us?" But the physical condition of the consulate dampened his hopes, bespeaking both the general tight-fistedness of the Republican administration in Washington and the lack of esteem congressional appropriations committees displayed toward the foreign service. For want of funds the consul, Charles Hathaway, had engaged offices on the second floor of a dingy building in one of the worst districts of Dublin. The dreariness of an early winter added to the gloom.

Somewhat apologetically Hathaway asked Henderson if he would mind beginning his work with a job that should have been done long before. The attic of the building contained piles of boxes filled with a century's worth of records. For some time the State Department had been asking him to sort and pack these records and ship them off to Washington for deposit in the archives, but between one thing and another he had never quite got around to it. With Henderson's addition to the staff, it now seemed possible to accomplish the task. Would he accept responsibility?

Of course Henderson had no choice. In the event, after digging through the decades of dust that covered the records, he found the duty, if not quite enjoyable, at least worthwhile. For two weeks he disappeared into the attic early each morning and emerged late each afternoon. In attempting to make sense of the mass of documents, he learned a great deal more than he ever had expected to about the effect of the potato blight of the 1840s on immigration to the United States and about the problems of British neutrality during the American Civil War. But he also absorbed considerable knowledge about the day-to-day affairs of a consulate.

Henderson so pleased Hathaway with his work on the records that when the vice consul finally reappeared from the attic, Hathaway announced that he had another equally important task, also unfortunately neglected. At the beginning of each administrative year departmental regulations required every consulate to order the official forms it would need during the coming twelve months. Although deficient in other resources, the consular service did not lack forms: in 1922 more than fifty varieties circulated. From lack of staffing, Hathaway explained, he had for the past few years been unable to comply with Washington's policy, instead borrowing forms from neighboring offices on a catch-as-catch-can basis. Since Henderson had such a flair for untangling messy business, would he take charge of setting the matter right?

Henderson again had no alternative, but once more he made the best of the situation. The forms covered all conceivable, and a few nearly inconceivable, functions of a consulate. In sorting the forms out and investigating how many of each the consulate had used in recent years, Henderson added to his insight into the kinds of things American consulates did. For better or worse, he gained a reputation among his colleagues as the resident expert on paperwork and a budding master of administration.

Fortunately for Henderson's sanity, Hathaway ran out of such chores, and his job grew more interesting. During the autumn of 1922 Irish republicans were in the process of writing a constitution and otherwise launching Eire's ship of state, recently unmoored from the United Kingdom. Dissidents rejected the cargo as deficient, since it did not include Ulster, and they were making their complaints by force of arms. By itself the struggle would not have involved the American consulate, beyond the naturally unsettling effects of occasional bombings and more-frequent gun battles. But American nationals of Irish descent had enmeshed themselves in the struggle. Money from America funded the rebels; American weapons armed them. More than a few Irish-Americans had returned to take up the fight directly.

While American officials could overlook the cash and contraband, leaving it to the Irish to stop the illicit traffic, they could not ignore the American guerrillas. In the midst of a war for its existence, the Irish government exercised less care for the legal rights of those caught with guns than the American constitution specified. In certain cases summary execution awaited those captured armed. By the time Henderson arrived in September 1922, Free State forces had not handled any American citizens with this full measure of vigor, at least not to the consulate's knowledge. Official forbearance, however, appeared to be thinning. For obvious political and military reasons the Dublin authorities could not allow themselves to be perceived as treating Americans more gently than they treated their own citizens. On the other hand, to countenance the execution of Americans without something approaching due process would jeopardize amicable U.S.–Irish relations, which Dublin and Washington both valued. In an effort to remedy the problem the Irish foreign office and Consul Hathaway had worked out an arrangement whereby the government simply detained without trial Americans captured in possession of arms. Needless to say, both parties kept the agreement secret.

Henderson soon landed squarely in the middle of this covert contract. Appointed by Hathaway to handle various general duties around the consulate, Henderson inherited responsibility for looking after the interests of American citizens in Ireland. When Americans got themselves arrested, as many did, friends and relatives inquired by letter and in person regarding what Henderson proposed to do to ensure that their loved ones received just treatment. Unable to reveal the consulate's arrangement with the Irish authorities, Henderson fielded the queries as best he could. He generally replied that although possession of weapons was a capital offense, no Americans had been executed to date and he saw no reason to expect the situation to change. He also suggested, without quite stating explicitly, that once the present unrest had ended, the Irish government would decide simply to deport all foreign nationals and forget about prosecuting. In the meantime he could only counsel patience.

By a peculiar twist of events, Henderson had occasion to meet one of the leaders of the rebels. A few years before the founding of the Free State, Irish nationalists had purchased bonds worth several million dollars in America. The bonds had been deposited in a bank in New York and were redeemable on receipt of an order signed by three of the nationalist leaders, of whom one was Eamon de Valera. A subsequent split in the nationalist movement had led to a falling out among the three co-signers. One joined the Free State government and led an attempt to redeem the bonds; de Valera, in a Free State jail, refused to sign. When the would-be redeemers sued in an American court, the judge directed the consulate in Dublin to take a deposition from de Valera.

As much from curiosity as for any other reason, Henderson tagged along when the senior vice consul, Harold Collins, went to record de Valera's testimony. Henderson found the nationalist leader an impressive individual. "There was nothing about de Valera's manner or well-groomed appearance," he wrote later, "that suggested he was being confined in a jail. He was composed and gave his answers without hesitation. After the testimony had been taken he shook hands with us and departed as though he had important business elsewhere."[5]

Most of Henderson's work was more mundane. At the beginning of 1923 he took charge of visa work in the consulate. From a slow start in the late winter of that year, business picked up when summer and settled sailing weather approached. As the first contacts most prospective immigrants to the United States had with the American government, Henderson and his associates bore a particular burden for separating those who would get in from those who would not. To Henderson fell the duty of determining whether applicants appeared to meet health requirements for admission, of verifying that they passed the literacy test Congress had mandated in 1917, and of recording applications and collecting fees. These duties filled most of his time, up to twelve hours at a stretch. On many days he issued more than one hundred visas.

The hardest part of the job was turning people away, but at first this did not happen often and did not entail inordinate anguish. Things changed following a 1924 reform of American immigration policy. For the first time Congress established ceilings on immigration, assigning quotas to each country. Implementation created an additional burden on American consulates overseas. Thousands of inquiries regarding provisions of the law flooded Henderson's office. Lacking staff and often information, Henderson could only respond with his best guesses. On one occasion he took it upon himself to interpret the literacy clause flexibly. A fair young lass, failing to decipher the text Henderson handed her, objected, "And faith, what do you expect? I have been out of school already more than six years." Henderson approved her application.[6]

For all the demands of the job, Henderson was beginning to settle into his existence as a vice consul, and despite the ongoing guerrilla war, which involved a tight curfew and some close brushes with violence, he found life in Dublin reasonably pleasant. Henderson did not socialize much; when opportunity presented itself he found relaxation in the Irish countryside. He would ride a train into the hinterland, where he would spend a day tramping the hills. But then, just as he was establishing a routine, he received a transfer to Cobh.

III

In Dublin the work of the consulate, aside from day-to-day matters of visas and the like, had reflected the inclinations of Hathaway, who preferred politics to economic and commercial matters. In Cobh, Henderson's new chief, John Gamon, enjoyed the traditional aspects of a consul's work. Consequently Henderson now spent more time investigating and writing about business conditions and opportunities. He forecast trends in the Irish economy and corresponded with American firms interested in commencing or expanding commercial relationships with Ireland.

Had Gamon not leaned toward economic affairs on his own, pressure from Washington at this time would have encouraged such an emphasis, for under the spur of competition from the Commerce Department and its forceful secretary, Herbert Hoover, the State Department was demanding an increase in economic reporting from American consulates around the world. Such activity ran against the grain of many of the diplomats. Even Henderson, who at this stage in his career detected silver linings in the most unlikely clouds, objected, albeit silently. Flogging American products hardly appealed to his sense of pursuing a higher calling.

The emphasis on economics, however, had a fortunate effect on Henderson's career chances. In the 1924 Rogers Act, Congress decided that American business opportunities abroad would best be served by combining the heretofore separate diplomatic and consular corps into a unified foreign service.

Henderson personally applauded the Rogers measure, but the first result he felt was a cut in pay. Prior to the passage of the Rogers statute, consuls, who were not expected to possess independent means, received higher salaries than diplomats—up to twice as much. In adjusting the scales for a unified service, the State Department tried for the most part to raise the diplomatic schedule. In some cases, though, the consuls lost ground. Additionally, acceptance of the notion that the foreign service might be a career led to the establishment of a retirement fund, financed in part by paycheck deductions. The two changes combined to leave Henderson short at the end of each month.

More discouraging than this decrease in buying power was the fact that in the amalgamation of the two services the diplomats gained preference on the ladder of seniority. Already old for his rank, Henderson now found himself classed beneath individuals several years his junior who had entered the diplomatic corps just prior to the passage of the Rogers Act.

On balance, however, he considered consolidation a stroke of good fortune. Whatever might be said regarding the significance of a consul's duties, they lacked the sophistication and intrinsic appeal of the diplomat's political work. Although Henderson had preferred the diplomatic service all along, he had chosen to become a consul because he knew he could not make a living on a diplomat's pay. Now a diplomatic career—one he could really devote his life to—opened before him.

IV

In September 1924 Henderson became eligible for home leave, having served abroad for two years. The following month he sailed for America, and after a brief stop in New York he traveled to Washington for consultation at the State Department. During a routine visit to the office of Assistant Secretary Wilbur Carr, he bumped into Evan Young, the commissioner to the Baltic states who had encouraged him to apply to the foreign service. Young by this time had advanced to the position of chief of the division of Eastern European affairs. He recalled Henderson's experience in the area and asked if a transfer would appeal to him. Henderson, not having become so immersed in the affairs of Ireland as to forget what had prompted his interest in the foreign service originally, replied that it certainly would. After a brief visit to Colorado he returned to Washington for a tour in the department.[7]

On his first day in the Eastern Europe office Henderson met Robert Kelley, Young's new assistant. For the next year and a half Henderson would continue to work for Young; after Young departed for the Dominican Republic in 1925, Henderson would report to Kelley. But even in the earlier period Henderson's outlook and style began to bear the Kelley impress. In the dozen years from 1925 to 1937 Kelley ran the Eastern Europe division. During that period—and in the decades beyond, when Henderson, George Kennan, Charles Bohlen, and others trained in the Kelley school of diplomacy moved into positions of responsibility—Kelley's stamp indelibly marked American foreign policy.

For Henderson, Kelley served as a role model and a bridge between Henderson's world and that of the department's gentry. "I had a tremendous admiration for him as a man of integrity, a scholar, and a diplomat," Henderson later remarked. Just one year older than Henderson, Kelley, like Henderson, had attended public schools rather than the eastern prep schools favored by the old-style diplomats and their younger protégés. Kelley had won a scholarship to Harvard, where he flourished academically; from Cambridge he had gone to Paris and the Sorbonne, specializing in Russian studies. The world war had interrupted Kelley's education, as it had Henderson's. The army sent him to Eastern Europe at the time Henderson was there, although the two did not meet. Kelley, like Henderson, had encountered Young, who convinced him to try for the consular service. After success on the entrance examination and a brief stint in India, Kelley joined Young on the Eastern Europe desk of the State Department.[8]

Kelley's passion was Russia; for the Soviet Union he had little use. When he began grooming his cadre of Russian specialists he constantly emphasized the importance of a firm grounding in the basics of Russian language and culture, sometimes to the neglect of Soviet affairs. At one point George Kennan, then studying Russian in Berlin, wrote to the department requesting permission to enroll in what were probably the best courses in the world on Soviet politics and economy. Kelley told Kennan to stick to his Russian.[9]

Kelley's character and intellect so influenced those who studied under him that almost without exception they adopted his belief that the leaders of the Soviet Union had usurped the Russian legacy. Kelley was not alone in this view, of course. His anti-Soviet opinions were reinforced by the Russian émigrés who staffed many of the universities and institutes where American diplomats studied. From working with these individuals Kelley, Henderson, Kennan, and Bohlen developed considerable sympathy for Russia's ancien régime. Bohlen summarized the feeling when he commented that "the Russian intelligentsia before the Revolution constituted one of the most remarkable collections of men that the world has seen." Concomitantly, the Americans inherited a distaste for what Kelley called the "Boles." Despite this distaste, and the better to know the enemy, Kelley began to collect data on the Soviet government, until by the late 1920s the Eastern Europe division and its extension office in Riga boasted the world's most comprehensive research collection on the Soviet Union. At one point, in fact, Soviet foreign minister Maxim Litvinov commented that Kelley possessed better records regarding Soviet foreign policy than did the Kremlin itself.[10]

Kelley's study convinced him that the Soviet Union could not make a fit partner for diplomatic intercourse with the United States. In most matters Kelley enjoyed a reputation for detachment and analysis of issues on their merits. Samuel Harper, a Russian specialist at the University of Chicago, a diplomatic part-timer, and later a frequent correspondent of Henderson's, described Kelley as "too legalistic." But on issues pertaining to the Soviet Union, an emotionalism—critics called it paranoia—supplemented the logical arguments he leveled against any accommodation of the Bolsheviks, and he found it difficult to credit the honesty of those who disagreed. In 1925 he wrote to Harper, regarding statements by President Coolidge on the administration's continuing non-recognition of Moscow: "I trust you noted the two interviews from the White House re our Russian policy. They could hardly have been more to the point—yet certain *dark forces* persisted in misconstruing them."[11]

Kelley devoted his energies to frustrating these "dark forces." He and his office marshaled arguments against recognition. Under Wilson the case against recognition had rested on two contentions: that the Bolshevik regime did not speak for the Russian people, and that as self-avowed world revolutionaries the Bolsheviks could not engage in civilized diplomatic relations. Following the Bolsheviks' victory in the Russian civil war and their consolidation of authority, the anti-recognitionists largely dropped the first contention. But they held tightly to the second. Kelley summarized this aspect of the argument in a letter responding to efforts by business groups to encourage recognition as a means of increasing trade and investment opportunities. Supporters of recognition, Kelley asserted, failed to understand the basic facts of the matter:

> (1) that the present regime in Russia is a regime dedicated to the promotion of "the world revolution," that is, the extension throughout the world of the political, economic, and social system set up in Russia in 1917;
> (2) that the fundamental purpose of the Bolshevik leaders precludes the observance by them of obligations ordinarily accepted as governing relations between nations; and
> (3) that, consequently, relations on a basis usual between friendly nations cannot be established with the present regime in Russia until its leaders have abandoned their world revolutionary aims and practices.[12]

Kelley's anti-Sovietism permeated the State Department when Henderson, now a "Foreign Service Officer, Unclassified," at an annual salary of $3000, took his desk on the third floor at the beginning of 1925. State's physical setting reflected the more deliberate pace of America's diplomatic life as it had developed in the nineteenth century. The department, previously housed in what had been the Washington City Orphan Asylum, occupied the rear portion of the State-War-Navy building on the block just west of the White House. Although the building's Gilded Age architecture appealed to some—whose number would increase as the years passed—many people considered it the ugliest and most pretentious structure in Washington. One pundit labeled it a fine specimen of "American ironic." A half-century's weather had had little effect on the elaborate stonework outside, but foot traffic within had worn depressions in the black and white marble tiles of the halls. In Washington's subtropical climate, heating came more easily than cooling, as evidenced by the high ceilings and the swinging saloon doors that allowed air to circulate into the offices. Scores of electric fans were more noticeable for their narcotic hum than for their ventilating effect.[13]

Except for brief visits, Henderson had never spent time in Washington, and he found the capital in the days of Coolidge interesting and more congenial than New York—a fact that owed at least as much to the more settled and directed circumstances of Henderson's life as to Washington's intrinsic merits. The taciturn Coolidge provided a certain entertainment. Henderson, low in rank, occupied an office facing an inner courtyard, but he often found himself in rooms overlooking the White House rose garden when visitors gathered for photographs with Coolidge. The president, of course, did not know the vast majority of people who wanted to be pictured with him, and he had no facility for small talk. Photographers would compose their portraits, leaving an empty space for Coolidge. When everyone was in place and the equipment ready, a White House attendant would signal the president, who would step out of his office and into the vacancy. In little more time than it took for the camera to click, and with not a word said, Silent Cal was back at his desk.[14]

To Henderson's gratification, the introduction the young diplomat received to the work of the department proved considerably more challenging than his initiation to consulate life in Ireland had been. Almost immediately Kelley set him to a systematic course of study of Eastern Europe and American policy toward the region. With customary organization and assiduousness, Henderson plowed through a list of books Kelley recommended. At the same time he pored over the division's files of telegrams, memoranda, and press releases concerning Russia and the neighboring countries.

Working under Kelley, Henderson realized he had found his calling. It was about time, since he would soon turn thirty-three. Henderson's duties in Ireland, however useful as training, were hardly what he had envisioned when he joined the foreign service. Now he was grappling with issues of real importance. Henderson was not an overtly moralistic type; many times in his later career he would go out of his way to avoid the appearance of moralizing on professional matters. But in dealing with the Soviet Union, whose policies so completely contravened what most Americans—Henderson included—deemed basic human values, he could not help feeling he was confronting fundamental issues of good and evil. This feeling would grow during succeeding years. Henderson doubtless would have prospered in the foreign service had chance, during this formative period, thrown him into another area of policy—Latin America, perhaps, or East Asia. But his outlook would have been different, and his views on America's proper role in the world almost certainly would never have acquired the edge they did.

The pupil shortly began to reflect the views of the mentor. Kelley assigned Henderson the duty of tracking Soviet connections to leftist organizations in the United States. An essential part of the anti-recognition case was the contention that Soviet leaders pulled the strings of the Communist International, or Comintern, and that directly or through the Comintern the Kremlin also manipulated such affiliated groups as the American Workers' Party, the Trade Union Educational League, and the Red International of Trade Unions. Henderson, utilizing research by American law-enforcement agencies, by the intelligence offices of the army and navy, and most importantly by the American legation at Riga, did not take long to agree with Kelley that these organizations were simply fronts for the Kremlin's subversive activities against the west. Moscow might disclaim responsibility, arguing that to the degree these groups followed the communist line they bowed only to the coercion of logic and experience. But Henderson soon convinced himself that the pipers in Red Square played the tune all the radicals danced to.[15]

3

Recognizing Reality

In the spring of 1927 Henderson received word of a possible transfer and promotion. Kelley told him of an opening at the American legation in Riga and asked if he would be interested. Henderson was. All along he had wanted to do diplomatic work overseas, but so far he had drawn only consulates and home duty in the department. The legation in Riga looked good. Moreover, the Baltic was territory he knew and appreciated. (That territory appreciated him in return: in 1924 the government of Estonia had decorated him for his Red Cross work.) Finally, Riga afforded an opportunity to study Russian. Despite his own and Kelley's continued opposition to recognition, it appeared unlikely that Washington would forever shun Moscow. Sooner or later the United States would need people to represent American interests within the Soviet Union. Henderson hoped to be ready when the chance arose.[1]

Arriving in Riga in the middle of September, Henderson was struck by the changes in the aspect of the city since his last visit in 1920. Then it had been struggling to gain its feet after a world war, even as revolution and civil conflict threatened to engulf it. Whether and how it would survive, no one could say. Now it was the capital of the Republic of Latvia, hiding its scars well and wearing its independence proudly, determined not merely to survive but to prosper. In 1927, Riga enjoyed a reputation as a cosmopolitan crossroads—the Paris of the Baltic, some called it—where on Henderson's daily walks he could hear Lettish, Yiddish, German, and Russian spoken; where he could buy newspapers in any of several languages; and where he could attend religious services conducted for vigorous Lutheran, Russian Orthodox, Roman Catholic, and Jewish communities. But stimulating as this variegation seemed to visitors, it was not what the predominant Letts had in mind, and in their nation-building enthusiasm they were attempting to impose a measure of cultural uniformity on the city. So far they had not conspicuously succeeded.[2]

Nor had they managed to overcome the housing and office shortage induced by the years of turmoil. Available buildings were few and dear. The American legation, entering a seller's market with the added handicap of congressional parsimony, was forced to make do with accommodations considerably less than satisfactory. The chancery, or business office of the legation, occupied what had been the house of a wealthy Latvian merchant. The merchant had fallen on hard times before he died; at present his widow, unable to afford repairs, was leasing it for whatever price it could command. Because it could not command much, the Americans got it.

But Henderson had encountered worse, and the depressing surroundings did little to diminish his excitement at beginning what seemed his real career. Henderson turned thirty-five just before leaving the United States; by anyone's standards he was getting a late start. Better late than not at all, however, and it appeared the best was yet to come. The cable from Undersecretary Grew assigning Henderson to Riga contained the explicit recommendation that he be detailed to the Russian section of the legation—the fast track for Soviet specialists. Grew remarked that the department had in mind "the desire of Mr. Henderson to prepare himself for possible future service in Russia"—this, six years before recognition—"and accordingly it expects insofar as it is practical that he will be afforded the opportunity to equip himself in this regard, especially to increase his knowledge of the Russian language." With such instructions from on high, Henderson had reason to feel excited.[3]

His great hopes proved premature, though, and he discovered that specialized training in Russian would have to wait. His administrative reputation had preceded him, and the American minister in Riga, Frederic Coleman, informed him that his skills were most needed in improving the legation's efficiency. As third secretary he would take charge of streamlining its operations. When he learned the ropes of the mission he would also assume responsibility for requesting additional funds and personnel from the State Department. As time permitted he might do some political reporting on conditions and developments in Latvia. For now he would not join the Russian section, but if after he completed his other tasks the opportunity arose to report on the Soviet Union, that would be fine. As for studying Russian, he would have to pick it up on his own.

Henderson had sufficient experience of the bureaucracy not to be crushed by this news, but it was disappointing nonetheless. He could have pushed paper and drawn organizational charts in Washington or Dublin, or Timbuktu for that matter; he had come to Riga to prepare for Moscow. And now he was not even being assigned to the Russian section. Once more, it seemed, his career was on hold. But there was nothing to do except make the best of the situation. Even had he been the type to object, he could see that third secretaries' complaints would not get them far.

So he set to his job with a will, and repeated the mistake he had made in Ireland. He demonstrated such capacity for administration that for the next several months he did little else. The 1920s were lean years in the State Department, as in much of the federal establishment. Convinced that the department had become bloated and profligate during the war, the Republican administrations and their supporters in Congress decided to reduce expenditures on diplomacy. In just three years, starting in 1923, outlays for the State Department and the foreign service fell by nearly 40 percent. Meanwhile the department's work was increasing, notwithstanding the rejection of the Versailles treaty.

The cost-cutting produced the situation Henderson confronted in Riga, of having to postpone specialized training in order to keep up with daily affairs. The sad condition of the legation provided further evidence of congressional budgetary priorities. So tight had budgets become, in fact, that Henderson discovered the Riga legation's economy squad recycling worn-out carbon paper by holding the used sheets over candles so the heat would melt and redistribute the blacking. Recalling that in Washington he had nearly gone blind trying to read faint cables from Riga, he decided that this practice had to end.

Henderson effected minor efficiencies by defining more precisely the duties of each officer and staff member at the legation and by laying off local employees who were pulling less than their weight; but he despaired of any serious improvements so long as the chancery remained in its inadequate and deteriorating quarters. At the beginning of March 1928, on a walk about town, he passed a new brick building under construction. He inquired whether the building had yet been leased. On learning it had not, he asked whether the owner might like to have the United States for a tenant. Indeed the owner might, he discovered. In Henderson's opinion the building represented an enormous improvement over present accommodations, in terms not only of morale but of efficiency. To document how the new location would facilitate the legation's work, he drew up a floor plan, which he submitted to Washington along with his recommendation that the department request funds for the move. The Coolidge White House approved the request, and Henderson and Riga got their new chancery.

Amid his administrative work, Henderson shoehorned in some political reporting. At first he concentrated on compiling dispatches explicating internal developments in Latvia. The country had calmed considerably since 1920, but its politics still reflected its ethnic and cultural antagonisms. As Henderson discovered, and as he explained to the State Department, local minorities, led by the Germans, had succeeded in forming a bloc in the parliament that often held the balance between two major parties. Chronic instability resulted, as the minorities repeatedly raised demands on their current allies until the latter balked, at which point the balance tipped to the opposing faction, often bringing down the government. In addition, although Latvian law prohibited the formation of a separate communist party, numerous Bolshevik sympathizers of undeniable energy and competence worked through political front groups. Like their comrades elsewhere in Europe, the Latvian communists aimed to keep local politics unsettled, in the hope of eventual revolution. Not surprisingly, considering the sensitive location of the Baltic states with respect to the Soviet Union, Moscow provided significant encouragement to their efforts.[4]

Interesting though the Latvian political scene was, Henderson still aspired to a transfer to the Russian section. To prepare himself he studied Russian on the side, but since his other tasks already occupied him overtime—and since in addition he served as local representative of the American Red Cross—he only succeeded in working himself sick. In the autumn of 1928 he began coughing up blood and showing other signs of strain. A visit to a doctor produced a preliminary diagnosis of a "catarrhal condition" of the throat and lungs. The physician feared something far more serious: tuberculosis. Taking no chances, he ordered Henderson to quit work and spend at least three months in a "dry, bracing climate," preferably in the mountains where the sun would be strong. The approaching winter on the Baltic promised to be bracing, but hardly dry or sunny. Consequently, but with great reluctance, Henderson requested permission to go to Switzerland.[5]

In recommending approval of Henderson's request for sick leave, Louis Sussdorff, the legation's counselor, filed a glowing account of the third secretary's "unusually efficient service." Sussdorf stated of Henderson, "He is an extraordinarily conscientious officer and has never spared himself in the service of the Department. His present condition is due to several years of overwork, culminating in an attempt to master the Russian language outside of office hours." Convinced, Washington approved the request, and Henderson headed south.[6]

Despite Sussdorf's warm send-off, Henderson viewed the future with foreboding and a sense of defeat.

> It seemed that my whole life structure was collapsing. Was it possible, after my family had sacrificed so much to assist my twin brother and me to obtain an education, that he should die in early manhood and that now I should also fall by the wayside? I was full of shame at the thought of leaving the Legation just when I was beginning to feel really useful. . . . I felt that I was not only letting down the Minister and my colleagues in the Legation but also those members of the Department of State who had had enough confidence in me to appoint me as a specialist in Eastern European affairs.

The next six months—which the original three stretched into when the doctors could not decide whether Henderson had tuberculosis or not—were the worst of Henderson's life. A "nightmare" was how he described it. The regimen at the health resort of Davos only made things worse. Following several batteries of tests, he was directed to spend at least eight hours daily, regardless of weather, in a fur-lined bag on the balcony of his room. On mild afternoons he might walk about for a short time, but only in a manner that did not raise his respiration rate. He must think pleasantly innocuous thoughts and must not read anything requiring concentration. He could not mingle with other patients, lest he infect them or they him.

After several weeks Henderson began to acquire the habits of the sanatorium.

Many of the patients had a morbid fear of the lower altitudes. They felt that any air other than that to be breathed on the mountaintop was mortally dangerous. In spite of my efforts to be optimistic I found that I also was beginning to fear the outside world. I had the feeling, furthermore, that I was degenerating into a living two-legged carrot.[7]

The worst of the experience was the uncertainty, and the pall his illness cast over his future. If he had tuberculosis he might well die. If the disease did not kill him it would probably end his career, since the lingering effects would unfit him for work in out-of-way locations. Even if he escaped uninfected, he still was losing precious time. At his age and place on the career ladder, every month counted. Early in 1928 Kelley had named Henderson one of a small group of Eastern Europe specialists. Naturally the assignment had pleased him, but since the department's grapevine indicated that only younger officers would be given leave for language training, he wondered what his new designation would mean. His worry on the subject, and the unrelenting burden of legation duties, had led to his efforts to learn Russian on his own—and to his sickness. The loss of another half year threw him further behind and nearly into despair. Now that the question mark of poor health hung over his head, the department would be even more reluctant to invest time and money into preparing him for service in the Soviet Union.[8]

Finally, it appeared that if the tuberculosis bacilli did not finish him off, the cost of the therapy might. His salary from the State Department would cease sixty days into 1929; after that he would be living off savings, which did not amount to much. He calculated that when May arrived he would be bankrupt.

By the middle of April he had had enough. He was still alive, although the doctors were not willing to pronounce him fit to leave. One insisted, in fact, that if Henderson returned to Riga before summer, he would be going to his death. Henderson decided to take the chance. He packed his bag, headed down the mountain and caught a train for Latvia.

Immediately on his arrival there he began to run a fever. Not long afterward he developed tonsilitis. But then things turned for the better. His health gradually improved, assisted by a change in his duties. The State Department, deciding it could not afford to grind officers like Henderson into the ground, lightened his responsibilities. Henderson did not cooperate, and soon he again was working nights and weekends. He did, however, take advantage of a respite from socializing. He received orders, which he did not dispute, to decline invitations to diplomatic receptions for a period of six months. He appreciated the break; it gave him time to catch up on his work.

Although this round of illness did not produce the spectacular psychological effects—especially the dream of Roy's death—of his earlier bout with typhus, it was no less revealing of the individual he was coming to be. As happened repeatedly during his life, he worked himself to the point of exhaustion—in this case recklessly close even to death. Ambitious careerists in any number of fields push themselves hard. And Henderson certainly desired to climb the career ladder. But as Henderson's previous experiences and his comments in this instance make clear, in climbing the ladder he felt he was carrying the weight of his father and Roy, as well Kelley and others who had had confidence in him, in addition to his own. It was quite a burden. That he perceived his responsibilities thus, and that he was "full of shame" at becoming ill and letting all these people down, indicates the extraordinary degree to which he had come to identify his personal worth with his work.

At a number of critical periods in his life Henderson would find himself under great physical or psychological strain. So far the physical pressure had been the more obvious, taking the form of overt sickness. Later, the psychological element would predominate, as when he came under political attack in 1943 for his tough line against the Russians, and again in 1946–48 for his opposition to American support for a Jewish state in Palestine. In greater measure each time, he emerged from the period of trial with his sense of duty enhanced, with the perception that he had suffered for his work and for what he believed in. The testing forged a strong character and a ferocious integrity, but not much flexibility.

II

About the time Henderson returned to Riga, Herbert Hoover replaced Calvin Coolidge in the White House. In choosing Hoover as Coolidge's successor, neither the Republican party nor the American people showed great interest in straying from the path Warren Harding had described eight years before when he said, "We seek no part in directing the destinies of the Old World. We do not mean to be entangled." Harding and Coolidge had consistently avoided entanglements. In 1924 Coolidge gave informal approval to a scheme designed by American banker Charles Dawes to reschedule European war debts, and in 1927 his secretary of state Frank Kellogg negotiated a pact with French foreign minister Aristide Briand to outlaw war; but neither seriously altered the essential unilateralism that informed American diplomacy during the 1920s. As American elections usually do, the 1928 election hinged on domestic issues—in this case on the prosperity that appeared to have no end. Americans believed they could have a chicken in every pot and a car in every garage, or if they did not believe it they still considered a dry Protestant preferable to a wet Catholic.[9]

Even so, Hoover's inauguration brought hints of a more active foreign policy.

In the first place, Hoover's personal experience of international relations surpassed that of either of his Republican predecessors, and although he, like they, counted on the private sector to secure the blessings of prosperity to America, his work in the Commerce Department had revealed an understanding of the potentially facilitative role of government. As an early indication of this understanding—one particularly appreciated by Henderson and his colleagues—Hoover requested of Congress an increase in appropriations for the State Department. Justifying his request, the president declared, "I know of no expenditure of public money from which a greater economic and moral return can come to us than by assuring the most effective conduct of our foreign relations."[10]

In the second place, Hoover's appointment of Henry Stimson as secretary of state guaranteed energy and acumen at the top of the State Department. The beginning of 1929 found Stimson midway through one of the most distinguished careers in the history of American public service. A product of Yale and Harvard Law, Stimson had busted trusts and scourged railroads under Theodore Roosevelt, headed the War Department under Taft, mediated various civil and international conflicts in Latin America under Coolidge, and, most recently, been governor of the Philippines. Although Stimson's style of management caused some consternation in the cloak rooms of the State Department—as an unreconstructed progressive he emphasized efficiency in a way that shook up the old boys—on the whole the career men sensed a return of stature to their work.

In the third place, Hoover early approved a new debt-rescheduling plan. Like the Dawes package, the Owen Young plan adjusted in the direction of reality—which was to say, considerably downward—the amount of reparations Germany was called upon to pay Britain and France. As in the Dawes case, the American role was unofficial—yet crucial, since London and Paris linked their receipts from Germany to their own payments to American banks.

Unfortunately for the Young plan, as well as for Hoover's incipient internationalism, the American stock market crashed just months later. Wall Street's collapse triggered the greatest depression in the country's history, which was part of a global economic meltdown. Americans turned more inward than ever. Seeking to preserve the domestic market in the facing of falling consumption, Congress passed the Hawley-Smoot tariff in the spring of 1930, effectively closing the American market to hundreds of imports. Hoover declined to buck the tide of protectionism and signed the bill, despite the misgivings of a substantial segment of the business community and dire predictions from a group of one thousand economists forecasting grave damage to the fragile system of international commerce, debt, and reparations.[11]

For once the economists—some of whom had predicted indefinite prosperity as recently as September 1929—got it right. Europe interpreted Hawley-Smoot as "a declaration of war, an economic blockade," in the words of an observer of the scene in France. Retaliation followed. Hoover's 1931 announcement of a moratorium on debt payments failed to stem the destruction. As the world economy slipped over the edge, almost everyone defaulted on nearly everything. Of America's debtors only frugal Finland survived to service its loan.[12]

The collapse of the tower of notes eventually helped open the way for American recognition of the Soviet Union, for when all the other countries stopped paying debts, Moscow's failure to honor czarist obligations—a failure anti-recognitionists cited as evidence of untrustworthiness—lost significance. But for the time being, Henderson and the other Eastern Europeanists at Riga could only view Russia from afar. (Or occasionally from a bit closer: Chip Bohlen frequented a particular beach in Estonia that once had served as the chief summer resort for St. Petersburg. Among its attractions was nude bathing for women. "I never could get over"—despite repeated tries—"the novelty of being greeted by a bevy of naked Estonian beauties," Bohlen remembered. The beach also lay near the border between Estonia and the Soviet Union. Once when an American friend visited Bohlen they decided to approach the line. "We walked along a dusty road until we came to the Estonian frontier post. Standing there, we looked into Russia. It was my first glimpse of the soil of that country.")[13]

For Henderson, the Soviet Union came a little nearer in the spring of 1929 when Minister Coleman informed him that he was being transferred to the legation's Russian section. In good conscience, Coleman said, he could no longer load Henderson down with administrative tasks, and if he was going to be doing political analysis it might as well involve Russia.

After an exceedingly dark winter this was wonderful news, and Henderson determined to make good at the assignment. The chief of the Russian section, David Macgowan, possessed a reputation as a prodigious writer of dispatches on developments in the Soviet Union. A pioneer Kremlinologist, Macgowan pored over whatever information Moscow deigned to release to the outside world and tried to make sense of it for the State Department. Under Macgowan's direction, Henderson forgot nearly everything he knew about brevity (not much, since most of what he had learned about writing he acquired in law school). Within a short time Henderson too began producing memos by the pound. Absent diplomatic relations with Moscow, Washington was in no position to act on most of the information; the bulk of Henderson's masterpieces were filed and forgotten. But two caught the attention of Kelley and the Eastern Europe desk.[14]

The first involved the triumph of Stalin over his political rivals. In 1924 Lenin had died intestate, at least concerning the succession, and for the next few years the self-proclaiming heirs fought for the mantle. Stalin, a narrow party man without the visibility or intellectual reach of stars like Trotsky, was by no means the obvious choice, but one by one he outmaneuvered his challengers—first Trotsky, then Zinoviev, and finally Bukharin. By the beginning of 1930 Henderson was able to report that the succession struggle was over. Stalin had emerged the clear victor. The dictator's recent fiftieth birthday, Henderson wrote, "was treated as an event of national and even world-wide importance by the Soviet press. It is doubtful if Lenin in the height of his power and popularity was ever given so much acclaim and praise." Henderson quoted a writer in *Izvestia:* "The whole world is thinking of Stalin, his enemies with hatred, his friends with love." Henderson considered it significant that the Kremlin, whether from egotism on Stalin's part or political calculation, did not attempt to disguise the fact that Stalin, as secretary-general of the Communist party, was also the effective ruler of the Soviet Union and the controlling director of external front groups like the Comintern and the Red International of Labor Unions. In other words, the Kremlin was now conceding what Kelley and the Eastern Europeanists had been saying all along: that at this stage in its development the world communist movement was for all practical purposes a monolith.[15]

Henderson's account of Stalin's victory drew praise from Counselor Sussdorff. Sussdorff declared that during fifteen years with the foreign service he had "never been associated with any junior officer who has possessed greater ability and energy as a political and economic reporter than Mr. Henderson." The State Department agreed, issuing Henderson a special commendation for his work.[16]

Henderson's second big project of this period, a study of the collectivization of Soviet agriculture, earned similar reviews. Requested by Macgowan to keep an eye on trends in this all-important sector of the Russian economy, Henderson decided he could not speak knowledgeably of current developments without understanding their historical context. When he proposed a comprehensive examination of farm policy from the revolution, Macgowan agreed with enthusiasm, assigning several translators and typists to assist with the task. For two months Henderson sweated over hundreds of documents generated by the Kremlin and lesser authorities. The end product was an opus dispatched to Washington in five fat parts. The work was a classic example of bureaucratic overkill; almost certainly no one in Washington then or after waded through all the thousands of pages of text, appendixes, and annexes. But the translated documents on which the analysis was based did prove useful, and for years they served as a reference text in the department. The principal profit from the exercise, however, accrued to Henderson personally. Having immersed himself in Soviet affairs and policy, he now spoke with a confidence and authority he might have acquired in no other way.[17]

Fortified by the favorable reception his work was receiving, Henderson wrote to the department requesting an opportunity to devote himself full time to the study of Russian. Henderson was not facile in languages, and his efforts to learn

Russian on his own had proved unavailing, as he explained to Kelley. He requested assignment to London, where he could attend the School of Oriental Languages. Kelley thought Henderson deserved the opportunity, and the Eastern Europe chief forwarded the application with the recommendation that it be approved.

Henderson, in this early part of his career, suffered from chronic bad timing. His request came just months too late. Before the stock market crash the State Department had commanded funds to sponsor the kind of advanced training he proposed; had he applied then he most likely would have received the go-ahead. But the onset of the depression had crimped the department's budget severely. Consequently, as Kelley explained in a regretful reply, not all qualified officers would get the further education they deserved. Under the circumstances, the department was obliged to focus its limited resources on younger men from whom the payback would last longer. Perhaps when the Bohlens and Kennans completed their study, or if the economy improved, the department could work something out for Henderson.[18]

For all the softening Kelley supplied, the rejection hit Henderson hard. Henderson's experience in producing the memos on Stalin and collectivization had demonstrated that a person could do creditable work in Soviet studies using translators; but anyone could see that facility in Russian would confer an enormous advantage on those who possessed it. In the kindest manner possible, Kelley was telling him he was too old to make the necessary investment worth the department's while. Now thirty-eight, Henderson had to admit Kelley was right. He also had to confront the fact that he would probably never become a genuine expert on the Soviet Union.

Henderson had been in Riga for three years, not counting his six months in Switzerland. He was due for rotation. Partly, no doubt, to ease his disappointment, Kelley offered him assignment to Washington again. Henderson replied that he would be "very agreeable" to such a move. By now second secretary, having moved up in October 1929, Henderson could console himself that if he was not on his way to becoming a true Russianist, at least he was making career progress in a more general sense. Besides, having lost as much time as he had, he was in no position to turn down any promising offer.[19]

Another factor weighed in his decision. In the spring of 1930 he met a young Latvian woman at a charity ball. The initial exchange did not go well: he addressed her in German; she, taking him for a Prussian who refused to learn the local language, replied with disdain in Latvian. He hastened to correct her mistake, and things proceeded more smoothly.

Elise Marie Heinrichson was the daughter of a once well-off Latvian landowner. The world war had brought a decline in the family's fortunes. During the half-decade after 1914, troops of the czar, the kaiser and the Bolsheviks destroyed the Heinrichsons' property and occupied their house. On numerous occasions Elise, who was in elementary school when the war began, saw her playgrounds become battlefields. Once she witnessed a group of boys, suspected of sniping, executed on her doorstep. Her mother died of malnourishment and disease.

But Elise survived, and as she grew older the bad memories diminished. They never disappeared, and to a significant but unmeasurable degree they became part of Henderson's mental baggage. If the values of the czarist ancien régime rubbed off on Kelley's protégés, Elise's Baltic background also shaped Henderson's outlook. At the end of World War I he had witnessed the hard lot of the unfortunate peoples fated to live too close to the Soviet Union. Later he would observe at first hand Stalin's reign of terror in Moscow. His marriage to Elise contributed in similar fashion to a personal understanding of the weight of Soviet oppression, the more so after the Kremlin swallowed the Baltic states in 1940. Henderson would have feared and distrusted the Russians had he never met Elise; now he had additional cause for suspicion and alarm.

At the time Henderson first encountered her, Elise had recently embarked on a singing career. Convinced that she needed to learn English, she decided to go to London. Henderson attempted to persuade her otherwise, proposing marriage just before she left. She accepted, although her acceptance did not prevent her departure.

With the marriage pending, Henderson had to find a way to deal with State Department regulations prohibiting an officer from serving in a spouse's native country. For once a coincidence of events worked to his advantage: Kelley's offer of a Washington job would solve the nationality problem while allowing Elise to continue her study of English.

At the end of November, Henderson left Riga for London, where he caught up with Elise. With the Latvian consul-general giving away the bride and the American consul-general acting as best man, the two were married. That evening they headed for the French Riviera and a honeymoon. At the beginning of 1931 they sailed for America.[20]

III

Henderson's return coincided with renewed efforts by supporters of recognition of the Soviet Union to get Washington to normalize relations with Moscow. The onset of the depression had reinforced the commercial argument for normalization, since the shrinking of domestic demand and the collapse of international markets made any sales opportunities more attractive than before. World political events of the early 1930s, especially Japan's 1931 seizure of Manchuria, added a strategic element to the recognition case. Washington disliked the idea of further Japanese expansion; so did did Moscow, whose hold on Siberia had not completely recovered from the revolution. By acting together the United States and the Soviet Union might curb Japan's appetite. But cooperation required, or at least strongly implied, recognition. Although the argument failed to sway Hoover, and still less the State Department's bureaucracy, anchored by Kelley and the Eastern Europeanists, the Republican president's successor Franklin Roosevelt indicated greater receptivity.

Roosevelt had learned the ways of bureaucrats in the most impenetrable bureau of all, the Navy Department, and he quickly demonstrated that he would leave the professional diplomats to their memos and position papers while he ran foreign policy from the White House. His handling of Soviet recognition was typical. During the first half of 1933 a diverse group called for normalization. Democratic elder statesman Al Smith told the Senate Finance Committee he saw no reason for continued ostracism. Citing the economic benefits recognition would bring, the former New York governor and presidential candidate declared, "There is no use trading with them under cover. . . . We might just as well be represented there and let them be represented here at Washington, and let us do business with them in the open." Republican governor Gifford Pinchot of Pennsylvania, a progressive in the mold of Theodore Roosevelt, asserted that normalization would be "in the best interests" of the United States. Senator William Borah, who had helped scuttle the Versailles treaty, described Russia as the "greatest undeveloped market in the world," adding that America "cannot avoid taking her into consideration."

Opponents of recognition, however, had not fled the field. Father Edmund Walsh of Georgetown University, an acknowledged authority on Soviet affairs, albeit a hostile one, stated that Washington's acceptance of the Kremlin's ambassador would amount to "a canonization of impudence." Walsh added, "You cannot make a treaty with that evil trinity of negations—anti-social, anti-Christian, anti-American." Labor leader William Green publicly worried about the depressing effects of Soviet low-wage competition on an American economy depressed enough already. Former secretary of state Bainbridge Colby announced that his views had not changed since his rejection of recognition in 1920, and, with touching immodesty, he asserted that for the Roosevelt administration to renounce his handiwork would "repudiate one of the historic achievements of the Democratic Party."[21]

Observing this debate from the sidelines, Henderson, Kelley, and the other members of the Eastern Europe division realized that the opponents of recognition needed help. Always ready to oblige in a worthy cause, the division sharpened its anti-recognition brief. During the last year of the Hoover administration, Henderson, at Kelley's direction, had resumed his study of Moscow's international commitments and the Kremlin's record in keeping them. The task required little originality, principally involving tracking down copies of treaties, contracts, and the like; editing and retyping them; and writing commentaries so that even the most harried and superficial congressional committee member could see that the Russians cheated on their pledges.

Early in the summer of 1933 the Eastern Europe division prepared a memo delineating, in the words of its achingly descriptive title, "Problems Pertaining to Russian-American Relations Which, in the Interests of Friendly Relations between the United States and Russia, Should be Settled Prior to the Recognition of the Soviet Government." Whether this paper served as an argument against recognition or a simply a list of recommendations designed to guarantee, as its authors declared, "that the United States may derive from the recognition of the Soviet government the benefits which normally follow the recognition of a foreign government," was a matter of interpretation. Roosevelt's designs were becoming clearer daily. Surrounding himself with such known supporters of recognition as legal scholar Felix Frankfurter, treasury-secretary-to-be Henry Morgenthau, and sometime diplomat William Bullitt, the president was moving quite evidently toward normalization. Under these circumstances, stonewalling would only result in the isolation of the State Department. So the Eastern Europeanists beat a strategic retreat. They opposed not recognition per se, but recognition uninsured by significant concessions on the Kremlin's part. If Moscow accepted the conditions, recognition might prove worthwhile. If the Russians refused, nothing would be lost.

The greatest concession required by the Kelley cohort, and the one least likely to meet Moscow's approval, entailed a basic change in the Soviet approach to international affairs. "It is obvious," the Eastern Europe paper asserted, "that so long as the Communist regime continues to carry on in other countries activities designed to bring about ultimately the overthrow of the Government and institutions of these countries, the establishment of genuinely friendly relations between Russia and those countries is out of the question." Such activities must cease.

The second concession was payment of outstanding debts. Moscow's failures in this regard transcended the monetary amounts involved. "The Soviet government," the Eastern Europe memo asserted, "has rejected international obligations which the experience of mankind has demonstrated are vital to the satisfactory development and maintenance of commerce and friendly intercourse between nations." As to numbers, the paper set the Soviet debt to the United States government at $192 million, to private American firms at $106 million, and to individual owners of confiscated property at $330 million, all exclusive of interest.

The third concession was an iron-clad guarantee of fair treatment of foreign nationals, whether engaged in business or otherwise. No one in the Eastern Europe division questioned the general principle that private foreigners were subject to the laws of the host country. And with most countries the principle worked satisfactorily. But not with the Soviet Union. The Soviet legal system was "so far removed" from western norms and the communist conception of justice was "so alien" to westerners that the only security for foreigners came from the willingness of their governments to engage in tit-for-tat reprisals against Russian nationals. Soviet authorities defined espionage broadly enough to place nearly all visitors at risk of stiff prison sentences; enforcement had become a matter of caprice. Without the strictest agreements in advance of recognition, differences bound to arise in this and related areas would lead to constant friction, not to mention injustice to Americans.[22]

This all amounted to a powerful case against hasty recognition, or so it seemed to Kelley and Henderson and the East Europeanists. But Roosevelt moved forward undeterred. Early in October the president instructed Henry Morgenthau and William Bullitt to enact a little play for Boris Skvirsky, head of the Soviet Information Bureau in Washington and the unofficial representative of the Kremlin. As per the president's stage directions, Morgenthau invited Skvirsky to his office, where he announced that a person from the State Department would arrive shortly with an important unsigned letter. "His face lit up with a big smile," Morgenthau recorded. Bullitt then entered with a flourish, brandishing the letter and announcing, "This document can be made into an invitation for your country to send representatives over here to discuss the relationship between our two countries. We wish you to telegraph the contents of this piece of paper by your most confidential code, and learn if it is acceptable to your people." If so, the president would sign the letter and make it public. If not, Skvirsky must give his word of honor never to divulge the contents of the proposal or even the occurrence of the present meeting. "Does this mean recognition?" asked Skvirsky. Bullitt sidestepped: "What more can you expect than to have your representative sit down with the president of the United States?"[23]

The letter in question had been drafted by Kelley and the Eastern Europe division. It was written with the aim of committing the Roosevelt administration to nothing—since no matter what Skvirsky might promise, the Russians would surely leak its contents if they considered such a move advantageous— while leading the Kremlin to send to Washington a top emissary, one who could actually negotiate for Stalin. In the letter Roosevelt described the current state of U.S.-Soviet relations as "abnormal" and "most regrettable." The difficulties responsible were serious but not insoluble. They might be removed by "frank, friendly conversations" at the highest level.[24]

The letter and its presentation produced the desired effect, and within the week Stalin replied that he would send Maxim Litvinov, commissar for foreign affairs, to Washington to engage in the discussions the president proposed.[25]

IV

Henderson had been preparing for this turn of events for several weeks. At the end of September he began a study of the implications of recognition, focusing especially on the economic aspects of normalization. Since a principal argument in favor of recognition was that the Soviet Union would provide a significant outlet for American manufactures, Henderson, at the request of Secretary of State Cordell Hull and Kelley, undertook to estimate the potential magnitude of Soviet-American trade. In part Henderson's survey constituted an exercise in foot-dragging: by exploding the notions of those talking as though recovery for the American economy awaited only recognition, Henderson's findings might slow what seemed reckless progress toward an exchange of ambassadors. At the same time Henderson would measure the Russians' ability to sell goods in the west, a matter crucial to a determination of the Kremlin's credit-worthiness. Hull in particular, hoping to use trade to reduce international strains, wanted to avoid the mistakes of the previous decade, when unrealistic demands on the economy of Germany had exacerbated world tension.

In the main, Henderson's study involved gathering and analyzing such statistics as existed regarding trade between the Soviet Union and the United States. Noting that a large majority of Russian exports to America consisted of just two dozen or so items, Henderson and his assistants projected future demand for these items. They weighed other factors as well, including the degree to which exports from the Soviet Union would face competition from other countries, and they attempted to identify additional products, beyond the core two dozen, that might find buyers in America. They finally produced an estimate of slightly less than $90 million for annual Soviet sales to the United States. Since this was three times what Americans had imported in 1930, Henderson considered the estimate generous—as it proved to be. Enthusiasts of recognition, however, who had been speaking of hundreds of millions in bilateral trade, deemed it more of the East Europeanists' sabotage.[26]

Roosevelt pressed on. The president ignored some of the opponents of recognition and sweet-talked others. Bringing Edmund Walsh to the White House, he declared, "Leave it to me, Father; I am a good horse dealer." With what Walsh later described as "that disarming assurance so characteristic of his technique in dealing with visitors," Roosevelt persuaded the priest to do what he could—which turned out to be a lot—to quiet Catholic criticism of recognition.[27]

Litvinov proved proficient in horse-dealing himself. The first phase of the negotiations regarding recognition took place at the State Department. Kelley believed that the American negotiators might most readily elicit desired guarantees from Litvinov by pointing to similar guarantees given by the Kremlin to other countries. Since Henderson had done work in precisely this area, Kelley gave him the job of hunting precedents, especially relating to Soviet interference in other countries' affairs. For two weeks Henderson drafted, discussed with Kelley, and redrafted one proposal after another. The objective, he later recalled, was

> to produce a commitment that (1) would on its face seem so reasonable that it would be difficult for Litvinov to refuse without leaving an inference that it was not the sincere intention of the Soviet Union to try to maintain genuinely friendly relations with the United States; (2) would unmistakably obligate the Soviet government not to interfere in our internal affairs as well as to not permit the Communist International and kindred international subversive organizations to continue to operate on Soviet territory; and (3) would be heavily interlarded with sentences and phrases borrowed from Soviet commitments to other countries.[28]

The effort produced a partial success. After ten days of negotiations Roosevelt and Litvinov exchanged letters expressing the mutual desire of their governments to establish normal diplomatic relations. Accompanying these letters—but following them, since Litvinov refused to allow attachment of any conditions to recognition—were several memos and statements declaring the intentions of each government toward the other. The first reflected the influence of Henderson and the Eastern Europe division. "It will be the fixed policy of the Government of the Union of Soviet Socialist Republics," the statement affirmed,

> 1. To respect scrupulously the indisputable right of the United States to order its own life within its own jurisdiction in its own way and to refrain from interfering in any manner in the internal affairs of the United States, its territories or possessions.
> 2. To refrain, and to restrain all persons in government service and all organizations of the Government or under its direct or indirect control, including the organizations in receipt of any financial assistance from it, from any act overt or covert liable in any way whatsoever to injure the tranquillity, prosperity, order, or security of the whole or any part of the United States. . . .
> 3. Not to permit the formation or residence on its territory of any organization or group—and to prevent the activity on its territory of any organization or group, or of representatives or officials of any organization or group—which makes claim to be the Government of, or makes attempt upon the territorial integrity of, the United States. . . .
> 4. Not to permit the formation [etc., of any group] which has as an aim the overthrow or the preparation for the overthrow of, or the bringing about by force of a change in, the political or social order of the whole or any part of the United States.

There was much more along these lines, with the result being a Soviet pledge of noninterference. ("No monkey business in our country" was how Henderson's associate Elbridge Durbrow put it.) In one respect, of course, the effort represented a waste of time. If the Kremlin wanted to continue to fund or otherwise encourage the Comintern, it would do so quietly and lie about its actions. On the other hand, Litvinov's public acceptance of these conditions marked a victory of sorts for the East Europeanists. When, as they fully anticipated it would, Moscow persisted in underwriting revolution, they could use this letter against the Soviets—and, just as important, against Soviet apologists in the United States.

On the issue of treatment of foreign nationals, American negotiators persuaded Litvinov to agree to a form of the most-favored-nation principle. Henderson's research had uncovered an article in a 1925 German-Soviet treaty that went further than any other pact in granting consuls the right to visit nationals arrested and detained by Soviet authorities. Litvinov accepted a statement that American officials and citizens would possess rights "not less favorable than those enjoyed in the Union of Soviet Socialist Republics by the nation most favored in this respect." The statement cited the Soviet-German treaty by name.

In the matter of debts, Roosevelt's eagerness to have done with recognition precluded a settlement. The Kremlin no longer dismissed the issue entirely, but Litvinov argued that he could offer no more than $75 million toward making amends. The president asserted that he could not get Congress to acquiesce in less than twice that amount. To prevent the talks from breaking down, Roosevelt agreed to leave the question unresolved.[29]

V

For all Henderson's suspicions of the Russians, the conclusion of the negotiations afforded a certain relief. It also presented an exciting prospect. However much he distrusted the Soviets, the idea of seeing for himself what he had only read and heard about held terrific appeal. Besides, maybe the Russians really *did* intend to act like normal citizens of the international community. It was a long shot, to be sure, but perhaps as the first generation of Bolsheviks, whose outlook on life often reflected years in czarist prisons, passed from the scene, a younger, less bitter cadre would come to power. Heaven knew Russia had plenty of domestic problems for ambitious types to tackle, without challenging the whole world. For Henderson, Kelley, and others of the Eastern Europe division to have had open minds on the subject would have required their forgetting all they had learned in the previous decade and a half; but one might leave the mind's door at least slightly ajar.

The day after Roosevelt exchanged notes with Litvinov, the president appointed William Bullitt to be America's first ambassador to Moscow. As Bullitt's assistant, the State Department named Earl Packer, Kelley's second in command, whom Henderson would replace on the Eastern Europe desk. By this time Henderson's skills at administration were well known in the department, and Kelley put him to work organizing the new Moscow mission.

Henderson's experience in the consular and diplomatic branches of the foreign service had convinced him that the traditional practice of maintaining both a consulate and an embassy in a foreign capital was anachronistic and wasteful, and would be so especially in Moscow where space was at a premium. Consequently he drew up plans for consolidating the functions of the two offices under one roof. Although this innovation raised hackles among the old birds in the department, Assistant Secretary of State Wilbur Carr approved it.

When word returned from Moscow of the space allotted to the Americans, the situation proved worse than expected. Henderson had requested 280 rooms for office and living quarters. The Kremlin granted 72. Upon receving this news Bullitt announced that unmarried personnel would receive first consideration for assignments and that married officers would leave spouses and children at home. In addition, to allow doubling and tripling up in living accommodations, the embassy staff would be almost exclusively male. On its face this last decision made sense, but Henderson and others feared the consequences of turning a large number of American bachelors loose in Moscow, where the only available female companionship would undoubtedly be vetted by the secret police.

Bullitt, like Roosevelt, distrusted the foreign-service bureaucracy. When selecting individuals to staff the embassy, he gave preference to young officers who had not yet acquired the rigidities of the diplomatic caste. His first choices included George Kennan, Charles Bohlen, and a third Russian-language specialist, Bertel Kuniholm. He passed over Henderson as too old and probably too set in his ways—and who, in any case, was slated for advancement in Washington.

Bullitt got his way in most of the appointments, but not all. When Earl Packer's assignment to Moscow fell through, for reasons only bureaucrats could appreciate, the State Department had to find a replacement. Kelley asked Henderson if he would take the job. Henderson replied that he understood that Bullitt wanted younger assistants. But if the department thought he would best fill the post, he certainly would go. Kelley told him to talk to Bullitt.

When Henderson repeated this reply to the new ambassador, Bullitt was not quite overwhelmed by Henderson's enthusiasm. Nor did Bullitt appear unduly taken by Henderson's qualifications. Henderson left the meeting expecting to stay in Washington. But he had made a better impression than he thought, and next day he received word that Bullitt had requested his assignment as second secretary of the Moscow embassy. He would leave on February 15, 1934.[30]

II

THE EASTERN FRONT

4

Red Tape

Henderson had mixed feelings regarding William Bullitt. America's first ambassador to the Soviet Union had long agitated for recognition of Moscow, from the time he and journalist Lincoln Steffens had conducted a secret mission to Moscow during the Paris peace conference of 1919 and negotiated what Bullitt hoped would be an end to the Russian civil war. Wilson had disowned the Bullitt mission, thereby turning Bullitt against Wilson but not against the Bolsheviks. In the eyes of Henderson, Kelley, and the Eastern Europeanists, Bullitt displayed far too much enthusiasm for the communist regime. Henderson recognized that he owed his job to Bullitt. Henderson had missed out on the specialized training Kennan and Bohlen had received; this deficiency might well limit his future. But for now Moscow represented a signal step forward. And again, maybe Bullitt and the optimists were at least partly right about the Russians. Perhaps to some degree the Kremlin's revolutionary talk was intended to keep up spirits on the home front. Besides, Bullitt was the boss, and in what surely would be a tough assignment he deserved a fair shake.

On the other hand, Henderson considered himself more of an authority on the Soviet Union than Bullitt, and in his soberest thinking Henderson doubted seriously that the Bolsheviks had junked the idea of world revolution. Accommodation with the west suited Moscow's purposes for the moment, but this would turn out to be nothing more than a Leninist zigzag. When the time came and the opportunity presented itself, the Kremlin would resume its campaign of subversion.

On the ship carrying the members of the Moscow mission across the Atlantic, Bullitt called a meeting that reinforced Henderson's reservations. Attempting to set the tone for business and life at the embassy, Bullitt began by saying he intended to ignore the requirements of foreign-service protocol. He had visited many American embassies and legations, he said, and found them stuffy and more concerned with etiquette than with managing diplomatic relations. He would dispense with rank and hierarchy in the Moscow embassy. He would observe no distinction between professional and nonprofessional staff, nor between career and temporary personnel. He would make himself equally available to all members of the mission. Ideas would be judged on their merits rather than on their source.

Bullitt went on to allude to the difference of opinion regarding the Soviet Union that existed between himself and certain members of the State Department. He described the gracious treatment Soviet officials had accorded him on his most recent visit, during which he had finalized arrangements for opening the mission, and he said he believed that this reception indicated a genuine wish to forge friendly relations between the two great powers. For his part he considered cordiality both necessary and possible. Consequently he desired that each member of the mission contribute wholeheartedly to the goal of Soviet-American friendship. The embassy's task was to promote the interests of the United States in relations with Moscow; an essential aspect of this task involved cultivating the good will of the Russian people. Perhaps, although he did not expect such an outcome, friendly relations would not develop. But if they did not, he wanted to know that the failure resulted from a refusal by the Russians to meet the United States halfway, rather than from any inadequacy or hostility on the part of the embassy.[1]

While Henderson harbored doubts about Bullitt, he possessed considerable confidence in other members of the mission. John Wiley served in the capacity of counselor of the embassy, the second-ranking post. Born into the foreign service in France in 1893, son of the American consul in Bordeaux, Wiley had grown up in Europe and was fluent in French, Spanish and German. Wiley did not marry until the age of forty, and in a bureaucracy that liked its junior officers unattached he moved ahead quickly. By 1934 he had put in nearly eighteen years at embassies throughout Europe and Latin America. As might have been expected of one who joined the consolidated foreign service from the diplomatic side, he preferred political reporting to what he considered the more pedestrian tasks of administering the embassy. In this, if not in enthusiasm for the Soviet experiment, his attitude matched Bullitt's. As a result, Henderson, third in command, inherited the burdens of day-to-day management.

Below Henderson in the hierarchy—which, despite Bullitt's design, reasserted itself before long—were three Russian-language specialists. The intellectual of the crew was George Kennan, whose studied aloofness, ascetic demeanor, and preference for Russian art and literature over Soviet politics made him seem even more profound than he was. Kennan's service with Henderson and Charles Bohlen in Riga had confirmed the suspicions of the Kremlin he had developed in Robert Kelley's finishing school. With Henderson, Kennan viewed Bullitt's high hopes with considerable skepticism. Bohlen provided a singular contrast, in terms of personality, to Kennan. As outgoing as Kennan was withdrawn, a connoisseur of life where Kennan dined on ideas, Bohlen brought to the embassy what Henderson described as "a pleasing personality, a keen sense of humor, a gift for amusing conversation, and a certain amount of spontaneous joyousness." Bertel Kuniholm rounded out the trio. A West Point graduate who had joined the foreign service after several years in the army, Kuniholm, like Bohlen, had taken his Russian studies in Paris. After Moscow he would serve in Switzerland and Iceland before accepting an assignment during World War II in Tabriz. As would Henderson, and at about the same time, Kuniholm would run afoul of those in the Roosevelt administration who placed a higher premium on unruffled relations with the Russians than on what to the Kelleyites seemed clear evidence of the Kremlin's malign intent.[2]

Other officers at the embassy included Elbridge Durbrow, who ran the consulate and shared an apartment with fellow bachelor Bohlen. Durbrow succeeded Henderson as head of the embassy's economic section during Henderson's extended stints as chargé d'affaires. Durbrow found the work so absorbing that he requested specialized training in economics, which he received at the University of Chicago. In 1944 he served as adviser to the American delegation at the Bretton Woods conference. Angus Ward, a Canadian by birth and a naturalized U.S. citizen, had met Henderson in Estonia in 1920, where Ward worked for Hoover's relief commission. A linguist of remarkable agility, Ward had learned English and Gaelic (the latter in a Scottish community in Canada) while growing up, Spanish and Basque during the war, Russian and Finnish a few years later, and Chinese and Mongolian after joining the foreign service in 1925. As a hobby he set to work compiling a Mongolian-English dictionary. Ward did not accompany the rest of the mission from America to Moscow, arriving instead from the east, stepping unannounced off the Trans-Siberian railway, wearing shorts in the subzero cold of a Russian March and sporting a fierce red beard.

More colorful still was Charles Thayer, another West Pointer who had decided against military life. Thayer encountered some trouble resigning his officer's commission, partly because he wanted out before the army had recouped its investment in him, and partly because his commanding officer, a cavalry colonel named George Patton, intended to assign the blue-blooded and horsey Thayer to his polo team. At Thayer's repeated insistence that the army would benefit from his transfer to the foreign service—he mentioned in particular a desire for a posting to the Soviet Union—Patton finally exclaimed in exasperation, "Go ahead, God-damn it! If you don't know what's good for you, I can't teach you. Damned cookie-pusher instead of good decent polo. And Russia on top of that! It's damned nonsense, I say. God-damned nonsense!"[3]

Thayer took to heart Bullitt's admonition to promote friendly U.S.-Soviet relations. He had no choice. Early in 1933, while not yet officially a member of the foreign service, Thayer had guessed that Roosevelt would soon normalize relations with the Kremlin. He hastened to Moscow to learn Russian and throw himself in the ambassador's path. Bullitt was taken by Thayer's brash good humor and hired him to help Kennan open the embassy. He stuck, and until 1937 when he normalized his own relations with the State Department by passing the foreign-service examination, he did assorted odd jobs around the embassy.

II

Starting up an embassy in any country is a major undertaking, but the Soviet Union in 1934 presented special problems. The country faced shortages, some severe, of all manner of goods and services. Although the diplomatic standing of the Americans, combined with the Kremlin's desire to commence relations on the right foot, put the embassy ahead of the locals on various waiting lists, accomplishing the most basic tasks sometimes required extraordinary efforts. Kennan and Thayer experienced myriad troubles arranging transport for the furniture and other equipment needed to outfit the embassy. Housing proved consistently vexing. Even as minor a matter as finding coathangers required a day-long expedition by Bohlen and Thayer, who hired a motorcycle and sidecar for commando raids on Moscow's shops and closets.[4]

While locating what the embassy needed was difficult, paying for the goods and services was more problematic still. Economic chaos abroad and government policy in Moscow had conspired to grossly overvalue the ruble. At times the official price exceeded the natural—that is, black market—figure by as much as 3000 percent. As part of the normalization agreement of 1933, the State Department had received what it considered a commitment by the Kremlin to find a way to bridge the gap between the two rates. At the time Henderson arrived, however, the Soviets had not made good their pledge, leaving the Americans in a quandary. To pay the posted price would prove impossibly expensive. Considering the modest amount Congress had appropriated for the mission, shutting down the embassy before it had fairly opened might turn out to be necessary. On the other hand, to do as the members of some missions did, with the tacit approval of the Russians, and buy rubles on the black market or purchase them abroad and slip them into the country under the cover of diplomatic privilege, would place the United States in a compromising position as a violator of Soviet law.

Acting for Bullitt, who was making his way slowly along the embassy trail in Europe, Henderson succeeded in deferring the problem for a time by negotiating a loan, in rubles, from the Soviet state bank. This provided the cash to get the embassy running and its staff settled. But it left open the questions of how and at what rate the United States would acquire the currency to redeem the note and how the mission would operate in the future.

When Bullitt arrived he rejected the notion of bending the law. "I am absolutely opposed to the smuggling of roubles in our diplomatic pouch, or to the purchase of roubles in the Black Bourse in Moscow," he wrote. Deeming the issue a test of Soviet good faith, Bullitt declared that the matter must be resolved in "an honorable and above-board manner."[5]

In this area, as in several others, the Kremlin failed Bullitt's test. Unwilling to admit publicly that the posted price of the ruble bore no relationship to its buying power, and disinclined to establish what might prove an embarrassing precedent by openly allowing the Americans to turn to extralegal means of acquiring currency, the Soviet leadership found one excuse after another for refusing to resolve the dispute. Eventually Litvinov hinted that the Soviet government would avert its gaze if the United States bought rubles in eastern Europe and quietly shipped them to Moscow. Bullitt did not like this solution, nor did Henderson. Beyond the impropriety involved and the potential for political extortion, it left the American mission and Henderson as chief administrator at the mercy of wild fluctuations in the market price of the Russian currency. But for the moment neither Henderson nor Bullitt could manage anything more satisfactory.[6]

Bullitt's hopes hit snags in other areas too. With the goal of moving the embassy to larger quarters, the ambassador found a location in the Lenin Hills overlooking the Moscow River. There he intended to reproduce Jefferson's Monticello as a symbol of America's world vision. Although the local Moscow Soviet had expressed reservations, claiming that a proposed canal would slice through the property, Stalin personally told Bullitt he could have the land. Later communications from Litvinov and the Soviet ambassador in the United States confirmed the deal—or so the Americans thought. In March 1934 it became evident that the Kremlin was reconsidering. A representative of the foreign ministry called to say—with great regret, he assured the ambassador—that there had been some confusion. It appeared that certain intragovernmental messages had been incorrectly interpreted and that the property available was not precisely the piece the Americans had in mind. Bullitt replied that the matter was settled, that he had assurances from Stalin himself. The ambassador added that it was not wise to create an impression among Americans that Stalin's promises were worthless.

The matter was settled, all right, though not as Bullitt desired. By the time he realized that the United States would not get the site, Bullitt had been forced, reluctantly, to conclude that Stalin's promise might indeed be worthless.[7]

The unresolved issue of debt repayment likewise disillusioned the ambassador. Litvinov attempted to link repayment to a new loan from the United States. Despite his—albeit decreasing—willingness to accord the Kremlin the benefit of doubts, Bullitt recognized this as a ruse for simply recycling, rather than repaying, the debt. He went around and around the issue with Litvinov for months. In October 1934 he finally exploded. He told the foreign minister that a loan was "impossible, had always been impossible, and always would be impossible." Bullitt added that Litvinov seemed determined "to kill all possibility of really close and friendly relations between our countries."[8]

Bullitt was not quite ready at this point to give up on the Soviets, although he realized profitable ties would come harder than he had anticipated. "We have not gotten nearly so far in our official relations as I hoped we might," he confessed to a friend. For a time he clung to the hope that the embassy's problems originated with Litvinov rather than Stalin. He wrote Kelley that while many Soviet officials seemed "inclined to be just as nasty as can be . . . Brother Stalin and his intimates in the Kremlin are extremely friendly." Bullitt asserted to an old acquaintance that despite Litvinov's obstructionism "the United States continues to be the most favored of all nations here."[9]

The ambassador tried to work around Litvinov, cultivating Karl Radek, Stalin's propaganda chief (a man, Bohlen commented, of "brilliant and cynical mind" and "surpassing ugliness"), Defense Commissar Klimenti Voroshilov, and Assistant Commissar for Foreign Affairs Lev Karakhan, the last especially known as an opponent of Litvinov. In an effort to promote closer ties to the military, Bullitt directed Charles Thayer to instruct the Red cavalry in the intricacies of polo ("polo for the proletariat," Thayer called it). He also brought baseball to the Soviet Union, importing balls, bats, gloves, and uniforms at his own expense. He had Thayer organize a party the likes of which Moscow had not seen since the days of the czars, if then. "Springtime" furnished the motif and Thayer supplied the props: tulips from Finland, roosters from a collective farm, baby mountain goats and bear cubs from the Moscow zoo—in addition to a jazz band from Czechoslovakia and a sword-dancer from Georgia.[10]

Bullitt's anti-Litvinov campaign succeeded slightly. In May 1935, Kelley, who disliked Litvinov even more than Bullitt did, thought the foreign commissar appeared "on the defensive in respect to his policy towards the United States." But Kelley had always believed that the underlying problem was systemic rather than personal, and he indicated that nothing had happened to change this view.[11]

Events gradually compelled Bullitt to agree with Kelley. By the summer of 1935 the ambassador had swung around to a position diametrically opposed to his earlier hopefulness. The coup de grace to Bullitt's optimism came with the seventh congress of the Comintern in Moscow. As the meeting approached, Bullitt reminded Litvinov of the Soviet government's pledge to avoid interference in American internal affairs. Litvinov's response, that he "could not promise anything," was discouraging. The event proved even worse. Soviet party officials feted attending American communists, who in turn hailed Stalin as their true leader. Although the congress called for a popular front against the fascists, it asserted that the main responsibility for world troubles lay with the American imperialists and their British accomplices. To Bullitt this seemed a "flagrant violation" of at least the spirit of the recognition agreements. No longer could he blame Litvinov for bad relations between the United States and the Soviet Union, since circumstances made it patently obvious that the "entire course of the Congress was dictated in advance by Stalin."[12]

Bullitt went so far as to consider recommending a pullout from Moscow, on grounds that "the Soviet Government has broken its pledged word to us and cannot be trusted." When he calmed down he decided that simple expedience advised against such a move. "As the Soviet Union grows in strength it will grow in arrogance and aggressiveness, and the maintenance of an organization in Moscow to measure and report on the increasingly noxious activities and breach of faith of the Soviet Union seems definitely in the interest of the American people." But Bullitt had no doubt regarding the Kremlin's intentions. "The aim of the Soviet Government is and will remain, to produce world revolution."[13]

III

Henderson might have told Bullitt as much all along. Yet while the scales were falling from the ambassador's eyes, the first secretary—Henderson had moved up again, in part as a result of reports by Bullitt that he was "really delighted" by Henderson's work—had his hands full coping with the physical and bureaucratic vicissitudes of daily life in Moscow. Housing and work space presented a continuing problem, with the most serious but not the only difficulty resulting from the initial fact of the Kremlin's allotting only a quarter of the rooms Henderson had requested. The location of the building in which these rooms were situated—nearly adjacent to Red Square—offset Henderson's disappointment to some degree, although it did not ease the crowding.[14]

As de facto chief administrative officer, Henderson sought to relieve the space shortage by all means possible. At first, three apartments in the building assigned to the Americans were reserved for Soviet officials. Lev Karakhan, the assistant foreign commissar occupied one; Dmitri Florinsky, director of protocol in the foreign office, had another; Pavel Mikhailsky, an approved journalist, lived in the third. A few months after the embassy opened, Florinsky disappeared under mysterious circumstances. Later, when the purges got into full swing, there would be nothing mysterious at all about such happenings, and no one would bother or dare to ask what had happened. In Florinsky's case the Americans only knew that two unidentified men had summoned him from a dinner at the British embassy and no one had seen him since. Various rumors indicated that he had been shot, sentenced to prison camp in Siberia, or exiled. Although Henderson considered Florinsky's fate unfortunate, as soon as it became evident that the protocol chief had committed some unforgivable faux pas and would not be returning, he went straight to the Soviet authorities and petitioned for annexation of Florinsky's flat. After several months of effort and repeated visits to assorted offices, he acquired one more apartment.

Henderson may have played an unintentional role in the vacating of Mikhailsky's rooms. For three years the members of the American embassy watched the elderly journalist come and go. He said little, keeping late hours and avoiding foreigners. But Mikhailsky, like Henderson, took his exercise in the form of daily walks, and one spring afternoon in 1937, when their paths crossed near the Kremlin, he asked Henderson to join him. The invitation surprised Henderson, what with the purges by now causing all but the most intrepid or foolhardy among Moscow's citizenry to shun members of the diplomatic community like the very death they often proved to be. For most of an hour Mikhailsky reminisced about the old days in the underground. He described Lenin, with whom he had reentered Russia in 1917 on the notorious sealed train. He spoke wistfully of his carefree days in exile in Prague, when he could work and study during the day and sleep easily at night. Now he dreaded going to bed, for fear of discovering next morning that friends and associates had been taken away.

Whether or not this conversation contributed to his demise, a few weeks later Mikhailsky disappeared. The only sign of his departure was an NKVD (secret police) seal on the door to his rooms. Later Henderson discovered that the journalist was one of the lucky ones, receiving a prison sentence instead of a bullet in the neck.

Once he concluded that Mikhailsky's removal was permanent, Henderson again visited the housing authorities. He asked to rent the apartment for embassy use. By this time most of Moscow lived in terror of the secret police, and with the mark of the NKVD on the door, no mere housing bureaucrat wanted to take responsibility for turning the apartment over to the Americans. For months Henderson's entreaties disappeared into the void. Spring became summer, then autumn and winter. Finally, during an early thaw, the occupants of the rooms just below Mikhailsky's came running frantically to Henderson to report a flood of water pouring through their rapidly eroding ceiling. Realizing what had happened, Henderson immediately sought a maintenance crew to shut off the water before the broken pipes in Mikhailsky's flat ruined half the building. Unfortunately, the weekend had just begun and no one was available. With no other choice, Henderson broke the NKVD seal himself and entered the apartment to stop the torrent.

Henderson guessed his action might stir trouble, and as soon as the appropriate office opened he personally delivered a note explaining the circumstances. The individual he spoke with looked at him in horror. "You have committed a grievous crime," he said. "A diplomat can be declared persona non grata for less." Henderson replied that he was ready to take the consequences for his actions, but he did not think the destruction of the American embassy would serve either Soviet or American interests. Henderson's interlocutor refused to accept Henderson's note, insisting that the entire matter be kept off the record.

Once the NKVD got wind of the affair, things moved quickly. The operatives of the secret police stripped the apartment of everything relating to Mikhailsky. Repairmen whisked through next. Henderson obliquely received word that although his transgression in this case would be overlooked, he must never, never try anything like it again. Finally, after a suitable period to let the American first secretary ponder the gravity of his offense, the Soviet bureaucracy released the apartment to the embassy.

The last of the three apartments became available in a similar, if less liquid, fashion shortly before Henderson left Moscow for Washington. At the end of 1937 Karakhan received a midnight summons. Soon afterward his execution was announced. By the following spring the Americans had claimed his flat.[15]

These additions eased the housing crunch only marginally, and Henderson had to deal with the problems that developed in the confined atmosphere of the Soviet capital. Through judicious, if unsolicited, advice to Washington he sought to steer to other posts individuals who had caused or might cause difficulties. The nonprofessional staff, whose work lacked much of the stimulation that sustained Henderson and the Soviet specialists, felt the shortage of space most acutely. "Although some of our clerical employees stand up remarkably well in the face of conditions here," he told George Messersmith, the assistant secretary for Eastern Europe, "others are beginning to show signs of strain, and in my opinion their transfer in the near future would be beneficial both to themselves and to the Government." Citing cramped quarters and inadequate health-care facilities, Henderson argued against sending couples with small children. When the department did so over his protest, he commented that he hoped the children were "robust and healthy." For a similar reason he advised against assigning to Moscow any officer who consistently required medical attention. Responding to rumors that such an officer was coming, he remarked, "I have a feeling that the assignment if executed may result in a tragedy."[16]

Certain individuals simply were not cut out for work in Russia. Describing a disbursing agent at the embassy, Henderson wrote to Messersmith, "He is an extremely capable man, an excellent accountant, an expert stenographer-typist, and a good file and code clerk. He is of a highly strung temperament, however, and the Moscow atmosphere is beginning to have its effects upon him." A vice consul possessed "considerable ability" but was prone to "outbursts of discontent" and tended to "sulk and even to agitate among the other clerks." The spouse of a doctor newly assigned to the embassy had made a bad showing immediately. "She was quite free in stating that at other posts State Department personnel had in general failed to accord her husband and herself the rank which they deserved and in hinting that they expected to be given better treatment here." Efforts at placation had succeeded momentarily, but Henderson was not especially hopeful. "Moscow is not the ideal spot for a woman like Mrs. Nelson, and it may be that in spite of our efforts we shall not be able to keep her satisfied."[17]

While the purges eased the pressure on personnel marginally, by making more apartments available, in other respects they made matters worse. The Russian employees at the embassy felt the effects first. After an initial experiment with a dozen Muscovites as porters, chauffers, typists, and translators, Henderson, with Bullitt's approval, had gradually increased the number of locals on the staff to more than thirty. Although hired primarily to save money, the Soviets brought an expertise not available readily, if at all, from the American side. Among the additions to the original group were various professionals, including a lawyer, a political scientist, a specialist in foreign trade, and an agricultural economist.

In September 1936, the last, Valentine Malitsky, failed to arrive for work one morning. His wife, an American citizen, came to the chancery, asking to speak with Henderson. Through her tears she explained that the secret police had taken her husband the previous day, and she implored Henderson to help her find him. Henderson raised the matter with the agency that handled relations between Soviet nationals and the foreign community and asked to know what had happened to Malitsky. The official Henderson spoke with evinced no desire to get involved with the secret police. The man demanded to know what business it was of the American embassy, since this was an internal matter unrelated to foreign affairs. To Henderson's reply that Malitsky was an employee of the American embassy and the husband of an American citizen, the official muttered that it was still highly irregular but he would find out what he could. On returning to the embassy, Henderson reported the matter to the State Department, which lodged a formal protest with the Soviet embassy in Washington. For three months Mrs. Malitsky's entreaties, Henderson's inquiries, and the department's complaints produced nothing. Finally, in December, Henderson received word that Malitsky was in a prison camp, serving a ten-year sentence for conspiracy to commit terrorism.[18]

Only in the definitiveness of its outcome did the Malitsky case differ from several others involving embassy employees. Normally Soviet authorities offered no information regarding the disappeared, beyond saying that the disappearances had nothing to do with the individuals' work for the Americans. This Henderson and others at the embassy found hard to credit. It seemed clear that the Kremlin was aiming to isolate foreigners by intimidating their Russian contacts. In November 1937 George Kennan reported "a deliberate anti-foreign campaign of almost unparalleled intensity." The Soviet strategy, Kennan explained, was to convince the Russian people that all foreign representatives were nothing more than "'accredited spies,' engaged chiefly in endeavoring to inveigle unsuspecting Soviet citizens into entanglements." Such propaganda, Kennan continued, contributed neither to the prestige of foreign governments nor to the peace of mind of the diplomats. This, of course, was precisely the point. Textbooks, periodicals, and other government organs repeatedly equated diplomats with espionage. "The regime has added considerable cogency to its arguments by seeing to it that most of the Soviet citizens who have had suffi- cient temerity to associate with foreign representatives—whether with official permission or not—eventually encounter misfortune." Those linked to the American embassy had been specially targeted. "Nearly every one of the Soviet nationals who could be said to have constituted the Embassy's important con- tacts with the Soviet world from 1934 to 1936 has since suffered at the hands of the Government." The anti-foreign campaign was working. "The Soviet public is very actively aware that most of those who have had relations with diplomats sooner or later come to a bad end."[19]

Henderson remarked similarly on the situation. At the beginning of 1938 he sent Washington a list of ten top-level foreign ministry officials with whom the American embassy had worked closely who had disappeared during the previ- ous several months. He described a "marked decrease" in the cooperativeness of those officials who remained. In light of the thoroughness of the houseclean- ing, Henderson suspected that the foreign commissariat might by now have felt the worst of the purges. Even so, he predicted it would be a long time "before the morale of the officials of that Commissariat will rise to such an extent that they will dare to show much energy in assisting foreign diplomatic missions."[20]

5

Moscow Trials

The isolation of the Americans in Moscow weighed on Henderson less than on most others associated with the embassy. He was not a person of broad interests. He neither read widely nor cultivated his tastes in the arts. Games and sports seemed somewhat frivolous; even had he felt otherwise, his stiff arm would have limited his activities. For exercise he continued to walk, weather permitting—which in Moscow it often did not. He engaged in such social activities as Thayer and the others could devise, but none ever mistook him for the life of the party. A serious man engaged in a serious task, Henderson asked for little more than the opportunity to pursue his work.

Of work there was no lack. Following the Comintern congress of 1935, Bullitt left Moscow in disgust. He returned for a brief stay at the beginning of the following year, but after he departed in April 1936 he never came back. The White House required several months to appoint a replacement, and between confirmation by the Senate and slow travel across the Atlantic, it was not until January 1937 that Joseph Davies arrived.

During Bullitt's lengthy absences and in the interregnum, Henderson served as acting chief of mission. In all, his twenty months in this capacity constituted almost half his term in the Soviet Union, and although he would not formally achieve the rank of ambassador for another decade, he acquired considerable experience serving in an ambassador's capacity in Moscow. Foremost among his tasks was explaining the Soviet Union to the Roosevelt administration, or at least to the State Department.

During the middle 1930s, probably more than at any time before or since, the Soviet Union needed explaining to the west. Even to the Russian intelligentsia, the purges that began with the arrest of the assassin of Sergei Kirov in December 1934 were a source of great, if muted, speculation; to the world outside they were a nearly impenetrable enigma. No one knew the magnitude of the shakedown, since the public tip of the iceberg bore an indeterminable relationship to the secret system of terror. Observers could only guess what lay behind the purges and what their aims were. The world watched and listened to that part of the drama staged for its benefit, drawing what conclusions it could.

In August 1936 Henderson attended the first of the great Moscow show trials. He was fascinated by the proceedings, which brought to the bar Zinoviev, Kamenev, and various smaller fry; and during and immediately after the trial he dispatched numerous lengthy telegrams describing the atmosphere of the courtroom, the demeanor of the defendants and prosecutors, the cogency and plausibility of the state's case, and the reactions of local and foreign observers. Reflecting on the matter during the several weeks following, he decided that he had not done the affair justice—if such was the appropriate term—and at the end of the year he filed an additional fifty-one page memo reconsidering and reinterpreting the whole business.

The attention of all observers at the trial focused on the three principals: the two major defendants and prosecutor Andrei Vishinsky. Henderson could not know what kind of pressures, physical and psychological, had been applied to Zinoviev and Kamenev, nor was he aware that the old Bolsheviks had received Stalin's pledge to spare their lives in exchange for public confessions. (Stalin's failure to keep his word would have surprised Henderson not at all.) In Zinoviev's case it was obvious that the prisoner had suffered during imprisonment and interrogation. Henderson remarked that Zinoviev appeared "disheveled" and "very tired" even at the beginning of the trial. By the end he seemed "crushed and unnerved." Kamenev had fared better: he looked and acted "as though he might be a distinguished scientist or a member of the Supreme Court of a western country rather than a world revolutionist." Kamenev's closing argument in his own defense moved to tears foreign observers and even several members of the otherwise hostile Soviet audience, as he described his three sons serving in the air force of the motherland and expressed his desire that they live to redeem the family name. The state's men, however, were conspicuously unimpressed. The judge yawned repeatedly while Vishinsky read a newspaper.[1]

Vishinsky, in Henderson's assessment, was the affair's "most active personality." Throughout, Vishinsky conducted himself in what Henderson labeled a "highly theatrical manner." Attempting to convey the impression Vishinsky made on those in the courtroom, Henderson could think of no better comparison than to Lionel Barrymore in one of his legal roles.[2]

The court convened in the former Noblemen's Gathering Place, lately converted to the House of Labor Unions. Seating accommodated some four hundred spectators. Since a major objective of the trial was to convince the world that the accused indeed were guilty, considerable space was set aside for foreign journalists and diplomats. During intermissions Soviet news officials spared no efforts to confirm to the press corps the validity of the testimony, while the commissariat for foreign affairs provided a similar service for the diplomats.

Assisting the Soviets in this endeavor were a number of present or former American citizens associated with the American Communist party or various western publications. Evidently by some oversight, the journalist Henderson considered the chief American apologist for the Kremlin, Louis Fischer ("an almost fanatical exponent of the Soviet experiment") was out of town, showing tourists the giant strides being made by Soviet agriculture.[3]

The trial itself, Henderson commented, was "beautifully staged." Behind tables on a raised platform, facing the courtroom, sat the uniformed judges. To the right, on a separate platform, this one surrounded by a fence and guarded by armed soldiers, were Zinoviev, Kamenev, and the fourteen other prisoners. Vishinsky as state's attorney, or procurer, confronted the prisoners from a third platform across the room. Despite Vishinsky's consistent vituperation and except for occasional sarcastic asides by the presiding judge, the court treated the accused with respect. As Henderson pointed out, nearly all the prisoners had distinguished themselves at least in part by their speaking ability. Now and again their rejoinders to the prosecution drew grudging grins from the bench and the prosecutor, and even laughter from the audience. On the whole, however, a solemn dignity marked the proceeding.

Notwithstanding the protestations by Soviet officials, it appeared obvious to Henderson that the entire affair was rigged.

> As I observed the performance day after day I began to feel that I was looking at a circus director putting a group of well-trained seals through a series of difficult acts. The manner in which the prisoners, while testifying, anxiously watched the face of the Procurer in order apparently to assure themselves from its expression that they were making no mistakes, and the way in which they hastened to correct themselves when they felt that they were not saying the proper things, were convincing evidence to me that they were under the absolute domination of the latter.

Not simply the mannerisms of the defendants but the substance of their testimony contributed to Henderson's conviction that the participants had carefully rehearsed their performances.

> The foreign correspondents and diplomats present who had never witnessed a Soviet trial sat astounded as the various prisoners eloquently endeavored to convince their hearers not only that they were guilty of conspiracy to murder Stalin, Kirov, Voroshilov, and other prominent Party leaders, but also that they themselves were despicable, irredeemable characters, that they had acted under the instructions of Trotski, who was a traitor to the working classes of the world and an ally of the German Fascists, that the German Fascists, with whom they had cooperated, were the vilest of reptiles, that Stalin had always been right and that the Soviet Union as a truly socialist state was making wonderful progress in the direction of bringing happiness and prosperity to its inhabitants.

The verdict—guilty, of course—and the sentence—death—interested Henderson less than the mechanisms by which the prosecution elicited such abject and absurd testimony from the defendants. "It is doubtful that the full explanation of its mysteries will ever be clearly known," he commented. Yet he did his best to guess. Although he judged it unlikely that any organized conspiracy had existed, he thought it plausible that Zinoviev, Kamenev, and some of the others had discussed the advisability of assassinating Stalin, who certainly deserved it. In part, therefore, the accused may have confessed to assuage uneasy consciences. In support of this argument, Henderson cited a centuries-old Russian belief that confession must precede repentance and ultimate forgiveness. Also, Henderson suggested, the defendants may have followed the prosecution's script in hopes of lightening their sentences. Henderson said he got the "strong impression" that at least some of them were trying to save their skins. Threats against family members may have played a role as well. Henderson believed the most likely explanation lay in some "judicious combination" of psychological and physical coercion probably tailored individually to the separate defendants.[4]

Equally intriguing and more significant from Henderson's perspective was the question of why the Kremlin chose to make such a spectacle of affair, and why just then. Again he could only guess, but he suggested several reasons. First, he thought Stalin might be trying to silence criticism from leftists regarding recent economic innovations.

> There is a growing fear that Stalin is leading them away from Communism in the direction of state capitalism. The growing differentiation of wages which is becoming more pronounced with the raising of production norms; the tendency to organize collective farms along capitalistic lines; the steady growth in influence of the so-called new intelligentsia, technicians, high state officials, and even persons connected with the former bourgeoisie; the encouragement by Party and State of patriotism; and similar trends have caused some alarm among the old or ideologically inclined Party members.

Second, Stalin may have chosen the trial of Zinoviev and the others to serve notice that a recent announcement of a new constitution did not mean that the party leadership intended to allow criticism from below. Third, the trial had the obvious aim of eliminating individuals Stalin distrusted. Fourth, by playing on the theme of sabotage Stalin might be preparing an excuse for the likely failure of the current five-year plan. Fifth, the Kremlin hoped to discredit Trotsky definitively among the international revolutionary movement by connecting him to the fascists in Germany. Finally, the trial would help fan anti-German feeling in the Soviet Union.[5]

II

Henderson left reporting of subsequent trials to junior members of the mission. In February 1937 Kennan described the second of the three big trials, featuring Vishinsky against Grigori Sokolnikov, Grigori Piatakov, and Karl Radek. Kennan believed that this round, at least as it applied to Radek, was more spontaneous than Henderson had judged the earlier trial. "Anyone," Kennan wrote, "who witnessed the magnificent verbal duel between the State's Attorney and Radek and who saw the repressed excitement which took possession of these two men when certain subjects were brought up, could not fail to realize that this was not the mere recitation by an intimidated prisoner of recantations which he had learned by heart, but that very real things were involved." All the same, Kennan conceded, Radek confessed to crimes he never could have committed. In explanation, Kennan offered reasons similar to those Henderson had proposed, adding a characteristically Kennanesque coda:

> Even if all the facts of the case were available, which they certainly are not and never will be, it is doubtful whether the western mind could ever fathom the question of guilt and innocence, of truth and fiction. The Russian mind, as Dostoevski has shown, knows no moderation; and it sometimes carries both truth and falsehood to such infinite extremes that they eventually meet in space, like parallel lines, and it is no longer possible to distinguish between them.[6]

Bohlen covered the last of the major trials in March 1938. Bohlen interpreted the ordeal of Nikolai Bukharin in much the same terms Henderson and Kennan had applied to the earlier prosecutions. Bohlen had met Bukharin shortly after the embassy opened in 1934 and had been impressed by the sincerity of his beliefs in the Bolshevik cause. Although Bukharin maintained his composure throughout the trial, Bohlen believed he must have found what the whole affair revealed about the transformation of the revolution thoroughly distressing— aside from his feelings regarding his personal stake in the outcome.

Bohlen described the chief prosecutor as being at the top of his form. "Anyone who saw him, as I did, mercilessly pursuing, mocking, and prodding defendants will never forget the ferretlike quality of Vishinsky," Bohlen wrote. Adjudging Vishinsky "one of the most unsavory products of Bolshevism" and pointing out that the procurer had originally been a Menshevik and was thus a late arrival to Lenin's party, Bohlen suggested that much of his "bloodthirsty career" could be traced to a desire to compensate for early sins.

As for the testimony, it was "fantastic." One witness asserted that defendant Genrikh Yagoda, former head of the secret police, had on his last day at work sprayed the NKVD headquarters with a toxic material designed to kill his successor. Three expert witnesses—Vishinsky's description, not Bohlen's— affirmed that the substance in question could have have led to the death of

Comrade Director Yezhov if it had not been detected in time. In the event Yezhov had in fact fallen ill. (Bohlen took this as a hint that Yezhov's days were numbered too. Sure enough, a few months later the purge claimed him, making room for a longer-lived successor, Lavrenti Beria.) More unbelievable still was the means by which Maxim Gorky was said to have been murdered in 1936. Yagoda, in all apparent seriousness, declared that his agents had got the old man drunk on a winter night in a room heated only by a fire. After he was suitably inebriated they drew him close to the hearth, so that his face and front side overheated, while his rear grew freezing cold. The resulting temperature gradient had induced an acute and fatal case of pneumonia.

At one point in the trial a defendant spoiled the show by recanting his confession. Bohlen surmised that the individual, a former vice-commissar for foreign affairs named Nikolai Krestinsky, was laboring under the belief that the Kremlin's top leadership was not fully aware of what the secret police were doing and that by making waves he might force Stalin to take notice. The incident created barely a ripple, however, and following a night of encouragement to ponder matters, Krestinsky recanted his recantation.

After nearly two weeks, at four o'clock on a March morning, the trial came to an end. The judge, who looked to Bohlen like a "sadistic pig," read the sentences. "With obvious relish," Bohlen wrote, "he intoned the names of the defendants, followed, in eighteen cases, with the refrain, 'To be shot, to be shot, to be shot.'"[7]

III

If Bohlen considered the testimony at the Bukharin trial fantastic, Henderson and the other career officers deemed Ambassador Davies' reportage only slightly more credible. The appointment of Davies, who qualified for the Moscow assignment on the strength of his longtime friendship with Roosevelt and his deep pockets, had provoked a near mutiny at the embassy. "He drew from the first instant our distrust and dislike," Kennan commented later.

> We doubted his seriousness. We doubted that he shared our own sense of the importance of the Soviet-American relationship. We saw every evidence that his motives in accepting the post were personal and political and ulterior to any sense of the solemnity of the task. We suspected him (and there could have been no more grievous failing in our eyes) of a readiness to bend both the mission and its function to the purposes of personal publicity at home. What mortified us most of all was the impression we received that the President himself knew nothing about, or cared nothing for, what we had accomplished in building up the embassy at Moscow; that the post, with all that it stood for, and all that depended on it, was for him only another political plum, to be handed out in return for campaign contributions.

Bohlen was more succinct, labeling Davies "sublimely ignorant of even the most elementary realities of the Soviet system and its ideology," and unwilling to learn. Elbridge Durbrow considered Davies the most "mentally dishonest" person he had ever worked with. When Davies finished his tour by writing a piece of puffery entitled *Mission to Moscow*, the career officers privately redubbed it *Submission to Moscow*.[8]

The Davies appointment placed Henderson in a difficult spot. Henderson, no less than the Soviet specialists, considered Davies the wrong man for an exceedingly important job, but as the top career officer at the embassy he felt a responsibility to give the new ambassador a chance to prove himself. In some respects, having anybody, even Davies, as ambassador was an improvement over a continued vacancy. Henderson had been able to decide day-to-day matters for the embassy, but certain questions required an ambassador. "The last six months . . . ," Henderson wrote to friend and confidant Arthur Bliss Lane, just after Davies' arrival, "have not been very easy for me, and I am glad that the Ambassador has appeared on the scene." Always willing to look on the bright side of issues relating to the State Department and the foreign service, Henderson pointed out that since Davies would be "starting without any hang-over prejudices" he might manage progress in some minor areas the disillusioned Bullitt had failed to make. Henderson added that although the new ambassador had not yet demonstrated much interest in the daily routine of the embassy, he was a "very active man" who had plunged into his job with "considerable vigor."[9]

Unfortunately, vigor could not compensate for what seemed to the career men a serious case of wishful thinking on Davies' part regarding the Soviet system. The ambassador's commentary on the Bukharin trial appalled Bohlen, who had been there, and Henderson and Kennan, who had witnessed the earlier affairs. That Davies could even entertain the notion that the trials represented a sincere effort to uncover the truth struck them as ludicrous or worse, but apparently he did. "If the charges are true," Davies reported after the opening session, "a terribly sordid picture of human nature at its worst is being unfolded." When the court returned what Henderson and the others considered a clearly cooked verdict, Davies argued that justice had been done. He considered Soviet jurisprudence somewhat deficient procedurally, but he contended that the prosecution had proved its case "beyond a reasonable doubt."[10]

Davies did not shut his eyes entirely to the realities of the Stalinist system. "I think it is safe to say," he wrote, "that the dictatorship here now is not a dictatorship *of* the proletariat but a dictatorship *over* the proletariat." Yet he insisted on balancing the bad with what he conceived to be the good. "This laboratory in political science over here is the most interesting thing of its kind that I have ever seen. It is full of contradictions. There are splendid indications of the finest altruistic aspirations for the benefit of mankind and at the same time indications of the most ruthless cruelty of method." Davies believed that the French revolution was replaying itself in slow motion, although he did not think the end would be the same. At present a group of "very strong men" had seized control of the government, but Davies said they appeared to possess "high motives" and aimed to create a society "founded upon the ideological concept of selfless industry."[11]

Neither Henderson nor the Russianists for one moment thought Stalin possessed "high motives." They wondered how Davies could, and they found the contrast between the later Bullitt and the new ambassador dismaying. By the time Bullitt left Moscow he had taken to describing the "so-called Soviet Government" as "a conspiracy to commit murder and nothing else." Henderson and the other career diplomats might not have phrased the matter quite so bluntly, but they considered Bullitt's conversion a commendable recognition of reality, in part the result of their own educative efforts. Now they would have to start all over again with Davies.[12]

Meanwhile there was the matter of reporting the truth from Moscow. With Davies providing what Bohlen called the "Pollyanna" interpretation, Henderson took to writing private letters to Kelley delivering the dissenting opinion of the majority. A few months after Davies' arrival, Henderson recounted the effects of the purges on Moscow society. The nightly visits of the secret police, he said, had pushed aside nearly all other topics of conversation.

> No one seems to know who will be the next to disappear. Since, during the last ten years, it has been almost impossible for anyone to survive in the Soviet Union unless he broke a law now and then, and since nearly every person at some time or another in an unguarded moment has made statements critical of certain policies of Soviet leaders, there is practically no one who does not have a guilty conscience.

When the disappearances began, many ordinary people evidently had taken "a sort of malicious pleasure" in watching the mighty fall. "Of late, however, as the arrests have grown in number and have commenced to include all categories of workers, faces are beginning to look more worried than usual."[13]

Henderson analyzed the impact of the purges on various segments of the population. Intellectuals had come under particular fire. He described the arrest of a prominent jurist thought to have played a lead role in writing the new Soviet constitution. The arrest and subsequent trial had shaken the foundations of law in the country. "It looks," Henderson wrote, "as though the Soviet legal theories which have prevailed from the time of the Revolution up to the present are being ripped wide open." When the Bolsheviks came to power, Henderson explained, they had reorganized the courts to reflect their belief that laws were "a sort of disgraceful and necessary heritage of bourgeois regimes which should be dispensed with as soon as relationships of a socialist society could replace the relationships of a bourgeois society." Like much of the early idealism, this notion was apparently being discarded. What would replace it remained unclear. The uncertainty had paralyzed the law schools, since no one wanted to guess wrong.

The secret police had intimidated writers and artists also. Several noted novelists and playwrights had been arrested and charged with espionage, "leftism," and Trotskyism. At the same time government-approved reviewers and critics had set about discrediting the unfortunates' work. Historians found themselves in a quandary similar to that of the law professors. After the revolution Bolshevik historians had denigrated the accomplishments of czarist Russia, but this approach was now changing, in line with a general effort to revive patriotism. Yet no one cared to risk going very far in rehabilitating the presocialist past, lest the pendulum swing back. Even dramatic circles, heretofore considered safe, were under siege. This part of the Stalinist offensive had only recently started, and reports from the front were fragmentary. But "almost fantastic stories" regarding the alleged crimes of actors and directors had already surfaced. Among the most egregious, yet not the least common, were charges that certain individuals had attempted to plant bombs in the boxes Stalin used when at the theater.

The NKVD had gone so far in some instances as to arrest foreign nationals. Germans especially felt pressure, and at the moment of Henderson's writing some thirty Germans awaited trial after up to half a year in jail. From the Kremlin's perspective the arrests fit into a general anti-German policy. As to why Berlin had not made more of an issue of the matter, Henderson suggested that Germany was keeping relatively quiet in hopes of having the individuals expelled rather than tried and perhaps executed. Besides, he added, "the German Government probably realizes that there is little which it can do to assist its citizens other than by sending an army to Moscow."

Particularly shocking to the foreign community was the arrest of Boris Steiger. One of the few Soviet regulars on the diplomatic dinner circuit, Steiger had served as the American embassy's primary contact with the Kremlin. He was respected because he returned respect. Henderson commented:

> One of the members of the Diplomatic Corps told me yesterday that Steiger had been the only Soviet official who had ever talked with him as though he were an intelligent human being. He said that the others continually told him things which only a person with a childish intellect could believe. Steiger, on the the other hand, usually gave his auditors credit for having fairly well developed powers of observation. He never, for instance, tried to convey the impression that Soviet officials always acted with altruistic motives for the benefit of the working masses.

For this reason Steiger's loss was felt the more acutely. The charge against him—homosexuality—had provoked "incredulity" among foreigners in Mos-

cow. The diplomatic corps, Henderson said, included a large number of "obser-
vant and gossipy" people. "It hardly seems possible that if Steiger was abnormal
as charged he could have successfully hidden his weakness during ten years of
close association with that Corps." Besides, in the case of another official
arrested on the same charge, the foreign community had spotted the individual's "peculiarities" years before. Henderson could only surmise that for some
reason yet unknown the Kremlin preferred to put Steiger away on a nonpolitical
charge.

Henderson thought Steiger's boss similarly was feeling the heat. "Litvinov is
rarely to be seen these days," Henderson wrote. "Apparently he is spending a
good portion of the time at his dacha." Relaying a rumor that might or might
not have bearing on the foreign minister's fate, Henderson reported that the
cognoscenti were saying Litvinov had separated from his wife and taken up
with his young foster daughter. But Henderson did not think the bell was tolling
yet. "I feel that despite his numerous enemies Litvinov may survive for at least
a year or two longer."

One explanation floated for Litvinov's uneasiness, as well as for many of the
arrests, stressed an incipient revival of Russian antisemitism. Henderson did not
agree. "I don't take very seriously the talk which has been going around for
some time that Stalin is anti-Jewish and that he is planning to ease the Jews out
of their commanding positions in the Government and Party apparatus." Henderson conceded that Stalin might well be "impatient and distrustful of persons
with the mentality possessed by so many Eastern European Jews." But as evidence against a new pogrom Henderson cited reports of continued appointments of Jews to midlevel positions throughout the party and government.

If antisemitism did not account for the purges, Henderson could not say with
confidence what did. "The situation is extremely confused. We in the Embassy
in trying to analyze it have endeavored to approach it from various directions.
I must confess, however, that none of us is prepared as yet to advance any final
explanation for the weird developments which are taking place."

Nevertheless Henderson offered an interim assessment. It seemed, he said,
as though the dictatorship intended to destroy persons who for a decade and a
half had filled key roles in running the affairs of the Soviet state. "It would
almost appear as though Stalin is trying to discredit some predecessor who has
been ruling the country unwisely for the last fifteen years." Since Stalin himself
had been in charge during most of this period, one could not avoid the impression that Stalin was attacking himself. Henderson suggested that the world
might be witnessing a turning point in the development of Soviet society.

> Perhaps in the future, historians may say that from 1926 until July 1936 Stalin ruled
> the Soviet Union not entirely as he wished; that either because of lack of full self-
> confidence in his own ability as a political theoretician or because he felt himself

not to be sufficiently strong alone to cope with other forces about him, he has been steadily compromising with those Party members who have insisted that the communist revolution could be achieved best through doctrinaire methods. Is it possible to speculate that Stalin has grown impatient with the intellectuals who have so circumscribed his freedom of action that he is now determined fully to rid himself of them and in the future to take the advice of men of action who will not confuse him with abstract theories?

As support for the latter suggestion, Henderson pointed out that Stalin was himself a "man of action" who "like Mohammed would be much more inclined to make converts at the point of the sword than to resort to the tedious process of trying to change human habits of acting and thinking by the application of psychological and sociological theories."

Henderson remarked that Stalin's course presented certain dangers for the dictator. If indeed he planned to turn to "men of action," he might find that they were more difficult to control than theoreticians. In addition, the purge had eroded the prestige of the party by bringing scores of its leading members into disgrace. Whatever effect this might have on Stalin's own position, it could not but weaken the unity of the country. Yet Henderson did not discount the dictator's shrewdness.

> In the past Stalin has demonstrated a marked ability in judging the distance which he may safely go in any single direction. It is probable, therefore, that he will call a halt before the game of running to cover Bolshevik intellectuals and theorists of the Party is carried so far as entirely to undermine respect for the Party and for the opinions of Party leaders.

Meanwhile Stalin was settling old scores and getting rid of corruption in the ranks, which surely existed. "Undoubtedly by this process many thoroughly rotten branches are being lopped off the Soviet tree." Where it would end, Henderson could not tell. "One is beginning to wonder . . . ," he concluded, "how much hacking and pruning will be necessary before good solid wood is to be found."[14]

For the next several months Henderson continued to report on the pruning. In June 1937 he stated that the officials of the foreign ministry, the group he knew best, were "so patently in abject terror that one must pity them. . . . They fear to talk on almost any subject and apparently dread meeting foreign visitors, particularly those from the foreign Diplomatic Corps." In recent weeks the army too had come under attack. Because the purge there had just started, Henderson declined to guess what effect it would have on Soviet military preparedness. But it did not bode well for Russian security that scores of generals and hundreds of officers of lower rank had been arrested or had simply disappeared.

Henderson took advantage of this latest development to refine his earlier analysis regarding the causes of the housecleaning. Considering the elevated position of some of those removed from power, it now seemed undeniable that the purge was Stalin's handiwork. A good deal of the explanation appeared to lie in the dictator's personality. Despite his calm demeanor and certain undeniable gifts of leadership, Henderson said, Stalin was reputed to be "extraordinarily vain, loving of flattery, vengeful and unforgetful of real or fancied slights, jealous, suspicious, crafty, hot-tempered and capricious." The one thing more than any other that brought out the worst in Stalin was the mention or mere thought of Trotsky.

> During the long conflict between Stalin and the clever and glib Trotski both before and after Lenin's death, Trotski easily showed himself to be Stalin's superior in abstract thinking and doctrinaire argumentation. On some occasions, it is said, Trotski succeeded in making his slower-witted opponent appear ridiculous. Although Stalin eventually defeated Trotski, his victory was not a sweet one since, according to those who knew Stalin, he had the consciousness that he had not triumphed through personal brilliance or as a result of intellectual prowess, but merely because of his superior organizational ability. The struggle with Trotski is said to have left two deep imprints on upon Stalin, namely, an intense hatred for his opponent mixed with fear which is said to be increasing as the years go by, and a suspicion and fear of persons who tend to think in the abstract and whose intellectual processes he is unable to follow and comprehend.

Recent efforts by Trotskyists outside the Soviet Union to establish a competitor to the Comintern apparently had alarmed Stalin sufficiently that he had abandoned nearly all restraint and had decided to humiliate his old foe by portraying him as an agent of Hitler. "The Zinoviev-Kamenev trial was primarily staged in order to discredit Trotsky once for all in the eyes of the world international revolutionaries and to place him in revolutionary history not as an exiled martyr but as a traitor and double-dealer." Not incidentally, the trial also provided an excuse for liquidating some old Bolsheviks who, although imprisoned, still claimed enough support to worry Stalin.

But the attack on Trotsky had backfired, serving primarily to increase the exile's visibility and prestige abroad and straining the credulity of many even in Russia. This unexpected consequence had infuriated Stalin, Henderson wrote.

> In a spirit of revenge he began to seek out all means in his control to hurt his rival. He caused one of Trotsky's sons who still lived in the Soviet Union to be arrested and, it is reported, had lists made of Trotsky's former allies and friends. One by one, the persons whose names appeared on these lists were disgraced and usually arrested. Not only were Trotsky's former friends disgraced, but friends of his former friends, and even friends of friends of his former friends.

Those of Stalin's advisers who recommended moving slowly found themselves charged with Trotskyism. Such advice soon ceased.

From an attack on Trotskyism to an assault on the intelligentsia was a small step. Members of this influential group exhibited what to Stalin seemed a disconcerting independence. "They are in general loyal to the State and, in theory at least, to the principles of world revolution," Henderson explained. "They are inclined nevertheless in varying degrees to consider themselves as adults who can think for themselves and to look with a somewhat amused and indulgent attitude upon the methods by which the Kremlin endeavors to control, and retain the support of, the masses of workers and peasants." Worse, although the intelligentsia was in some measure Stalin's own creation, at least in the sense that he had allowed its development, the intellectuals looked down on their former patron. "Every now and then some of the new intelligentsia are said, without reproof from their fellows, to have made remarks which might be interpreted as ridicule of the dictator." When word reached Stalin, as most things in the Soviet Union eventually did, the result was predictable.

In addition to the Trotskyists and intellectuals, Stalin feared the "leftists": persons who demanded more-rapid progress toward communism within the Soviet Union and more-active pursuit of revolution abroad. Some but not all of these individuals were old Bolsheviks. Newer recruits simply took the old teachings seriously and suspected Stalin of abandoning socialism for state capitalism, and revolution for Russian patriotism. Although the leftists had created no formal organization, they apparently represented a growing body of opinion.

Whatever the reasons for the purge, its effects spread daily.

The bureaucracy, engineers, technicians, and so forth, are, for the most part, in a panic. . . . They are so frightened that in general they do not want to attract attention to themselves. Accordingly they dodge responsibility and endeavor to pass along to one another the duty of making decisions. They are faced from below with a lack of respect and discipline on the part of workers and from above with officials who themselves would not hesitate to pass on to them any blame which may arise from shortcomings in work.

In the universities and intermediate schools, professors and instructors in many instances are afraid to face their classes. Their histories have been condemned, their philosophical textbooks have been found wanting, their legal textbooks have been banned. Students are inclined to show their teachers less respect than hitherto, and in some cases to take advantage of old grudges to report statements made in lecture rooms which may be interpreted as treacherous.

The theaters are just as brilliant as ever, but it is understood that when Stalin and his suite appear at any performance everyone connected with it, including the playwrights, the directors, the actors, and the managers are in a fright until a verdict has been rendered.

The members of the new as well as the old intelligentsia are both frightened and resentful. Many of them not only fear for their own personal safety, but are resentful at having been deprived of the right of freedom of discussion on almost any subject. . . .

The prestige of the Red Army has suffered tremendously. Only a few months ago it was considered as the bulwark of Soviet power. Its officers and men were accorded special treatment. Discipline was good. It was said that it was freer of internal dissension than the armies of capitalist countries. In a comparatively brief period officers and men have been given to understand that the superiors in whom they had confidence are untrustworthy. The feeling is prevalent among all ranks that the new political commissars are being sent out to spy on the officers. The officers have lost confidence in themselves and the full respect of their men. . . . A considerable period must elapse before the Army can recover from the blow which it has received.[15]

6

Minuet for Dictators

In November 1936 Henderson took the occasion of the third anniversary of normalization to assess the nature and development of U.S.–Soviet relations since 1933. The prevailing motif, he declared, was mutual disappointment, and Moscow's arose principally from American reluctance to collaborate against the fascists. Recently enacted neutrality legislation, by which the American Congress attempted to prevent the kinds of entanglement that had drawn the United States into the world war, had upset the Soviets. In the Kremlin's view, Americans and Russians ought to be able to set aside whatever differences they had regarding internal politics to put a halt to the depredations of the Germans and Japanese. But Washington, by adopting a policy of knee-jerk neutrality, was, according to Henderson's Soviet interlocutors, "failing to assume its share of responsibility for the maintenance of world peace."

Henderson went on to describe other sources of dissatisfaction. The Russians had hoped normalization would lead to enhanced prestige for their country. "American recognition meant that the last of the Great Powers of the world had finally decided to admit the Soviet Union as a full-fledged member into the family of nations." The Kremlin's leaders also hoped normalization would result in economic aid and most-favored-nation status, and they had thought that to some degree it signaled a willingness on Washington's part to cease harping on Moscow's connections to the Comintern. Washington had disappointed Moscow—largely because Moscow had disappointed Washington. The United States had watched with dismay the Kremlin's continued stonewalling on the debt issue and the failure of a larger Soviet market for American goods to materialize. As for the Comintern, it continued to grate on the Americans precisely because of Moscow's connections.

The fact that expectations on both sides had not been met, Henderson continued, had created "considerable strain" between the two countries. A variety of lesser irritations contributed to the tension. The smaller annoyances owed primarily to "the shock of the contact between the American and Soviet systems." The numerous ways in which Soviet officials made life difficult for the embassy staff—refusal to grant adequate housing and office space, imposition of arbitrary exchange rates for currency, attempts to isolate the Americans from Soviet society—could be traced "to the fact that representatives of foreign governments, regardless of what might be their feelings towards the Soviet regime, or the extent to which they may be in the good favor of the Soviet Government, are nevertheless products of what is deemed to be a hostile system and therefore automatically subjected to suspicion and restriction." Henderson expected some sources of irritation, such as the housing shortage and the currency problem, to ease with improvements in the Soviet Union's economic situation. But others were so deeply embedded in Russian practice that they would last indefinitely. Foreign diplomats, he noted, had been describing Moscow as a hardship post since the days of Ivan the Terrible.

Further barriers to better relations dated only from the Bolshevik revolution. Of these the regime's revolutionary messianism constituted the most serious obstacle. "I am convinced . . . ," Henderson wrote, "that the establishment of a Union of World Soviet Socialist Republics is still the ultimate objective of Soviet foreign policy." Henderson admitted that this objective was "somewhat dimmer" than it had been during the pre-Stalin era, and he conceded that short-term exigencies of foreign affairs might require setting it aside momentarily. Nevertheless, he argued, "this objective is a real one at the present time and is a factor not to be ignored in any discussion of Soviet-American relations." Although Russian diplomats habitually claimed that tranquility between the two countries was "being sabotaged by American officials who fail to present a true picture of the situation to their superiors," it was Moscow's abiding radical vision, rather than any personal factors, that lay at the base of U.S.–Soviet problems. American monetary claims created difficulties because the Kremlin refused to waive "the right of revolutionary governments to repudiate debts." Despite Americans' best arguments, Soviet leaders insisted on nationalization of property as a "basic revolutionary principle." Regarding the Comintern, Henderson wrote, "I am convinced that unless the whole system of the Soviet State and the ideology of its leaders should undergo a complete change—and this does not seem likely in the foreseeable future—the Kremlin will continue to express certain of its foreign policies through the Communist International or similar organizations."

The doctrinaire philosophy of the Soviets gave rise to a characteristic bargaining style. Unwilling to grant any legitimacy to the position of opponents, the Soviets accepted concessions as their due and immediately raised their demands. To a greater degree than most governments, the Kremlin had laid out a series of definite objectives, and officials in the Soviet bureaucracy rose or fell depending on their success in achieving these goals. To be sure, the Soviets did not entirely lack flexibility. Firm believers that ends justified means, they could and did adopt interim tactics apparently at odds with long-range strategy.

The popular-front policy advocated by the Comintern's seventh congress fell into this category. For the moment the Kremlin was soft-pedaling revolution in favor of opposition to fascism, which threatened the homeland, and therefore existence, of the revolution.[1]

Henderson's report elicited enthusiastic reviews from the State Department, with Washington sending congratulations on an "excellent dispatch" demonstrating an "objective and scholarly" understanding of the vexing issues of U.S.–Soviet relations. Encouraged, Henderson supplemented this memo a few months later with a more thorough analysis of the Kremlin's foreign policy aims. For several years, he wrote, the Soviets had been preoccupied with the activities of the Germans. Hitler's rise to power had increased Moscow's wariness, so that now the Kremlin read sinister motives into everything Berlin did.

> When Goering goes hunting in Poland it is announced in Moscow that Beck has sold out to Germany. When Schacht goes to Belgium it is pointed out that he is engaged in maneuvers for the purpose of obtaining foreign credits for Germany. When Germany and Yugoslavia sign a trade treaty, Moscow finds that Germany is making progress in its aim to break down the Little Entente. When the countries of the so-called Oslo bloc, namely Belgium, Luxemburg, Holland, Denmark, Norway, Sweden and Finland, engage in conferences, the Soviet press insists that Germany and Great Britain were quarrelling for economic ascendency in those countries.

Article after article in Soviet publications, Henderson reported, declared that Hitler was preparing for war. The authors differed regarding where the first blow would fall. Some said to the west, others to the east. Many predicted that Czechoslovakia would provide a *casus belli*. But all agreed that so long as the Nazis ruled Germany the peace of the continent was at grave risk.[2]

As for Moscow's views on the other countries of Europe, these views depended largely on the Kremlin's interpretation of how those countries were preparing to stand up to Hitler. The British came in for considerable criticism for failure to announce themselves definitely on the side of peace. Likewise Poland, whose foreign minister, Joseph Beck, Henderson described as Moscow's choice for "the most unpopular European statesman." Beck was charged not only with passivity in the face of Hitler's bullying but with active skulduggery against Russia. The Polish diplomat's recent trip to Romania, for example, was being branded an effort to revive the "border state front" against the Soviet Union. France had fallen into Moscow's bad graces as a consequence of the habit of French officials, when referring to the "peace-loving democracies," of including France, Britain and the United States but omitting the Soviet Union. In addition France's apparent efforts to find a supplement to or substitute for the 1935 Franco-Soviet pact as an instrument of European security had the Kremlin on edge.

This last point, in combination with Stalin's capacity for suiting means to ends, caused Henderson to warn against ruling out a Soviet about-face regarding Germany. For some time rumors had circulated in Moscow of a thaw in relations between the Kremlin and the Nazis. To a certain degree these stories might represent nothing more than a Soviet attempt to smoke out the west. On the other hand, the Red army appeared sincerely interested, for obvious reasons, in not having to fight the rapidly improving Reichswehr. In any event Henderson stated that a sudden rapprochement, despite the current bitter criticism on both sides, was "not at all impossible."

In Asia, Japan occupied a position in Soviet thinking analogous to that of Germany in Europe, except that Moscow seemed to fear the Japanese less than the Germans. Soviet militarization in the Far East, Henderson said, had made "tremendous progress," boosting Russia's confidence vis-à-vis Tokyo. "The Kremlin apparently feels itself to be no longer under the necessity of mincing words or stalling for time in dealing with the Japanese." Almost until the present, Soviet officials had declined to protest when Japanese troops in China had taken liberties with the frontiers of Siberia and Mongolia. Such toleration would soon end. "It seems to me that Moscow is likely to become progressively more curt in its treatment of Japan."

Relations with China tracked those with Japan, but inversely. According to reliable reports, Soviet agents were working for a reconciliation between Chiang Kai-shek's Kuomintang party and the Chinese Communists, the objective being a common front against the Japanese. In recent negotiations between the two parties, the hand of the Kremlin was evident. "It is quite obvious that Chow [Zhou Enlai] is taking his directions from Moscow and that his conversations with representatives of Nanking [the Kuomintang government] form an integral part of Soviet-Chinese relations." In the interests of thwarting the Japanese, Russian officials were not hesitating to counsel the postponement, perhaps cancellation, of the revolution in China. "It seems that Moscow is willing to consent to the Chinese red army and the Chinese soviets losing their identity as separate entities if in return Nanking is willing to follow a line more in conformity with the policies of Moscow and at the same time to offer stiffer resistance to the Japanese."

Should Chiang refuse the offer, however, Stalin could flip-flop in the east as quickly as Henderson suspected he might in the west. If the Russian dictator could get no one else to stand up to Japan, it was not inconceivable that he would agree to some partition of China between Japan and the Soviet Union.

As for Stalin's plans regarding the United States, Henderson reported a variety of opinions circulating in Moscow, particularly in the Soviet press. Yet the different writers tended to concur on several points. In the first place, the consensus held, Washington's response to events in Europe would probably swing the balance on the continent. If the Americans stayed out of continental politics, Hitler might well run rampant. If the Americans came in, he could be stopped. Second, despite the current wave of isolationist passion, "influential circles" in the United States, centered on the White House and the State Department, would eventually bring American power to bear on world affairs. Regardless of the wishes of the masses, the capitalist class in the United States could not afford to abandon its overseas interests and ambitions. In this respect a recent proposal by the Roosevelt administration to amend American neutrality legislation to allow cash-and-carry trade indicated that a shift in favor of Britain and France had already begun, since those two countries more than others possessed the dollars and bottoms to make good use of a pay-and-take-it policy. Third, any de facto alliance among the capitalists would not outlast the threat from Germany and Japan. The imperialists, inveterate rogues at heart, eventually would fall out as they found themselves competing for the same markets.[3]

II

The demands of the American foreign-policy bureaucracy interrupted Henderson's reporting of Soviet reactions to world events. As the scorn the Russianists heaped on Davies' appointment as ambassador demonstrated, the Roosevelt administration was split into two warring camps regarding the Soviet Union. On one side, the Eastern Europe division of the State Department, headed in Washington by Kelley and represented in Moscow by Henderson and the other career officials, looked askance at Moscow's motives and read revolution, if sometimes disguised and delayed, into the Kremlin's every action. The opposition consisted of various agents of the White House, including Davies, Undersecretary of State Sumner Welles, Eleanor Roosevelt, and other "left-leaning New Dealers"—to use Henderson's loaded phrase—who held that Stalin was neither the revolutionist Lenin had been nor the world-swallowing ogre his current critics made him out to be. These two schools of thought were not completely irreconcilable: the Kelleyites admitted that even a Kremlin bent on revolutionizing the world might first be helpful in eliminating the Nazis, while the president's men and woman could not ignore the possibility that an alliance with the United States might only represent a decision to defer the day of confrontation with the foremost of the capitalist powers. But in the natural course of intramural tussling, the proponents of the different views personalized and polarized the issues, while subsuming them in the larger suspicions the careerists and the politicals felt toward each other. As a consequence, those close to the president treated Kelley and his crew as reactionary obstructionists, while the East Europeanists sniffed at what they called the "hands across the caviar" naïveté of the first lady and her allies.[4]

By the spring of 1937 the acrimony had reached the point of bureaucricide.

Even as Henderson quelled talk of a mutiny in Moscow on Davies' arrival, Sumner Welles moved to liquidate his opponents in Washington. Henderson once described the undersecretary as "extremely able and ambitious" and "impatient with any person or thing that might restrict his activities or block his advancement." In June, Welles vented his impatience on the East Europeanists. He summoned Kelley and told him Eastern Europe no longer existed on the divisional charts of the department, that his job was being abolished, and that he was being transferred to Turkey.[5]

Henderson found this news distressing, though not entirely unexpected; but he hardly had time to grieve for Kelley before he learned that the shakeup might reach Moscow. The embassy there was spending a lot of money. The department wanted to know if it all was necessary, and Washington sent a special inspector to find out. As chief administrative officer and principal representative of the Kelley clan, Henderson could not help feeling uneasy.

The demeanor and personal style of J. Klahr Huddle did little to assuage Henderson's worries. For two weeks, while Huddle pored over all manner of records, from invoices for stationery to cable traffic, and interviewed each member of the mission's staff, the inspector refused invitations to join the social life of the embassy. Following a final four-hour interview with Henderson, during which Huddle kept strictly to business and gave no indication of satisfaction or displeasure, the inspector retired to a borrowed office for three days to write up his report.

Finally, with the report completed and in the diplomatic pouch, Huddle broke a smile. He congratulated Henderson on the good work he was doing, in terms of efficiency of operation and astuteness of reporting. Needless to say, Henderson was tremendously relieved. Having seen Kelley exiled, he had imagined a similar fate or worse for himself. But surviving the test gave him a surge of confidence. He later commented that he considered Huddle's favorable evaluation one of the "turning points" of his career.[6]

III

With bureaucratic matters under control for the moment, Henderson returned to his observations of international affairs. He found most intriguing the manner in which Stalin and Hitler seemed to be sizing each other up, and despite the uncompromising hostility emanating from their respective propaganda departments, Henderson considered rapprochement increasingly likely. To be sure, such a volte-face would not come easily. On the Russian side, the spring of 1937 witnessed a wave of especially vitriolic anti-German articles in the Moscow press. But one could never tell just what such inspired journalism meant. Perhaps the significance lay right at the surface, and the Kremlin was aiming simply to rally the party and the reading public against the fascist threat from the west.

On the other hand the articles might be intended for German consumption, motivated, as Henderson put it, by a "desire to impress upon Germany the fact that the Soviet Union should not be overlooked as a factor in European affairs and that Germany might therefore find it advantageous to come to a working understanding with the Soviet Union." Perhaps the propaganda campaign was designed to serve both purposes. This seemed to Henderson the most likely explanation. He remarked that the Kremlin was being careful "not to burn its bridges behind it" so far as Germany was concerned.[7]

Henderson reflected on the preconditions and probable shape of a Soviet-German detente. If Great Britain and Japan came to an understanding in the Far East, one that failed to guarantee Soviet interests in the area; if political shifts in France brought to power a government opposed to cooperation with Moscow; if the League of Nations continued to fail as a keeper of the peace— then, Henderson believed, the Kremlin would strongly consider a revival in some form of the 1922 Rapallo agreement with Germany, despite the fact that the fascism of the Nazis had supplanted the republicanism of Weimar. "The Soviet Union in the past," Henderson remarked, "has not been particularly squeamish regarding the non-socialistic standards of its friends." A few years earlier Moscow had sought cooperation with Japan in the face of the militarism already in evidence in Tokyo. Until Mussolini's Abyssinian adventure the Kremlin had experienced little difficulty getting along with the duce. "I can see no reason why it should not be willing to work with Germany merely because the German Government might be reactionary."

Should the Kremlin decide to collaborate with Berlin, Henderson thought, Stalin would give away even fewer of his principles than he had in relations with the United States. "If Hitler has any expectations that the Soviet Union will formally drop its international revolutionary ideas ... I think that he will be disappointed." Yet Stalin might well rein in German communists and order the Comintern and other front organizations to tone down their attacks on Berlin. In return the Nazis would have to adopt a friendlier policy toward the Soviet Union and, to some degree, lay off the Jews. Because this latter estimate might seem surprising, Henderson elaborated.

> Please don't gather the impression that I feel that the Soviet Union is dominated by the Jews. Nevertheless, I am convinced that the Soviet Union realizes that, by and large, international Jewry is an important supporter of it in international affairs, and I further believe that it would prefer, at the present time at least, to continue to have bad relations with Germany rather than to allow the internationally-minded Jews of the world to feel that it has left them in the lurch.

Henderson cautioned that his thoughts on these matters did not constitute firm predictions. They were simply "in the realm of poor speculation." On the other hand, he would not have raised the issues had he not had some confidence in his analysis. "I think it quite possible that there may be extremely interesting developments in Soviet-German relations during the next two years," he concluded. "We might find it profitable to keep this fact in mind."[8]

Russia's external policy, of course, bore a close connection to what was happening within the country. Henderson continued to devote considerable effort to reviewing developments on the domestic front. In the late winter of 1937 he took a trip through the hinterland beyond Moscow. He had traveled outside the capital earlier and found the rural regions generally bleak. This time he acquired a different opinion. For all the paralysis the purges had induced in Moscow and among the bureaucracy, life on the farms seemed to be improving. Living conditions for workers and peasants remained spartan, to say the least, with a wide variety of consumption goods taken for granted in the west either unavailable or exorbitantly priced. But the acute shortages common just recently had eased. The peasants had benefited from what Henderson deemed a "genuine effort" by the Kremlin to remedy rural deficiencies. The payoff was psychological as well as material. Henderson described a "new vitality" animating village life. He recognized that his guided tour was precisely that, but even at the appropriate discount he considered significant the upbeat mood of the communities and people he encountered.

> In every one of the twenty villages through which we passed, we noticed that extensive repairs of houses and clubs were being carried on; in some instances we observed that new houses much superior in construction to the old peasant shacks were being built. In these villages Sundays rather than the rest days are still observed and along the road we met peasants out for strolls dressed apparently in their best clothes. I was impressed by the fact that at least the Sunday clothes of the peasants in that part of the country visited were just as good as the clothes generally worn by the people in Moscow on free days.
>
> Not only was there much construction going on in the villages but there was a hustle and bustle in the whole countryside which I didn't note in my trips last year. Although this part of the Soviet Union is not so suitable for large-scale farming as are the prairies of the Ukraine, we saw tractors in almost every village.

Henderson had returned to Moscow by a road following the new Volga-Moscow canal. The magnitude alone of the project struck him as noteworthy. He commented that the amount of material and labor involved must have been "enormous." The latter resource came cheap; reports indicated that up to 200,000 convicts had dug, hauled, blasted, and poured for hundreds of millions of man-days. He found these reports quite credible, as he personally had seen "village after village" made up entirely of prisoners. They obviously worked hard. "At 9:00 o'clock in the evening we passed one group of some 20 convicts lying on the road with shovels and picks in hand sound asleep with the guard standing over them. They seemed to be too tired to give any heed to the coldness of the ground and the noise of the passing traffic." If the canal lived up to expectations—Henderson had doubts, since although the canal was set to open officially within days the amount of work obviously remaining would require another year or two to complete—it would not only link Moscow with the Caspian Sea but provide a large supplement to the city's water supply.

In the defense industry as well, the Soviets seemed to be making big strides. Here Henderson had to rely on the accounts by the military attaché to the embassy, Colonel Philip Faymonville. Henderson was skeptical regarding Faymonville's reliability, not least since Faymonville came to Moscow with Eleanor Roosevelt's seal of approval. Henderson considered the colonel too quick to credit the Soviet interpretation of debatable issues. But he conceded that Faymonville had cultivated informed sources, and he thought the attaché's observations deserved passing along to the State Department. Faymonville had recently visited an installation near the Urals devoted to the production of engines for warplanes. He returned quite taken. "He said the factory was by far the best plant that he had seen in the Soviet Union and in some respects superior to any American airplane motor factories." The operation, which was nearly self-contained, with casting, milling, and molding done on site, employed 12,000 persons, each working twelve hours per day. Henderson found the figures impressive. Other observers apparently did too.

> It is rather startling to realize that a single factory in the Soviet Union is turning out motors sufficient in number to equip almost 4,000 war planes a year. One can understand why the Japs and the Germans, as well as the Poles and the people of the Baltic States, are inclined to look with some concern upon the increasing industrial and military power of the Soviet Union.

Henderson noted reports that in striving for these gains the Kremlin had rolled back some of the principles of socialism, but he argued against inferring any significant liberalizing tendencies from such measures. Indeed, although the Soviet government had resorted to "practically all the artifices employed in the so-called capitalistic countries" to improve efficiency and labor productivity—differential pay scales, bonuses, and the like—the party bosses kept the pace and nature of change entirely under control, and they seemed little inclined to go much further. "I don't believe that Stalin or the leaders around him have any intention to permit the Soviet Union to develop into a capitalistic economy."[9]

Henderson described reforms in the political arena. During the spring of 1937 the Moscow press made much of recent announcements of forthcoming elections for representatives to positions in the Communist party and the various soviets. The voting in this case received special attention due to the fact that it would be the first held under the universal suffrage provisions of the new constitution and would be by secret ballot. Did the Kremlin intend to democratize Soviet political practices?, Henderson asked rhetorically in a letter to the State Department. The question, he replied, was "extremely difficult." In certain respects, he believed, the answer was yes. The different groups authorized to select candidates—unions, cooperative societies, youth organizations, cultural societies, and others—would probably nominate individuals of their choice without overt interference by the state. But Stalin need not worry. "Since no society can publicly exist for a day unless it obeys the orders of the Party, there is little likelihood that persons will be nominated for offices if the Party has any objections." Yet a certain substance lay beneath the reforms. Stalin was "anxious for the office holders, particularly those in the provinces, to have a little competition." The dictator recognized the tendency of bureaucrats to avoid risks and look first to their own security. The election plan should stir things up. "Stalin wants the Government to be manned by real leaders, and through the new electoral system he hopes to be able partially to gauge their real leading ability and at the same time to put them on their mettle."

Henderson did not envy the candidates in such a setting, for now they had to answer not only to the party above but to the people below. "If they meticulously carry out orders received from their superiors which are sure to be unpopular in the villages, their own popularity will certainly suffer. On the other hand, if they fail fully to carry out such orders, they will be reprimanded by their superiors for failing in the performance of their duties." The Kremlin had recognized the problem, as indicated by the fact that the central government had relieved elected officials of some of their most obnoxious duties, such as collecting grain and taxes. These chores now rested with appointed agents.[10]

When the balloting, after a great deal of preparation, took place at year's end, Henderson analyzed the outcome. The elections, he said, were "a gigantic dumbshow in which the voters were not permitted to have any voice whatsoever," and represented "a farce from the American or Western European conception of what elections are." The pre-election publicity campaign "was in fact nothing more than a pure propaganda drive for the purpose of popularizing Stalin and his adherents and consolidating their power, and for persuading, at times by intimidation, the electorate to go to the polls and register their approval of the Stalinist regime." All the same, Henderson believed that the exercise had planted "the seeds of a primitive democracy." Whether the seeds would sprout and mature was another matter, of course. Stalin evidently had no such intention. Nonetheless, a country that had never known democracy had to start somewhere, and Stalin's heirs might take a different view.

To the reasons he earlier had advanced for the Kremlin's decision to hold elections, Henderson now added two others. Aside from the introduction of the secret ballot, which for all the limitations placed on its use Henderson deemed an important precedent, the authorities had adopted nonpartisan registration for government positions. Thus voters saw no distinction on their voting forms between Communist party candidates and nonparty nominees. This erosion of distinctions between those in the party and those outside complemented similar trends in other aspects of Soviet life and seemed to Henderson part of a larger effort by Stalin to weaken the Communist party and create a ruling group "supremely loyal" to him.

The second consideration Henderson perceived affecting the Kremlin's decision to put on some of the trappings of democracy involved the continuing fear of war. In the view of Moscow's leaders, a showdown was shaping up between the fascist and the nonfascist states. Henderson still considered a Nazi-Soviet deal possible, but the Kremlin apparently deemed such an expedient a last resort. Meanwhile it sought to secure defense arrangements with the nonfascists, and Stalin was doing his best, within the narrow confines of a totalitarian order, to make the Soviet Union appealing to the democracies of the west. Some foreign observers in Moscow, Henderson said, considered this concern for international appearances the "deciding influence" leading to the elections. Henderson himself did not go so far as this, but he conceded that "there is little doubt that the Soviet desire to ingratiate the Soviet Government with the believers in democracy of the western countries played a role in causing the Soviet leaders to have the elections conducted somewhat along the lines of elections in democratic countries."[11]

IV

The spring of 1938 marked the end of Henderson's fourth year in Moscow. By ordinary standards of State Department and foreign-service practice, he was due for reassignment. In January, George Messersmith had notified him he might expect a transfer to Washington in the not-distant future.

Henderson was in no hurry to depart. He told Messersmith, "I personally would have preferred to continue for a time in field work." When the department indicated that his transfer might come through by the beginning of summer, Henderson replied, "There is no reason in so far as I am personally concerned for it to be speeded up." He offered to stay on in Moscow until the end of the year.[12]

Henderson's reluctance to leave is significant, especially in light of the fact that others at the mission considered Moscow an increasingly difficult and decreasingly rewarding post. The social isolation of the embassy bothered Henderson scarcely more than before. Elise had joined him after the initial settling in; with her company and that of the rest of the foreign community he could put up with the lack of Russian contacts. More to the point, he felt he was doing the most important work of his life. Nearly forty-six, he held a position of large responsibility in a field in which he had trained for a decade and a half. He recognized his limitations—his lack of Russian, for example—but he compensated for these by making good use of persons like Kennan and Bohlen who did command the language and by putting in more effort than anyone else. He was a meticulous drafter of dispatches, revising and editing endlessly. If the resulting prose often lacked grace, it rarely fell short in clarity. Washington valued his reportage, as his numerous commendations testified. He felt he knew Stalin's Russia about as well as any American—and he was probably right. Although he did not possess some of the historical and cultural perspective of the Russianists, he was an astute observer of contemporary politics. From Kelley he had acquired the habit of extreme skepticism regarding everything Soviet leaders did. Later this inclination to assume the worst and most devious would result in false, or at least incomplete, conclusions. But in dealing with Stalin in the 1930s it was an entirely appropriate attitude. In Moscow, Henderson had found his niche. It had taken him a long time and a great deal of physical, mental, and emotional effort to do so. Naturally he hated to leave.

Washington insisted, however; and in July Henderson paid a farewell visit to the Soviet foreign ministry. Litvinov was his usual self, which Henderson found thoroughly disagreeable. On hearing that Henderson was being detailed to the State Department, the foreign commissar frowned and remarked on the "reactionary" nature of that organization. Most officials there, he asserted, had fascist friends, were sympathetic to the fascist cause, and were antipathetic to the Soviet Union. It was unfortunate that Henderson could not go to work for President Roosevelt, a "great liberal" who was anxious to further the cause of democracy and liberalism throughout the world. That the United States and the Soviet Union had not managed better to resolve their differences owed not to any deficiency on the part of the president but to the perfidy of the "reactionaries in the State Department," who attempted to distort the policies laid down by the president and give them an anti-Soviet and pro-fascist bias. Litvinov commented that he knew all this not simply from the Soviet embassy in Washington; American visitors to Moscow, some of whom were "particularly well informed with respect to what was going on in Washington," concurred.

Henderson objected to this caricaturization of the State Department, although he recognized the basis of truth from which it was drawn. He told Litvinov he was mistaken regarding America's career professionals. From his own experience in Washington, he said, he knew them to be "conscientious, painstaking and able public servants" who were doing their best to carry out the president's directives. If the foreign commissar had heard otherwise, the discrepancy no doubt resulted from the existence in America of opportunity for vigorous debate on all manner of subjects. The American government even tolerated the activities of individuals and groups who were more interested "in furthering one or more international movements than they were in advancing the interests of the United States." The persons in question—Henderson was too diplomatic to name names—were "extremely vociferous" and prone to personal attacks on the integrity and loyalty of the State Department. Evidently the foreign commissar's informants fell into this category, which accounted for his mistaken impressions.[13]

After leaving Moscow and en route to Washington, Henderson stopped in the Baltic republics and in Finland. In each country he discovered in heightened form the same war fears he had encountered in Moscow. Poland, which he visited next, felt even more pressure. Samuel Harper, Henderson's Chicago friend, after a trip to Warsaw two years earlier had sent Henderson an account of his journey, concluding with the question, "Can Poland hold out against the two mountains on its two sides, from which pour down on it violently competing doctrines?" In 1936 Harper had thought Poland could. By the middle of 1938 Henderson was not so sure. The Poles were relying on the League of Nations and on the west to protect them from Germany and the Soviet Union. Considering the record for irresolution of both the League and the west, such reliance seemed misplaced, although Henderson had to admit he could not see what else Poland might do.[14]

A visit to Berlin proved no more encouraging. Henderson watched Hitler's goose-steppers on parade; he heard Nazi orators thunder their denunciations of Versailles and demand the respect due the thousand-year reich; he consulted with American military attachés who stated that already the Germans had outstripped the rest of Europe in fighting strength; and he shuddered for the future.

Discussions with William Bullitt in Paris improved his spirits slightly. Bullitt, now ambassador to France, was in one of his optimistic moods, although his comments hardly supported his optimism. He said Franco-American relations were improving, without adducing any persuasive evidence. He admitted that Hitler was a source of legitimate worry. Just what the Nazi leader's game was, he could not say, but it surely boded ill for Germany's neighbors.

In London, British officials were the gloomiest of all. Some of Henderson's Whitehall contacts confided that Britain was in no condition to fight. Should war come soon, they declared, the empire might well collapse.

At the beginning of August, Henderson and Elise landed in New York. He checked in with the State Department, and then the two set off on a cross-country tour. Elise had seen parts of the Atlantic seaboard but had never penetrated the interior. Henderson showed her where he had spent his youth in Arkansas and Ohio, where he had attended college in Illinois, and where he had taught in Colorado. He introduced her to relatives in Colorado Springs and Albuquerque.

An urgent message from Washington cut the vacation short. Hitler had demanded part of Czechoslovakia. The British and French were pondering their response. Chamberlain and Daladier had arranged a meeting with Hitler in Munich. Whether war would come, none could tell. But in any event the new man on the Eastern Europe desk had better get back to Washington right away.[15]

7

Between the Devil and the Deep Blue Sea

The State Department to which Henderson returned in October 1938 retained most of the somnolence it had possessed when he first crossed its marble threshold in 1922. With the country still mired in depression and the president nursing bruises from his fight to pack the Supreme Court with cooperative justices, the Roosevelt administration evinced no desire to challenge the isolationist mood of the electorate. Following the Japanese attack on China in 1937, which led to a conveniently—for FDR—undeclared war, the president declined to invoke the neutrality legislation, thereby allowing American weapons and other supplies to flow to the Chinese. But when he attempted to prick what remained of the nation's Wilsonian conscience with a call for a "quarantine" against the "epidemic of world lawlessness," declaring that "there must be positive efforts for peace," the angry response from isolationists in Congress and the press caused him to pull back. Franklin Roosevelt, unlike cousin Theodore, had little use for bully pulpiteering. No immersionist, he preferred to let converts to internationalism get used to the chilly waters gradually. With luck the shock would not wake them until after the election of 1940.[1]

Even had Roosevelt adopted a bolder approach, State Department officials would have been the last to know, for the president regularly jumped the chain of command in favor of his own personal agents. Secretary of State Hull often learned what his nominal subordinate, Undersecretary Welles, was doing by reading about it in the paper; and if special envoys like Harry Hopkins deigned to inform the careerists what foreign leaders had told him, the information was provided more as a favor than a right.

On the other hand, the lack of White House interest allowed the department, especially Henderson and others working on Eastern Europe, a certain amount of autonomy. Although Kelley's division, like some bureaucratic Constantinople, had fallen beneath the assaults of the liberal infidels, the Kelley spirit lived among the refugees. The Russian collection had vanished into the basement of the Library of Congress, and many of the division's irreplaceable files, as Henderson subsequently recalled, had been "mercilessly destroyed." Yet the pillagers overlooked one loyalist who, as Henderson related with a thin note of triumph, had managed "secretly to salvage some of them and to keep them hidden until the mania to destroy had subsided." More important, Henderson's immediate boss and the director for European affairs, Jay Pierrepont Moffat, had plenty to worry about on the rest of the continent and was happy to leave the mysterious east to Henderson and his staff, who were happy to make the most of their modicum of freedom.[2]

In the immediate aftermath of the Sudeten deal, Soviet leaders roundly condemned the western democracies. The Kremlin branded the Munich accord a betrayal of Czechoslovakia, a blow to the principle of collective security and the safety of smaller countries, an encouragement to aggression, and a general threat to world peace. With such hopes as they had rested in Britain and France now destroyed, the Russians felt more vulnerable than ever. Consequently they indicated a desire to improve relations with the United States.

In the opinion of the Moscow embassy, this desire raised the possibility of resolution, on terms favorable to America, of the outstanding difficulties between the governments of the two countries. At the same time, the analysts there—Alexander Kirk, Bohlen, and Stuart Grummon—warned against thinking that current Soviet cooperativeness represented a change of heart. "It may be assumed that the Kremlin does not envisage cordial relations with the capitalist governments on any permanent basis but rather as a temporary expedient dictated by the more immediate objectives of Soviet policy," they wrote.[3]

Divining Moscow's intentions occupied most of Henderson's time during the latter part of 1938 and the beginning of 1939. In March 1939 Kirk and his colleagues in Moscow reported a major foreign-policy speech by Stalin, in which the Soviet leader asserted that the capitalist states of the west were endeavoring to turn the fascists against Russia. As a complement to their policy of appeasement, Britain and France were attempting to incite Russia against the Germans—"to poison the atmosphere and provoke a conflict with Germany without any visible grounds." In Kirk's view this reference to no "visible grounds" for conflict between Moscow and Berlin, which flew in the face of years of propaganda, seemed intended as a clear hint to the Germans of interest in a nonaggression pact.[4]

A short while later, following Hitler's invasion of Czechoslovakia and Lithuania's forced cession of Memel to Germany, Kirk reported further indications of a desire for rapprochement. The Moscow press had dropped its anti-Hitler campaign, and Stalin had stated that the Soviet Union would refrain from opposing any country that did not directly threaten its vital interests. The Kremlin, in the opinion of the embassy in Moscow, was trying to reinsure itself.

Although Stalin's regime had been pouring money and energy into building Soviet defenses, considerable question remained regarding Russia's ability to fend off a German attack, especially when Japan's continued adventurism in Asia made a two-front war a genuine possibility. Summarizing Stalin's position, Kirk wrote, "He is exercising and will continue to exercise extreme caution in his relations with all foreign countries."[5]

During the spring of 1939 Stalin concentrated his attention on securing Russia's western marches against Germany. In June, Henderson described to the State Department's hierarchy efforts by Britain and France to negotiate an anti-German pact with Moscow. Discussions had stalled on the question of guarantees of the borders of the countries forming the buffer between Germany and the Soviet Union. Britain had declined to extend to the Baltic republics the assurance it already had granted to Poland and Romania, namely to defend them in the event of German attack. The Baltic states, as Henderson put it, formed "the weak link in the chain" Stalin was attempting to forge around Germany. From the Kremlin's point of view, this link needed hardening more than any other, especially because the Baltic countries appeared to prefer, if the choice came to that, German to Soviet control. Under the circumstances, a German ultimatum might suffice to bring the small countries into Berlin's sphere. What the Soviets wanted, failing a guarantee by the British and French of Baltic independence, was the right to preempt a German blow by a move of their own—either sponsored revolution in the Baltics or military occupation.[6]

However necessary such a policy might appear to Stalin and his generals, the Russians surely knew, as Henderson explained, that Britain would never agree to such an arrangement. "The question, therefore, presents itself as to whether the demands are made merely for the purpose of gaining greater security or in order to effect a breakdown in the negotiations which would result in the Soviet Union being able, for an indefinite period, to play off the so-called democratic bloc against the Axis."[7]

The British in fact refused Stalin's terms, lending support to Henderson's suspicions that the Soviets were not negotiating in good faith and that playing one side against the other was indeed the Kremlin's game. The likely winner of such a game, he believed, would be Hitler. The losers would be the rest of the world. In the second week of July, Henderson wrote to a friend:

> Until very recently I have been hopeful that a war might be averted this summer and fall. The situation at the present time, however, looks extremely black. I know there are many people who feel that a war would be a wonderful thing and that one who deplores the possibility of war is lacking in liberalism.
>
> I have seen close up enough of the aftermath of war to realize what it would mean to Europe, because I believe that if a war does take place the world will be a rather bleak place in which to live, at least during the remainder of our lifetimes.

It seems, however, that nothing except war will put a stop to Hitler's acquisitive tendencies and of course such tendencies cannot be allowed to develop forever.[8]

Two weeks later, following articles in American magazines and newspapers by authors he described as having "little knowledge or understanding of the mentality of the present rulers of Russia," Henderson felt obliged to spell out for the State Department the objectives of Moscow's diplomatic maneuvering. "The present rulers of Russia," he explained, "are still dominated by a spirit of aggressiveness; that is, they have not departed from the ultimate aim to enlarge the Soviet Union and to include under the Soviet system additional peoples and territories." Under current circumstances, however, expansionism took a back seat to holding what socialism already claimed. Since 1917 Soviet leaders had expected an attack from the capitalist countries of the west. Greed, they believed, specifically a desire to expand markets and investment opportunities, would provoke the imperialists to attack. Or the imperialists might fear that if they did not destroy socialism first, it would destroy them. But from one cause or another, war would come.

In recent years, Henderson continued, the Kremlin had drawn two particularly worrisome scenarios. The first involved a joint attack by Germany and Poland and other border states, perhaps aided by Japan. The second envisaged a transformation of the 1925 Locarno bloc—especially Britain, France, Germany, and Italy—into an explicitly anti-Soviet alliance. Although both possibilities seriously threatened Moscow, the former had usually seemed the more dangerous, and so long as the Germans and Poles had been speaking to each other, the Kremlin, in hopes of lining up the Western European powers on its side, had championed the cause of collective security. But the recent falling out between Berlin and Warsaw, succeeded by British guarantees of Polish security, had completely changed the view from Red Square. "For the first time the Soviet leaders are in no immediate dread of either a German-Polish combination or of a great four-Power European settlement."

With Europe quiet for the moment, Moscow was turning its attention to Asia, where the prospects were also improving. The Japanese, it appeared, had overreached themselves in attacking China, and no end of the conflict there could be seen. As a result the Kremlin was "in a position to create numerous incidents in the Far East and in general to make matters unpleasant for Japan." Even if Moscow went too far and found itself at war with the Japanese, such a conflict would take place, Henderson said, under the "most favorable circumstances for the Soviet Union." The Japanese could expect little help from other countries. "Most of the civilized world, with the exception of the totalitarian powers, would in general sympathize with the Soviet Union."[9]

II

"Where are we now?" asked Samuel Harper a month later, following reports of a commercial credit agreement between Germany and the Soviet Union. Henderson replied that it was "absolutely impossible" to guess the future, although in another letter, to John Wiley in Estonia, he remarked that the situation "looks extremely black." The guessing grew easier three days later when Berlin and Moscow announced a nonaggression pact. When word leaked out that the Molotov-Ribbentrop deal included a secret protocol effectively dividing Poland and the Baltic states between the Nazis and the Soviets, guessing became unnecessary.[10]

The German-Soviet treaty did not stun the State Department, since Henderson and others had been warning of the possibility for some time. In addition, an anti-Hitler diplomat in the German embassy in Moscow had been passing information to Bohlen for several months. But until the event occurred, it might have gone the other way. In any case, the pact and the war it touched off pushed the American foreign-policy machinery into overdrive. Henderson likened the department to a "madhouse" as he and other administration officials sought to keep the European conflict European.[11]

For the moment, even a fight confined to the continent created plenty of problems for American diplomats. Leaving Germany to his colleagues, Henderson confronted the fall-out from Stalin's cynical, if understandable, reversal of course. In some respects the Molotov-Ribbentrop pact eased Henderson's difficulties by knocking what he considered the blinders off those who had refused to understand the true nature of Stalinism. As Henderson explained, the Russians' "great flop from the camp of collective security into a working partnership with the aggressors" had laid bare the Kremlin's designs. And if the Soviet absorption of eastern Poland shocked most of those who had still hoped socialism might provide a peaceful alternative to fascism, the subsequent Soviet invasion of Finland alienated them further. "The attack on Finland," Henderson told Laurence Steinhardt, the new ambassador in Moscow, "has raised resentment in this country to an all-time high." Moscow's contention that it was merely reclaiming land lost to Finland during the chaos of war and revolution two decades earlier satisfied few critics. Because Russia took the whole winter to subdue the Finns, American resentment had time to fester and grow.[12]

Henderson found the swing of the popular mood simultaneously gratifying and dismaying. "The American public is certainly fickle," he commented to Harper. "It will be condemning us for opposing something one day, and on the next day for opposing the undoing of what it insisted on having done." Henderson could not help enjoying the discomfiture of "the radical New Dealers" who had lionized Soviet diplomats and slandered their American counterparts.

But he doubted that the edifying effects of Stalin's recent actions would prove more than fleeting. However justified it might be, Henderson considered the anti-Soviet reaction largely a case of "hysteria" that would "undoubtedly die down in time."[13]

Emotions aside, the fighting in Eastern Europe required a response, and American officials had to figure out what such a response should include. Few thought the semi-alliance between Stalin and Hitler would last. "It reverses the processes of men's minds in a way which I do not believe can be permanent," Assistant Secretary of State Adolf Berle said, adding, "Winning or losing, that combination must eventually break up." Naturally, it would serve American interests to promote such a dissolution.[14]

In the winter of 1939–40, however, any attempt to draw the Soviet Union away from Germany entailed considerable risk. Isolationism remained a potent force in the United States, and many Americans still considered Hitler preferable to Stalin. After all, serious fighting between Germany and the British and French had not begun. Perhaps a settlement might yet be achieved. Meanwhile Russia was beating up on poor—but not too poor to repay its debts—Finland. In addition, Roosevelt had the election of 1940 to think about.

Even so, the Roosevelt administration counted on a rupture between the fascist and socialist camps, and the president chose to treat the Soviets more leniently than the Germans. After the German invasion of Poland the White House had ordered an immediate American embargo against Germany; the Soviet occupation of the rest of Poland and the war against Finland prompted only a "moral embargo" against Moscow. The anomalies of the administration's policy became less noticeable in the spring of 1940 when fighting in the east ended, with Finland's defeat taking that country off the front pages, and when Hitler launched his blitz against France. At the same time, the escalation in the west supported the administration's premise that Hitler posed a greater challenge than Stalin.[15]

Henderson agreed with this premise, at least in the short term. The emergence of a militant Hitler hardly changed his mind about the Kremlin, but he did not think the United States had much to gain at present by slapping Stalin, despite the public outcry against the Russians. To Steinhardt he wrote in December:

> All kinds of pressure have been brought upon the Administration and upon the Department to take concrete steps in order to show our feelings. These suggestions, which have come from all kinds of sources, range from the severance of economic relations to the withdrawal of our Ambassador and even to the breaking of diplomatic relations. I am glad to say that in spite of the outraged feelings of the highest officials of the Government it has been possible to make them realize that nothing constructive could be accomplished just now by the adoption of any of these suggestions.

Henderson went on to say that the administration might stiffen the moral embargo if conditions warranted. The Russians would probably complain and might retaliate against the American embassy in Moscow—which was one reason Henderson was writing Steinhardt—but in this case Washington held the trumps. "If the Soviet Government tries to retaliate by taking repressive measures against our representation in Moscow, I think the President without any hesitation will sever diplomatic relations. If the Soviet Government, therefore, wants to continue to maintain relations with this country, it must watch its step."[16]

Watching Moscow's step was Henderson's job through the first half of 1940. The work, combined with developments on the continent, often proved discouraging. "Things in Europe seem to proceed from bad to worse," he told Harper in February. "It is difficult to be an optimist these days." Once Hitler's panzers began rolling through the west, the outlook grew grimmer still. "The international situation is too dismal at the present time to bear discussion," Henderson wrote. He could not say that current trends came as any surprise, at least not to someone who had been listening to the Russians. "I must admit that it has developed precisely as our Soviet friends said that it would several years ago . . . when they insisted that Germany's real target was not Eastern Europe but the Western Democracies."[17]

The problem of dealing with Russia encapsulated the Roosevelt administration's difficulties regarding the war generally. Diplomatic considerations required keeping open the channels to Moscow, in preparation for the likely day when Hitler would turn on Stalin. But domestic factors, especially the fast-approaching presidential election, precluded actions that might increase the risk of involving America in the fighting. Consequently the best that Henderson and the Russianists could hope for was to keep U.S.–Soviet relations from getting worse.

During the spring and summer of 1940, Soviet officials consistently complained of discriminatory treatment of the Soviet Union in trade with the United States. They contended that the moral embargo amounted to government encouragement to American firms to breach their lawful contracts with the Soviet government. When the State Department declared that certain strategic materials were in short supply and therefore unavailable, the Soviets dismissed this assertion as a subterfuge. They protested the barring of Soviet engineers and technicians from American munitions factories, and they claimed that the administration was manipulating commercial shipping regulations to prevent Moscow from chartering American vessels.[18]

Henderson bore the brunt of the complaints, which suited him well enough, since he thought Moscow was getting no more trouble than it deserved. But even he had to admit that this period was not especially enjoyable. An April memo characterized a conversation with Andrei Gromyko, the counselor of the

Soviet embassy, as "an hour's bickering." Nor did it help matters that the Soviets refused to recognize the political constraints under which the department and the administration were laboring. In a conversation with the Soviet ambassador in June, at a moment when France was collapsing, when Britain's expeditionary force was fleeing for its life and when the United States really was trying to do what it could to keep the Soviet Union from falling into Berlin's orbit, Henderson attempted to explain the situation.

> I told Mr. Oumansky that I realized that it was difficult for a person in my position to talk to him in a personal rather than in an official manner. I would, nevertheless, like to make a remark which I had no authority to make, provided he would be willing to regard it as a personal comment from myself. He replied that he would be glad to hear what I had to say and would consider it as personal rather than official.
>
> I then said, "For a number of months you have been talking to me about discrimination against the Soviet Union, and I have been endeavoring to reply to you. I feel it is terribly unfortunate, in the light of the present world situation, that your Government cannot be made to understand or to take an attitude which will allow us to know that it understands that what we are doing may eventually be of benefit not only to the American Government and the American people, but also to the Soviet Government and to the people of the Soviet Union."[19]

Henderson's arguments gained support from Hitler's quick victories in the west, which evidently surprised Stalin as much as they did American observers. The fall of France, together with renewed rumblings from the Japanese in Asia, inclined Moscow toward greater cooperation with the United States. When Sumner Welles proposed increased efforts to settle outstanding differences, the Soviet ambassador responded that he was "very much relieved" to hear the undersecretary's offer, adding that such efforts would be in the "highest interests" of the two countries."[20]

Henderson agreed that collaboration could serve American interests, but he feared that Welles would get carried away. Henderson particularly worried when Welles began speaking of a "friendly spirit" that informed U.S.–Soviet relations. However cooperative the Russians might become, Henderson refused to believe that friendliness motivated their actions. Cold calculation, not friendship, moved Moscow. Evaluating a recent suggestion by Soviet foreign minister Molotov that the Soviet Union and the United States combine forces against Japan, Henderson wrote, "After one of its periodic analyses of the world situation, Moscow has come to the conclusion that the time is ripe for it again to resort to the game of endeavoring to play off the United States and Japan against each other." He continued, "For some time the Soviet Union has been trying to persuade Japan that we are doing our best to promote friction between the two countries. Now, while still endeavoring to reach an understanding with Japan, the Russians refer to the Japanese as 'our common foe.'" Perhaps the

Soviets would prove useful in Asia, and in Europe as well. Still, the administration ought to guard the store and let the Kremlin come knocking. Henderson concluded, "At no time since 1917 has the Soviet Union had such an urgent need for good relations with the United States as now."[21]

Nonetheless Roosevelt agreed to negotiations between Welles and Oumansky, designed to resolve the points of contention between the two sides. In preparation, Henderson drew up a list of particulars. Trade in machine tools, he explained, for which the Soviets since the early 1930s had been America's best customer, and which they now sought with greater energy than ever, was in chaos. The situation was "critical," having deteriorated to a point where the American navy had taken to seizing tools already loaded onto Soviet ships awaiting clearance for sailing. The Soviet embassy continued to complain about discrimination in chartering vessels, and the Kremlin was protesting an American decision to freeze the assets of Latvia, Lithuania, and Estonia following the annexation of those countries by the Soviet Union. For its part the United States had signaled concern for the rights and property of Americans in the territories occupied by the Russian army, especially in Poland, and it still sought improvement of the conditions under which the American embassy in Moscow had to operate.[22]

The Welles-Oumansky talks began in August and proceeded slowly. Henderson expected no more. In fact he considered the lack of progress a hopeful sign, since it meant that Welles was not granting too many concessions. In this view Henderson gained backing from Steinhardt in Moscow, who wrote to encourage Henderson and the department to stand firm. The ambassador fretted that the discussions had attracted excessive attention and assumed exaggerated significance. This, he said, was having "a very bad effect in the Kremlin." The noises about friendship that Welles and other accommodationists were making would surely backfire. "I do not need to labor the point with you that this is the wrong approach to these people. They are realists, if ever there are any realists this world." Stalin was simply trying to get the United States to fight Russia's enemies, especially, at the moment, in Asia. "In the Far East, it seems to me that the Soviet objective must be war between the United States and Japan. Nothing would be more to their liking." Knowing that Henderson fully shared his skepticism, Steinhardt commiserated. "I can imagine what you are up against in trying to get this point of view across."[23]

Henderson felt he could use the sympathy. Welles recognized that the East Europeanists would gladly see the talks fail, and so he cut them out of the policy loop, leading Henderson to complain to Steinhardt, "I have not been informed of the real purpose of the conversations." But Henderson could guess.

They were undertaken as a result of strong feelings in certain Governmental circles that we must during the present time take energetic steps for the purpose of getting on a friendly basis with the Soviet Union so that we can be able more frankly to discuss with the Soviet Union matters relating to Germany on the one hand and to Japan on the other. I believe that there was the particular hope that through making certain concessions to the Soviet Government it might be possible to dissuade the Soviet Government from entering into certain commitments with Japan and Germany which would result in strengthening of the latter two countries.

Henderson took a certain relief in telling Steinhardt that until now the administration had offered "no great concessions" to the Kremlin. But he feared that following a scheduled visit by Molotov "more sweeping concessions may be made as a sort of reward" to the Russians. "One has to admire the success of the Soviet policy," he commented. The Kremlin knew how to manipulate sympathetic Americans. "By remaining non-committal and whispering here and there and by backdoor methods, it is able to get concession after concession from all quarters without sacrificing anything." Henderson admitted he was concerned. "I personally have some grave doubts that our policy of so-called appeasement will get us any place." There was little he could do about the situation, though. He must follow orders. "As long as that is our policy I am endeavoring loyally to cooperate in carrying it out."[24]

A month later Welles gave Oumansky just the sort of reward Henderson predicted, in the form of a lifting of the moral embargo. By this time the United States, with the help of a contact in the German foreign ministry, had obtained persuasive evidence that Hitler intended to invade the Soviet Union. The information contributed to the Roosevelt administration's decision to loosen trade. At the beginning of March, Henderson drafted a letter to Steinhardt instructing the ambassador to tell the Kremlin to expect a Nazi attack.[25]

Despite the previous reliability of the German informant, Henderson and others harbored doubts. "In so far as Russian matters are concerned," he told Steinhardt at the end of March, "it is hard to find two people in the Department, or anywhere else, who fully agree." Naturally, recently growing coolness between Berlin and Moscow deserved close attention. But Henderson thought Hitler might simply be trying to put a scare into the Soviets, and he argued against wishful thinking as a guide to American policy, especially as it involved relations between the Russians and the Germans. "I hope that I am being too much of a pessimist but it is still difficult for me to believe that after all that has happened, they will entirely terminate cooperation, let alone enter into a conflict with each other."[26]

A short while later, Henderson discovered that at least one Soviet official took the portents of war with Germany quite seriously. At a reception hosted by the Russian embassy, Oumansky buttonholed Henderson for a lengthy talk. "The Soviet ambassador came up to me," Henderson recorded the next day, "and after sighing rather heavily said that things appeared to look rather gloomy." What things?, Henderson asked. Soviet-American relations, Oumansky replied. What aspect of relations?, Henderson queried. "The whole general picture," the ambassador responded. Oumansky complained that despite the growing threat to the Soviet Union, the American government continued to block machinery and other items essential to Russian security. Referring specifically to a recent announcement of export controls on a variety of strategic materials, the ambassador remarked that once goods fell into the controlled category they invariably stopped flowing to the Soviet Union. What made the matter so incomprehensible was the fact that "without doubt the Soviet Union and the United States would eventually be on the same side."

This statement grabbed Henderson's attention. He told Oumansky he found his remarks "extremely interesting." He added that it would be "very helpful," if the Soviet government held such views, for Moscow to explicate them fully and frankly to the American government.

Oumansky retreated a step, perhaps thinking he had said too much. The ambassador assured Henderson that his comments reflected only his personal opinions.[27]

III

When the predicted attack came in June, Nazi forces sent the Red army reeling, and the great double-cross laid the foundations for the grand alliance. At once Roosevelt moved to include the Soviet Union in America's lend-lease program. Henderson and his colleagues on the Eastern Europe desk did not begrudge the aid to Moscow, but as before they sought to prevent the administration from going overboard. On June 21, only hours before Washington learned of the German attack, they drafted a memorandum outlining a preferred American policy in the event Hitler did what he was about to do. In the first place, they said, the American government should wait until the Soviets asked for help before making any statement regarding assistance. Second, if the United States did make aid available it should do so only to the extent such a program did not hamper American preparations for war. Third, the administration should not offer any commitments regarding the future. In particular, it should not tie itself to an exile Russian regime in case the Germans succeeded in overrunning the country. Although the reason for this last recommendation remained unwritten, it did not take much between-lines reading to perceive the East Europeanists' thinking. If Hitler did the west the favor of ousting Stalin, America should not oblige itself to reinstate him. Finally and most importantly: "We should steadfastly adhere to the line that the fact that the Soviet Union is fighting Germany does not mean that it is defending, struggling for, or adhering to, the principles in international relations which we are supporting."[28]

American officials could only guess how Stalin viewed the idea of collaboration with the United States. But some of the Kremlin's men found the prospect most heartening. "The Ambassador seemed to be in an extremely cheerful and optimistic mood," Henderson wrote after a dinner meeting with Oumansky on July 1. The Soviet diplomat expressed confidence in the fighting ability of the Russian army, which he said had laid careful plans for resisting a German attack. "He was sure that the world would be surprised at the results." Although the United States was not yet in the war, Oumansky went on to say he hoped there would develop the closest cooperation between the leaders of the two countries. He believed that American officials, especially military officers, were poorly informed about the morale and technical proficiency of the Red army. He trusted that the two governments could remedy this deficiency.

Henderson did not aim to spoil the party, which took place at a small country restaurant outside Washington, but he pointed out that there were good reasons why American officials knew little about Russian readiness, as the ambassador was in an "excellent position" to understand. If the Kremlin sought to promote genuine cooperation it might begin by opening up a bit more to American observers. Oumansky responded that he hoped Americans would become thoroughly familiar with the operations of Soviet army units.[29]

Henderson could not determine how much of Oumansky's optimism reflected honest feeling and how much an effort to put a bright face on what was proving to be a severe challenge to the Soviet military and implicitly to Stalin himself. Henderson remarked to Samuel Harper, "Now is certainly the test of the firmness of the Soviet regime and the extent to which it has been able to implant its roots firmly in Russian soil." Yet however skillfully or ineptly Stalin and his associates handled the crisis, the outcome would occasion disputes in the west and produce no more consensus than already existed. "Certain groups will maintain if the Soviet Union falls, that the fall was not due to the weakness of the Soviet Union, but to the sabotage of the democratic powers. Other groups, no matter how well the Soviet Union might fight, will find reasons for stating that the results of the campaign have demonstrated that they were right in insisting upon the weakness of the Soviet system."[30]

Within a few weeks it looked as though the German army would fail to achieve the quick triumph Hitler hoped for. While taking appalling losses, the Soviet Union proved not to be another France. The Red army, Henderson remarked, was "putting up a really good scrap." He added, "I certainly hope that they continue to do so." By the middle of August, Henderson could write that "it looks as though Hitler made what may be a fatal mistake in attacking the Soviet Union." Henderson did not think the Russians were quite out of the woods; Hitler might still have his way in the east. But "even though he may win a final victory, it looks as though the cost in men and supplies may almost fatally weaken Germany." Even so, the whole situation was most confusing. The fog of war compounded the Russians' refusal to take Americans into their confidence. "No one seems to know anything about anything."[31]

Soviet sympathizers and other critics of American policy alleged administration foot-dragging in getting help to the Soviets. As usual, the charges of anti-Sovietism focused on the career diplomats. Henderson had expected as much. "Attacks on the State Department," he wrote Harper, "are always much applauded and usually the charges are believed." The attacks were misguided, as Henderson pointed out. "The fact is that the State Department has been doing its utmost to obtain maximum assistance for the Russians in this particular period." The Soviets' failure to acquire the equipment they wanted resulted not from obstruction among the diplomats but from honest shortages and from a disinclination at the War Department to release anything American forces might conceivably require.[32]

The reluctance of the generals made it easy for the State Department to reject a request by Moscow at the end of July to share military secrets. Reviewing a list of matériel the Soviets had asked for, Henderson and Assistant Secretary Berle reflected that the items on the list indicated that the "real work," as Berle put it, of Russian engineers visiting American factories had not been limited to the nominal scope of their contracts, but in fact demonstrated "very efficient espionage." Berle and Henderson agreed that while the United States might safely provide hardware to the Russians, Soviet visitors ought not to be allowed into American factories on a regular basis. For one thing, the American government did not accord such privileges even to the British. For another, some of the secrets in question belonged as much to third countries, especially Britain, as to the United States. Finally, noting that extra-Soviet Communist parties, for example in Mexico, still preached anti-Americanism, Berle and Henderson contended that Washington should go slowly until it ascertained the precise nature of the Kremlin's overall policy toward the United States. "At least for the time being," they asserted, "we are not too clear about the Russian policy."[33]

8

Comrades in Arms

The Japanese attack on Pearl Harbor helped clarify matters. "We are certainly in it now," Henderson wrote to Samuel Harper. "Our initial losses, of course, are a tragedy," he continued, but he added that "from a point of view of unification of the country the Japanese could not have done a better job." Conceding that the United States would be fighting uphill for awhile, he declared that neither he nor anyone else in the department had "any doubt whatsoever" as to the ultimate outcome of the war.[1]

Henderson wrote on December 10, the day before Hitler solved the riddle of whether war with Japan meant war with Germany by declaring against the United States. This last news, on top of all the events of the previous months, heaped Henderson's plate higher than ever. Conscientious to a fault in the slowest of times, Henderson increased his workload to such a degree that at the end of December he had to tell Harper his eyesight could not stand the strain of some extracurricular reading the latter had suggested. "I am sure that I could learn a lot from these books but am compelled to use my eyes and my time on what seems most worthwhile." A few weeks later, Henderson's exhaustion and the general state of tension induced an illness that knocked him out of action for several days.[2]

Through the first part of 1942, while the Japanese overran Southeast Asia and the western Pacific, expanding their empire until it stretched from the borders of India to the Aleutian islands; while the Germans continued to push east into the Soviet Union, establishing a line from Leningrad to Crimea; and while the United States considered how to bring its weight to bear in the conflict, Henderson did his best to keep the Russians in the war. His job and that of the Eastern Europe desk was to facilitate delivery of American aid to the eastern front. In doing so, however, Henderson refused to be carried along by the tide of sudden good feeling toward America's new collaborators. He fully understood the danger posed by Nazism, and he completely agreed that smashing Hitler required helping Stalin. But as his letter to Harper just after the Japanese attack indicated, he had no doubt that the United States and its allies would win the war, and he thought it a grave mistake to forget a quarter-century of Soviet misconduct in the interest of what doubtless would prove a passing alliance. Others would become disillusioned with Moscow in the aftermath of the war. Henderson had no illusions to lose when the war began, and he had no intention of acquiring any during the conflict.

Henderson made his position plain in April 1942. The State Department had received a request to allow aliens residing in territory under the control of the Soviet Union to immigrate to the United States. The administration's stance before Pearl Harbor had been that such immigration should be discouraged if permitted at all. As Henderson explained to Welles, the Soviet government habitually enlisted many such immigrants as Soviet agents, enforcing cooperation by pressure on relatives remaining behind. The American government, Henderson pointed out, had no obligation to facilitate the Kremlin's efforts at infiltration. Now that the United States and the Soviet Union were allies, the question again arose, and some argued that Washington should loosen its restrictions. Henderson opposed any change, despite the appearance of distrust opposition entailed. As he argued to Welles,

> The attitude of the United States authorities in this regard is by no means unreasonable. Their policies with regard to persons desiring to come to the United States from the Soviet Union must be based not upon the relations existing between the Soviet Union and Germany but upon the Soviet attitude toward the United States.
> . . .
> Soviet authorities have not to any appreciable extent relaxed the close supervision and control which for years they have exercised over American citizens in the Soviet Union. They continue to prevent American officials in the Soviet Union from traveling throughout the country or from maintaining contact with the local population. They make it clear that they do not in general desire the presence of American citizens in the Soviet Union.
> Although little is heard of the Communist International at the present time we have no information which would cause us to believe that it is not continuing quietly to function with headquarters in the Soviet Union. The American Communist Party, the organ of the Communist International in this country, is supporting the war effort to the extent that such an effort might be helpful to the Soviet Union. It has not, however, ceased to work for the eventual overthrow by force of this Government and for the establishment of a Communist dictatorship.
> In these circumstances the American authorities would be derelict if they should fail to take adequate precautions to prevent secret Soviet or Communist International agents from entering, and carrying on activities in, the United States.[3]

Henderson insisted that concrete self-interest should inform American policy toward the Soviet Union. He had no patience for those who hoped wartime collaboration might lead to a lessening of philosophical distance between the American and Soviet systems. In some respects, while he remained at heart an idealist regarding American diplomacy—in that he believed American policy should be, and in most cases was, motivated by principles of self-determination, democracy and the like—his experience dealing with the Russians had made him the ultimate pragmatist in relations with communists. Convinced that a careful calculation of what would serve the Soviet Union motivated Moscow, he demanded that the United States exercise similar care guarding American interests.

Yet Henderson could not entirely ignore the waves of pro-Russian sentiment that swept America during the war. Engaged in what they judged a moral cause, Americans liked to think their allies were too. Official and unofficial propaganda re-created the image of the Soviet Union. The menacing Stalin of the Moscow trials became the kindly Uncle Joe of the fight against fascism; the sullen, submissive Russian peasants became the stubborn heroes of the battle of Stalingrad.

Henderson had to admit that the enthusiasm for things Russian made relations between the two countries easier in the short term. Of Joseph Davies' bestselling account of his experiences in the Soviet Union, *Mission to Moscow*, Henderson commented that the book had "certainly made a tremendous hit." Henderson added, "He couldn't have timed its appearance at a better moment."[4]

From a personal perspective, on the other hand, the ascendancy of the softliners caused Henderson discomfort. "These are interesting times here in Washington," he told Harper, "but you may be sure of one thing—it is not the most agreeable situation in which to carry on." Of late, Henderson had come under attack in the organs of the left. The communist *New Masses*, in an article entitled "Appeaser's Who's Who," singled Henderson out for membership in what the journal described as Washington's "Cliveden set." Two characteristics identified the group's members: "their hatred for President Roosevelt and their admiration for the devious ways of fascism." In Henderson's case, these opinions were said to be evidenced by his hobnobbing with the minister of fascist-controlled Finland and by his consistent ability to "find excuses not to do 'too much' for the Soviet Union." The article's author also described the alleged fifth-columnists as antisemitic, but here he did not refer to Henderson by name.[5]

This treatment was Henderson's first brush with notoriety, and as similar experiences would later, this one convinced him he was on the right track. It may be significant that these early public attacks came from sources clearly outside the American mainstream, sources easily dismissed. Subsequent critics would be located closer to the middle of American opinion, but by then Henderson would be in the habit of discounting criticism as politically motivated. In this instance, as often later, he considered the very fact of criticism, coming from the source it did, as vindicating the correctness of his position. It encouraged him to stand fast.

"I have known for some time," he told Harper, "that the communists would much prefer that my desk be occupied by someone much more sympathetic to them." But regardless of what the leftists might say and the uninformed think, he would not permit himself to be "terrorized by attacks from any source." "As long as I am in the Foreign Service I shall continue to do my duty as I see it."[6]

Yet Henderson was not—quite—so set in his thinking that he did not occasionally question his position; and at times the contradictions of fighting with the communists against the fascists made it difficult to determine just what duty demanded. "It is so hard these days to know what is the best thing to do," Henderson wrote in March. By mail and during occasional conversations he had conducted a running debate with Harper on whether "realism" was the proper attitude for conducting foreign affairs. Harper generally took the affirmative, while Henderson, although realist to the core in handling the Russians, preferred the negative. But the war was causing him to doubt the merits of his case.

> There are so many various sides to so many questions that I am not inclined to be critical of opinions expressed by people whose opinions I value, and I certainly value yours. It is possible that before this war is over we shall be compelled like various other nations to think only of what we can do to win rather than to bother too much about principles. I do not believe that we have yet reached that stage. Arguments can well be advanced that we should reach it without delay.
>
> Naturally, many of us who would like to have the United States come through the war not only as a victor but also with the reputation of upholding certain principles which eventually must be upheld if there is to be anything like normal international intercourse, would hate to see us compelled to scrap these principles. I am sure that you and I feel precisely the same way about this point.
>
> I am inclined to believe anyway that such terms as "realism," "international morality," "adherence to principles," et cetera, are likely to lead to confusion. There are times when the most realistic policy is that of the highest international morality and adherence to established principles and times when it is not. . . .
>
> I certainly do not have a closed mind about any of these problems which are confronting us, but I think we must admit that they are problems which are not so easily solved.

In any case, Henderson concluded, the problems were not his to solve. Rather they rested in the hands and on the consciences of "the highest authorities of our Government."[7]

II

The highest authority of all decided in August 1942 to send W. Averell Harriman to Moscow, to keep Winston Churchill company. With Stalingrad and Leningrad under siege and Moscow threatened, the Soviets were demanding that the Anglo-Americans open a second front against the Germans to lessen the pressure in the east. American war planners favored an invasion of France. Churchill adamantly opposed the idea, advocating an assault on North Africa. Roosevelt reluctantly acquiesced in the British prime minister's view.

Whether Stalin would go along was another matter. The State Department did not worry excessively about Stalin reversing course again; Henderson wrote, "There is no one in this Department who has for one moment considered seriously any danger of a separate peace." But the White House was not so sure. More than a year after the formation of the grand alliance, the president asked William Standley, the American ambassador in Moscow, regarding Stalin, "What do you think, Bill, will he make a separate peace with Hitler?" Yet even if the Soviets kept fighting, continued delay in opening a second front in Europe might have lasting ill effects. Standley predicted that if a second front did not open soon, Soviet suspicions of the United States and Britain would cause "inestimable damage" to the allied cause.[8]

When Churchill volunteered to break the news of no imminent second front to Stalin personally, the president sent Averell Harriman along to facilitate the business—and to keep an eye on the prime minister. Churchill was already on his way when Harriman got the order from the White House to proceed to Egypt at once to catch him.

In Cairo, Harriman ran into Henderson, who also was heading for Moscow to inspect the American embassy. Harriman knew of Henderson's experience in the Soviet Union and asked him to come along.[9]

On the last leg of the journey, from Tehran to Moscow, the Soviets insisted that a Russian navigator guide the American pilot. Henderson found himself pressed into service as a translator. Not having spoken Russian for several years and unfamiliar with the terminology of navigation, he was not sure how helpful he would be. In the event, the group arrived safely, though not without some trepidation. Churchill, on approaching what he later described as "this sullen, sinister Bolshevik State I had tried so hard to strangle at its birth," felt as though he were "carrying a large lump of ice to the North Pole."[10]

The talks went about as well as could be expected in light of the disappointing information Churchill and Harriman were bringing. "The atmosphere was tense," Harriman reported to Roosevelt. "Stalin took issue at every point with bluntness almost to the point of insult with such remarks as you can't win wars if you aren't willing to take risks and you mustn't be so afraid of the Germans." When Churchill described allied plans for a bombing campaign against Germany, the atmosphere began to relax. "A certain understanding of common purpose began to grow," Harriman said. "Between the two of them they soon destroyed most of the important industrial cities of Germany." Upon the conversation's returning to the relative merits of European and African operations, Churchill drew a picture of a crocodile and asserted that success could just as well follow a blow at the belly as a strike at the snout.[11]

Stalin was not convinced. He handed Harriman a letter for Roosevelt stating that the opening of a second front in Europe in 1942 had been agreed upon during a previous visit by Molotov to England. He pointed out that the high command of the Red army had built its plans for the year on the premise of such a second front, and he argued that "the most favorable conditions" for an invasion now existed. Whether they would persist could not be known. "We are of the opinion, therefore, that it is particularly in 1942 that the creation of a second front in Europe is possible and should be effected."[12]

But the decision had been made. Roosevelt replied that the American government well understood that the Soviet Union was absorbing the brunt of the fighting and losses in the war against Germany. "We greatly admire the magnificent resistance which your country has exhibited," Roosevelt said. He concurred that it was necessary "at the earliest possible moment" to bring the combined forces of all the allies to bear against Hitler. "Just as soon as it is humanly possible to assemble the transportation you may be sure that this will be done." In the meantime the United States would send more than one thousand tanks and various other strategic items, including aircraft, to assist the Soviet army in its noble struggle.[13]

The alliance survived this failure of the Anglo-Americans to open a front in the west in 1942, as the alliance would survive a similar failure the following year. But as Standley was predicting, the failure left a legacy of mistrust—or, more accurately, it confirmed the suspicions that already obtained. Stalin publicly took the line, as he declared in February 1943, that "since there is no second front in Europe the Red Army bears the whole burden of the war." Stalin exaggerated: his remarks came after the invasion of North Africa and the initiation of heavy bombing raids from England. But his words marked out a position the Soviets would take into negotiations for a postwar settlement. Perhaps more importantly, the slowness in opening a western front created a feeling in certain quarters in the United States that the Americans owed something to the Russians, who suffered greatly at a time when American soldiers had hardly begun to fight the Germans.[14]

But the end of the war remained a distant prospect in the summer of 1942, when Harriman and Churchill departed Moscow, leaving Henderson behind. Through the autumn of that year Henderson served as chief political officer at the embassy, backing up Standley. The ambassador, a navy admiral and former chief of naval operations, had been tapped for the Moscow post to demonstrate the administration's desire for military collaboration rather than to comment on the Soviet political scene. "I had no illusions as to my qualifications to be a diplomat," Standley remarked. "I knew that I was just an ordinary, run-of-the-mine Naval officer as far as diplomatic experience went." Consequently, following an inspection of the embassy, Henderson stayed on to lend a hand.[15]

Among his first chores was coping with the visit of Wendell Willkie, the

Republican presidential nominee of 1940 and aspirant for 1944. Willkie stopped in the Soviet Union on a global tour that would give rise to his hit book *One World* the following year. To Henderson and others in the American embassy, Willkie seemed the epitome of all that could be wrong in American political diplomacy. He operated out of channels. Following a conversation with Stalin, when Standley asked what the two men had talked about, Willkie replied that the discussion had dealt with matters so sensitive he could not confide them to the ambassador. He seemed naïve and irresponsible. After two days in the Soviet Union he announced to Andrei Vishinsky and Solomon Lozovsky, a for-eign-ministry official, "There's no great difference between the American and Soviet viewpoints. In a few years, the social systems of the two countries will be quite similar." Lozovsky replied, "I'm not so sure of that, Mr. Willkie." To repeated queries from factory workers and assorted other "ordinary people" as to when the Americans and the British would open a second front, he responded, "Sooner than you think." He breached etiquette and would have ignored entirely the dean of the diplomatic corps, the ambassador from Afghan-istan, had he not received repeated prods from embassy officials. He was per-sonally rude, proposing a drinking bout between Standley and the most capa-cious officer the Soviet military could produce. He was a political grandstander, seizing the least excuse for a photo opportunity.[16]

Standley decided not to take such treatment. Immediately after Willkie's departure the ambassador met with Henderson, and the two agreed that Stan-dley should request leave to return to Washington, where he would seek assur-ances from the president that the ambassador, not informal envoys, would rep-resent the United States in the Soviet Union. If satisfied with the response, he would come back. If not, he would resign.[17]

The State Department approved the plan, directing Henderson to take charge of the embassy in Standley's absence. Actually, Henderson took charge of what amounted to two embassies, for at the height of the German offensive toward Moscow the Kremlin had officially evacuated the capital, relocating the government to Kuibyshev. Formerly Samara, this Volga trading port had been renamed in honor of one of Lenin's comrades, and it received a flood of a half-million or more refugees fleeing the German advance. The influx strained the city's already burdened infrastructure, and problems of provisioning the Amer-ican embassy matched the worst of those Henderson had encountered in Mos-cow in 1934. Meanwhile Stalin and other prominent Soviet leaders remained in the Kremlin. Between communicating with them and dealing with the lower-downs in Kuibyshev, Henderson spent much of his time shuttling back and forth.

The months of Henderson's wartime service as chargé d'affaires marked the beginning of the period of the closest relations the United States and the Soviet Union have ever enjoyed. With the Anglo-American landings in North Africa in November 1942, the three countries were really in the war together, and the lifting of the siege of Stalingrad at the beginning of 1943 diminished the irritating effect of the delay in establishing a front in the west. In fact, when General Patrick Hurley arrived in Moscow in November to talk military strategy with Stalin, the Soviet dictator commented that while a western front must be opened "eventually," for the moment he was content for the Americans and British to exploit their advantage in Africa and for the Americans to continue sending aid to Russia.

At this meeting, the question also arose of an appropriate policy toward Japan. Hurley had come from the Pacific theater, where he had listened to arguments advanced by Australians, New Zealanders, and some Americans that prudence required the destruction of Japan's fighting strength before the Japanese managed to exploit the tremendous resources of Indochina, Indonesia, and other territories they had conquered. Stalin dismissed this line of reasoning. Tokyo lacked trained personnel to bring these resources into production, he said; the Japanese would require a year or eighteen months before they gained significant advantage from their booty. From Hurley's assertion that the German-Japanese alliance meant little to Tokyo, that the Japanese were simply waiting to take advantage of whatever success the Nazis achieved in the west, Stalin likewise dissented. Cooperation between Germany and Japan was complete, and the Japanese depended on the Germans for essential items of armament, especially airplane engines. Japan, he said, could not sustain itself without Germany, and the collapse of the former would follow necessarily defeat of the latter.[18]

Roosevelt agreed. In a letter sent to Henderson to pass along to Stalin, the president declared that "the menace of Japan can most effectively be met by destroying the Nazis first." He added that Churchill shared his view. In the same note he suggested a meeting of the three allied leaders, since what the Americans and British did next in the Mediterranean would have a bearing on Soviet strategy, and vice versa. Roosevelt concluded, "I do not have to tell you to keep up the good work. You are doing that, and I honestly feel that things everywhere look brighter."[19]

Through the next several weeks Henderson played the role of intermediary between Roosevelt and Stalin. He conveyed Stalin's congratulations on the operations in the Mediterranean, which were "developing so auspiciously" and which promised to "change the whole military situation in Europe." In this message Stalin ducked Roosevelt's invitation to meet, suggesting instead consultation among staff officers of the three allies. In a letter Henderson delivered at the beginning of December, Roosevelt reiterated his earlier proposal. "I am very anxious to have a talk with you," the president said, pointing out that military officials were not in a position to make the decisions required in the event of a German collapse. Roosevelt recommended some secure place in

Africa during the middle of January. "I hope that you will consider this proposal favorably because I can see no other way of reaching the vital strategic decisions which should be made soon by all of us together. If the right decision is reached, we may, and I believe will, knock Germany out of the war much sooner than we anticipated."[20]

A short while later Stalin repeated his *nyet*. "I cannot leave the Soviet Union," the dictator declared. "I must say that things are now so hot that it is impossible for me to absent myself for even a single day." When Henderson relayed a revised invitation from the White House for a meeting in North Africa in March, Stalin again begged off. Affairs connected to the fighting in his country required his "constant presence." He inquired regarding exactly what Roosevelt and Churchill wanted to talk about. "Would it not be possible to give consideration to these questions through an exchange of correspondence? . . . I presume there would be no disagreement among us."[21]

III

At this point, Stalin presumed more or less correctly. While the end of the war remained at some distance, the Americans, British, and Soviets managed to get along well. Differences still cropped up regarding where and when to press the attack against the Germans, but as long as Hitler remained dangerous he provided the glue that held the grand alliance together.

Standley arrived back in Kuibyshev in January, having convinced Roosevelt to grant him what he hoped would be greater control of American affairs in the Soviet Union; and Henderson returned to Washington. Resuming his position on the Eastern Europe desk, Henderson soon discovered that if the big three were willing to put off consideration of what would follow the war, some of the lesser powers were not. In March he received a visit from the Polish ambassador, who described the fears of his government regarding Soviet designs on Poland. In the Polish view the Russians intended to annex Poland. They would establish a communist provisional government for Poland, initially basing it in Soviet-controlled territory but transferring it to Poland with the Soviet army. The ambassador warned that unless the United States and Britain resisted Soviet plans, the west—and Poland—would confront at war's end a fait accompli: a communist Poland under Stalin's thumb. He appealed for American help, pointing out that Soviet pressure on Poland flew in the face of the Atlantic charter and the declaration of the United Nations. Poland therefore provided a test case of the willingness of the west to stand up for the ideals embodied in those documents. What was required was a "firm reassertion" by the United States government that it would not sacrifice its principles to Soviet imperialism.[22]

Henderson sympathized with the Polish diplomat. His sensitivity to the Soviets' cynical treatment of the peoples on their western marches had increased with the Kremlin's annexation of the Baltic republics. (The Soviet action even more enraged Henderson's wife. Elise's antipathy toward Moscow took the form of a vigorous verbal and nearly physical assault on the spouse of Russian ambassador Oumansky at a dinner party in the home of Joseph Davies.) Henderson found it easy to appreciate the worries of the Poles—which proved all too accurate.[23]

When the opportunity came, Henderson would not hesitate to urge a "firm reassertion" of American principles in the face of Soviet pressure. But at present he could see he would only be wasting his breath. Roosevelt quite obviously had no desire to disrupt relations with Moscow over Poland, whose fate was to him a peripheral issue. The president would worry about the Poles once Hitler's end was guaranteed.

In areas where he thought his efforts would do some good, however, Henderson expressed himself forcefully. In pursuance of claims to Polish territory, as well as to that of the Baltic countries, the Kremlin had demanded the right to dispose of property in the United States owned by individuals residing in those regions. Henderson argued that the American government should take a tough line of opposition on this issue. "We cannot accept suggestions from the Soviet Government with regard to the administering or disposition of property of persons in occupied Soviet territory except that territory which we recognize as Soviet." To do otherwise would be tantamount to accepting Soviet control as legitimate. In words he might have taken out of the mouth of the Polish ambassador—or vice versa—he concluded, "If we show the slightest weakness and equivocation in this regard the Soviet Government will at once bring tremendous pressure on us and in the end our relations will be more unfavorably affected than they would be if we display firmness at the outset."[24]

Predictably, such advice did not sit well with those in the administration who hoped to foster friendly U.S.–Soviet relations. Nor did it help matters that Henderson's old antagonist from Moscow, Litvinov, was now Soviet ambassador in Washington. If Litvinov had forgotten Henderson's distrust of the Kremlin's motives, his memory was refreshed by Welles and others in the Roosevelt administration who considered what remained of the Kelley influence detrimental to the war effort. During the summer of 1942, when Henderson had applied for a visa to the Soviet Union in preparation for his inspection tour, the ambassador had displayed visible annoyance. "He was quite rude," Henderson remembered later. "He said he did not see why I should go to the Soviet Union." At that time, though, the ambassador chose not to make a major issue of the matter.

Six months later, when Standley was preparing to reassume direction of the embassy in Kuibyshev, Litvinov asked if it was true that Henderson was returning to the Eastern Europe desk in Washington. To Standley's reply that such indeed was the case, Litvinov responded that the Soviet Union and the United States would never have good relations while Henderson held that position. Subsequently the Soviet ambassador made the same argument to Welles and Eleanor Roosevelt.

When Standley told Henderson of this exchange, Henderson wrote Secretary Hull volunteering for reassignment. At first Hull refused the offer, asserting that he would make his own decisions about staffing. A few months later, however, under pressure from the White House, Hull reversed course. Motioning across the street, the secretary explained to Henderson: "The people over there want a change."

But the secretary insisted that Henderson receive a promotion. He told Henderson he did not want Litvinov to think he could damage the career of a foreign service officer because that officer displeased him, or for other officers in the future to fear taking actions that might excite Soviet displeasure. Hull instructed Henderson to look around for an available position overseas as chief of mission: ambassador or minister.

Henderson felt uneasy about the promotion, since it would kick him two steps up the career ladder and past some of his seniors. He also felt disappointment, and considerable resentment, at being shifted away from his area of expertise. He had devoted two decades of his life—nearly his entire career—to study of the Soviet Union and its neighbors. He had served diligently and well. No one had acted out of greater devotion to duty; few had accomplished more in terms of revealing the nature of the Stalinist system to those in Washington with eyes to read and ears to listen. Now all that effort, all those years, were to be cast aside—at the insistence of Litvinov and his allies in the White House.

Yet in a curious way the reassignment afforded Henderson some satisfaction. He felt, justifiably, that he was being persecuted for duty's sake. He had spoken what he believed to be true, and he was paying for his candor. That his principal antagonist in this instance was the Russian ambassador increased his perception of being on the side of the angels; with enemies like Litvinov, who needed friends? Obviously Stalin's chief agent in Washington wanted Henderson out of the way because Henderson knew too much. The Kremlin could not stand the truth being told about its activities and plans. Henderson had no difficulty shrugging off the opposition of Welles and Eleanor Roosevelt. The former he considered a political maneuverer who would sacrifice principle to expediency; the latter simply knew nothing of the way the world worked.

In consequence, Henderson prepared to leave the field of policy toward the Soviet Union with mixed emotions. In fact, as matters developed, he left more in body than in spirit, for he soon demonstrated that it was easier to take Henderson out of Russia than Russia out of Henderson.

At Hull's direction, Henderson inquired of the personnel office regarding vacancies as chief of mission. Learning that Iraq was open and did not have a long line of applicants, he told Hull that Baghdad was where he wanted to go. The White House approved, judging Iraq sufficiently removed from the arena of Soviet-American relations that Henderson could not make more trouble. In the early autumn of 1943 Henderson headed for the Middle East.[25]

III

EXILE AND RETURN

9

"Head net, mosquito bar, sun helmet. . ."[1]

In 1937 an observer in *Harper's* magazine surveyed the activities of the State Department's various divisions and offices. Arriving at the bureau assigned to watch over the Middle East, the author wrote that the Division of Near Eastern Affairs, to use its formal title, was "responsible for our dealings with such peoples as those of Albania, Bulgaria, Greece, Iran, Iraq, Palestine, and Trans-Jordan, and the divers haunts of Turks, Arabs, and Syrians. . . . Our relations with these peoples are not important: their economic exploitation has been preempted by other empires. Now and then an American contractor aims to build a bridge or a dam." The following year, when Evan Wilson, who would serve as Henderson's strong right arm in the fight over Palestine, drew assignment to Cairo, a friend in the European office commiserated, "The Near East! Nothing ever happens there."[2]

This remark still fairly summarized popular American attitudes when Henderson joined the Near Eastern division in 1943. Although the peoples of the Middle East probably had no fewer troubles then than they would during the next half-century, the perturbations these difficulties generated in the American consciousness scarcely registered amid the turbulence of the world war. Among students of the region, however—and Henderson immediately became one, throwing himself with customary assiduousness into his new assignment—two issues, soon to be of surpassing general interest, stood out.

The first involved oil. In the early 1940s the United States remained a net petroleum exporter, indeed the largest in the world, despite the fact that it was also the world's biggest consumer. But the trend lines of American production and consumption indicated that this favorable balance would not always exist, and American firms increasingly looked to the Middle East for the big strikes they previously had hit at home. The attention of the American government was not far behind. As an indication of things to come, the Roosevelt administration included Saudi Arabia in America's lend-lease program.

The second issue was Zionism. For the most part the American government had stayed out of the problems of administering Britain's Palestine mandate. Woodrow Wilson had offered qualified support to the Balfour declaration, which advocated the creation of a Jewish homeland in Palestine, but the State Department had never considered the matter a priority and subsequent presidents had found other affairs to occupy their time. American Zionists had tried to raise the issue in American politics, with little success. When Henderson arrived on the Middle Eastern scene in 1943, Washington still exhibited a hands-off attitude regarding Zionism. But it was becoming obvious that this luxury would go the way of America's oil glut, and probably sooner. Although the full enormity of Hitler's treatment of European Jews was not yet common knowledge, what *was* known had raised pressure in America for a Jewish refuge in Palestine. While the war continued, Roosevelt could keep a lid on the Zionist issue in the United States, as the British were doing with less success on the ground in Palestine. In both cases, though, the likely effect of such compression appeared to be a more violent explosion over when the lid finally blew off.

II

Iraq in 1943 lay at the center of Britain's Middle Eastern sphere of influence, which stretched from Egypt to India. Occupied by British forces during World War I and mandated to Britain by the League of Nations in 1920, Iraq had gained independence in 1932. Nearly a decade of political instability followed, culminating in the accession by coup of Rashid Ali in 1941. While previous regimes had cooperated more or less in guaranteeing Britain's property and strategic interests in Iraq, Rashid Ali adopted a defiantly nationalist approach, making overtures to Berlin to offset British influence and even encouraging a German invasion of the country. Unwilling to tolerate this challenge, London sent troops, reoccupying the country, replacing Rashid Ali with the pro-British Nuri al-Said, and reestablishing thorough if informal British control.[3]

Henderson arrived in Baghdad with no specialized knowledge of the Middle East generally or Iraq in particular, but he set himself at once to the task of self-education. For a first project he analyzed the nature of British influence in the country. Partly from force of habit and partly because the model fit, he approached the subject in much the same way he earlier had studied the influence of the Comintern in the west. Throughout his career Henderson saw wheels within wheels in the activities of foreign governments—not surprisingly, for he acquired the tendency investigating the activities of the two governments, Moscow and London, that in the 1930s and 1940s surpassed all others in the art of diplomatic intrigue. His examination of Britain in Iraq resulted in an opus reminiscent of his earlier treatises on the Soviet Union. Complete with painstakingly drawn organizational charts showing who answered to whom, directly or otherwise, Henderson's study described a system whereby British civilian and military officials had intruded themselves thoroughly into Iraq's politics and society. The purpose of this elaborate apparatus, he said, was "to safeguard British interests in practically every field of Iraqi national life and to direct the trends of Iraqi internal and external policies into channels which will serve the well-being of the British Empire." Henderson located the nerve center of this paragovernmental network at the British embassy in Baghdad, which dwarfed in size and expenditure all the other foreign missions. The British ambassador—the only diplomat in Iraq so ranked, since London insisted that the chiefs of mission of other nations stand no higher than minister—was nothing less than a proconsul, the most powerful figure in the country.[4]

As in many places where Britain's informal empire exceeded its statutory reach, the imperial agents spent much of their time preempting possible competitors. Britain currently shared harness with the United States against the fascists, and British officials recognized the need for cooperation. But they also understood the threat American influence posed to their hegemony in the Middle East, should the Americans ever decide to contest Britain's primacy. Relations between British and American representatives in Iraq, Henderson pointed out, reflected this ambivalence. "On the one hand, British officialdom is anxious to prevent American economic and commercial interests from infiltrating into the country, while on the other hand they do not wish to take any action in time of war which might be considered as unfriendly toward Great Britain's most important ally." From the standpoint of personal dealings, Henderson found nothing in British conduct to complain about. His colleagues in the British embassy displayed a disposition to be "genuinely friendly." Even so, they were waging a diplomatic guerrilla war against the United States, if primarily a war of omission. The British were "not inclined to use their prestige and authority to suppress political or propaganda activities which tend to create a lack of confidence in the intentions of the United States toward the Arab world." In particular they appeared "not at all displeased" that the Arabs suspected that Washington would ultimately back the Zionists in the Zionists' quest for a state. Nor did the British make any effort to disabuse Iraqis of the notion that America preferred the house of Saud to the local Hashemites in the dynastic struggles of the region.

The British were paying for their influence, though. Henderson noted that they had alienated Iraqi nationalists, choosing to collaborate with an aristocratic class known for enriching itself at public expense. While the war continued, the British and their local allies might control the situation by jailing dissenters in the name of security. But when the fighting ended, change of one sort or another would surely come. Whether Britain would adapt, whether London could adjust itself to the new reality the war was creating in Iraq, as elsewhere in the colonial world, Henderson refused to predict. It was possible, he said, that Britain would "interest itself in problems of public morality as well as those of political reliability." But he had his doubts, and he would not be holding his breath.[5]

The key to Britain's position in the country, and a man with whom Henderson would have repeated dealings until the prime minister's murder in a 1958 revolution, was Nuri al-Said. Nuri had four years on Henderson in age, having been born in Baghdad in 1888 to a family of middle rank in the Ottoman hierarchy. Educated at a military academy and later at the imperial staff college in Istanbul, Nuri was one of the few Iraqi officers to desert the Ottomans in favor of the Arab nationalist movement prior to World War I. At this time he likewise opposed British influence in Iraq, and he fought to break Britain's hold on Basra near the Persian Gulf. He was captured in combat and spent a year in an Indian jail. When the war came the British released him to join the Arab revolt against the Turks. Early on, he hitched his star to that of Faisal, and when the latter became king of Iraq, Nuri received his reward in the form of successively more responsible offices. He served first as army chief of staff, then as minister of defense, and finally as prime minister. Through most of the 1930s he bounced in and out of office; in 1941 he landed back as premier, supported now by the British.[6]

As much as the British were gaining from Nuri's cooperation, he was profiting from theirs. He was as shrewd a collaborator as Britain ever encountered. "His skill in negotiation and intrigue is very great," one British diplomat commented, while another described him as "a devious intriguer with a passion for having a finger in every pie." London was hardly surprised when Nuri attempted to lessen his dependence on Britain by branching out and improving contacts with Britain's two wartime allies. Whether they would have been surprised to learn that he was also flirting with the Nazis is unclear.[7]

Almost as soon as Henderson arrived in Baghdad, the American minister found himself involved in Nuri's game. The prime minister summoned Henderson to the palace to announce that his government was pondering the establishment of diplomatic relations with the Soviet Union. Nuri assured Henderson he had no ulterior motives in considering such a move, that he simply desired to put Iraqi-Russian affairs on a more businesslike footing. Although Iraq, he said, like the United States, had long distrusted the Soviets for meddling in the affairs of other countries, in recent years the Russians had acted more civilly. In particular they had dropped their opposition to Arab unification, a principal objective of Iraq's diplomacy. Under the circumstances, normalization was warranted. Nuri commented that just before the war the Iraqi and Soviet governments had discussed an exchange of representatives. But before the arrangement could be consummated, Rashid Ali had taken power and, anxious to improve relations with Germany, had scotched the deal. Now Nuri was picking up the thread again, as were the Russians, who, Nuri confided to Henderson, were pressing "rather insistently" for normalization. Nuri said Iraq would not be rushed; he had told the Soviets he preferred to let the international situation clear a bit before committing himself and his country.

Although he saw no insuperable obstacles to recognition, one minor stumbling block remained. Nuri told Henderson that before he agreed to normalization he wanted to ascertain the validity of reports that Moscow was encouraging a separatist movement among the Kurds of Iran and Turkey. The prime minister shrugged off the potential for similar unrest among Iraq's Kurdish population. Iraq's Kurds, he said, were not a major problem. But he wished to avoid charges from Tehran and Ankara that Soviet diplomats in Baghdad were fomenting trouble across Iraq's borders. He personally had no information to substantiate the rumors, and once he convinced himself that the Kremlin was not trying to create a "puppet Kurdistan," he would consent to opening relations. In light of Russia's growing importance in world affairs, he hoped to accomplish the task soon. As something of an afterthought, he added that rapprochement would have mostly political significance. He saw little prospect of meaningful economic ties between the two countries.[8]

Henderson did not require much imagination to perceive that Nuri had a purpose in telling him all this, although, as was usually the case with Nuri, deciphering that purpose was not easy. Undoubtedly Nuri wished Washington to know that Iraq had other suitors besides Britain and the United States. And he probably desired to find out if the Americans had major objections to his opening to Moscow. But what else he had in mind, only time would reveal. Commenting on the conversation in his report to Washington, Henderson noted that Nuri had not told the whole story regarding his relations with Moscow. After King Faisal had died suddenly in 1933, Faisal's son Ghazi had assumed the throne but proved unequal to the task of fending off numerous contenders. Ghazi's sudden death in a car wreck in 1939 had put the crown on the head of Ghazi's young son and placed Nuri at the center of Iraqi politics—a circumstance that caused Nuri's many rivals to detect his hand in the accident. Distrusted at home, Nuri also had made enemies throughout the Arab world by pressing the Hashemite claim for primacy against those emanating from Saudi Arabia, Egypt, and Syria. As an official at the British embassy in Baghdad commented, Nuri was "suspect in Egypt as an opponent of Egypt's leadership in the Arab World, and in Syria and Saudi Arabia as a Hashimite henchman." Henderson believed that the opening to Moscow was designed to increase Nuri's credibility both in Iraq and in the region around. Regarding the latter arena, Henderson pointed out that Egypt was in the process of improving relations with the Russians. Nuri evidently thought Iraq could not afford to fall behind. Henderson commented that the prime minister did "not wish Egypt to have active relations with any great Power which is not represented in Iraq."[9]

Nuri also had understated the seriousness of the Kurdish revolt in the mountains of Iraq's northeast. In fact the Kurds had become so troublesome that by the beginning of 1944 the government was forced to invite the leaders of the rebellion to Baghdad to parley for a ceasefire. Henderson described one of the rebel captains, Mulla Mustapha, as "a small but impressive and handsome figure with walnut-colored skin, an aquiline nose, and heavy black eyebrows and mustache" who customarily traveled with "daggers and revolver" tucked in his sash. Henderson would have liked very much to talk with this formidable individual, but discretion forbade it. When Mustapha requested an interview, Henderson declined, not wishing to foster the appearance that the United States favored Mustapha's objectives.[10]

Henderson thought Nuri was deliberately downplaying the economic significance of normalization between Iraq and the Soviet Union. The economies of the two countries, Henderson argued, might complement each other nicely, especially when the war ended and the Soviets retooled their industrial plant to produce more consumer items. If Moscow made a conscious effort to strengthen its standing in the Middle East, as it would have every reason to do, the effect would be even more pronounced.

From an ideological perspective as well, Henderson detected greater opportunities for Soviet penetration than Nuri allowed. The prime minister had declared that communism held few attractions for Iraqis, but Henderson had discovered that Russian military successes during the war had made a "tremendous impression" on the popular mind in Baghdad, with pictures of Stalin and Molotov appearing frequently in public places. The war had shaken Iraqi society to its roots. "As the younger generation here, particularly the students, are becoming emancipated, they are inclined to look for radical cures for the economic ills of the country." The most attractive of these radical cures was socialism, which naturally directed the attention of its adherents toward Moscow. At the same time, pro-western statesmen like Nuri appeared increasingly anachronistic and were encountering "intense bitterness and dissatisfaction" on the part of the poorer classes.[11]

Nuri complemented his opening to Moscow with an effort to enhance relations with the United States. In February 1944 Nuri called Henderson in for a chat about what would follow the war. The prime minister observed that his government had great expectations for the United States in the Middle East, adding that he was "particularly anxious" to establish U.S.–Iraqi relations on a sound basis. As a first step toward this goal, he suggested direct conversations at the highest level—perhaps a visit by the regent Abdul Illah and himself to Washington to "assist in making more Americans conscious of the existence of Iraq, of the friendly feelings of the Iraqi people for the United States, and of Iraqi aspirations and hopes."[12]

The White House refused to issue Nuri an invitation, preferring not to upset American Zionists and the Saudis. The prime minister took the rejection hard.

"Nuri was obviously deeply disappointed," Henderson reported. Nuri tried by one means and then another to answer the administration's objections to his visit, but when Henderson made clear that the subject was not open to negotiation, and when he explained the political difficulties Nuri's coming would create for President Roosevelt, the prime minister, speaking as a politician himself, said he understood.[13]

In the event, developments in Iraq mooted the issue. Nuri's grip on power slipped, and in June his "makeshift, tottering cabinet," as Henderson described it, collapsed. Henderson attributed the fall to a variety of factors, including increased separatist activities among the Kurds, complaints of inattention by army officers, and religious tension between the dominant Sunnis and the challenging Shiites. Most important, nationalists who saw in Nuri nothing but continued foreign domination conspired to bring him down.[14]

A short while later Henderson had a discussion with one of these nationalists. This individual represented what Henderson characterized as a group of "youngish Iraqi intellectuals who hold positions of secondary rank in the Iraqi Government and whose influence in the country is gradually increasing." The man declared his ideological affinity with the United States and Britain in their struggle against fascism, but the merits of the allied case appealed less than its promise for Arab liberation. "Like many Iraqis," he said, "especially those educated in America and Britain, I believe that the cause of the Arabs is a just and human cause, and that it is directly connected, for political and cultural and international reasons, with that of the Allies." An allied victory would serve the purpose of the Arabs, who were "mainly ignorant, poor, diseased masses" yet who were awakening and becoming "more and more conscious of their status as Arabs and as human beings entitled to justice and human rights." He went on to identify four groups competing for the destiny of Iraq. The first comprised the "corrupt relics" of the Ottoman empire; the second, the feudal and backward-looking tribal leaders; the third, "elements who are mainly non-Arab or anti-Arab and who want to follow blindly the path of Moscow"; and the fourth, young Arab nationalists "who are both democratic and socialistic and who have social and economic reconstruction along the lines of the democratic countries at heart."

While Nuri wanted to use the Americans against his rivals, Henderson's visitor sought to use the Americans against Nuri. The future, the visitor argued, lay with the Arab nationalists, and the Americans would do well to take their part. President Roosevelt himself had made the nationalists' case in the Atlantic charter; the United States must not go back on its word. Neither should the Americans sell the Arabs out to the Zionists. "It is my sincere hope," he said, "that the Atlantic Charter will have its true spirit applied to the Arabs, and I pray that the Allies will not repeat the mistakes committed during the last war of promising Palestine to the Jews." Far better for the United States and for the world that the American government assist in the establishment of an Arab federation unified along "democratic and socialistic lines," a nation that would be "a great addition to the families of nations in a reconstructed world."[15]

Neither Henderson nor Roosevelt had any intention of sponsoring a socialist pan-Arab federation, but the appeal to the Atlantic charter struck a chord in the American minister. Some days after this conversation, Henderson spoke with the Iraqi foreign minister, Arshad al-Umari. Arshad expressed his hope that the United States would apply the principle of self-determination to Iraq in its relations with Britain. The British doubtless would press for a postwar settlement based on spheres of influence, but the United States, having taken a "high moral position," must not agree to any such arrangement. Flying in the face of American promises, this would amount to nothing less than "traffic in the fate and liberties of small countries or undeveloped peoples." Although Henderson at this time was in no position to act on such advice, beyond forwarding it to Washington, the sincerity of the foreign minister's presentation impressed him. During the next few years Henderson would make similar arguments himself.[16]

III

After a dozen months in his new field, Henderson's picture of the Middle East was coming into focus. Considering his background, it was unavoidable that he should interpret developments there in the light of what he judged to be the ongoing struggle between the principles America represented and those promoted by the Soviet Union. Henderson spent the dozen years from 1943 to 1955 concentrating on the region that stretched from Greece to India, but he never became an expert on the area. Who could, in such a short time? The diversity of languages, cultures, religions, and political systems would have overwhelmed anyone. Henderson, whose linguistic incapacity was demonstrated by the fact that he acquired no fluency in Russian during twenty years in the Soviet field, did not stand a chance. But in Henderson's case the problem lay less in ability or lack of it than in inclination. Henderson had been transferred to the Middle East, but his head and heart remained in the area of Soviet-American relations. That was where he had found his calling; that was the field he had sweated, bled, and nearly died over. Besides, from a purely objective standpoint, odds were that the Soviet-American contest would shape the world in the near future far more significantly than American relations with Iraq or any other country of the Middle East.

As a consequence, Henderson always viewed events in the Middle East in terms of their impact on the ideological and political balance of power between the United States and the Soviet Union. He eyed the British warily, as potentially poisoning the well of Middle Eastern opinion against the west. (The British watched Henderson with similar skepticism: an official in the British embassy in Baghdad suspected Washington of plotting "a long term drive to promote American interests in this ancient and backward land" and identified Henderson as an "energetic representative" of this competing influence.) In particular Henderson feared that the British would alienate the representatives of Arab nationalism, which he correctly identified as one of the decisive influences in regional politics for the next generation. Henderson's appreciation of the strength of the Arab nationalist movement demonstrated that he was not immune to developments close at hand. But for Henderson, Arab nationalism never mattered as much for its own sake as for what it represented in the conflict between the United States and the Soviet Union.[17]

He took much the same attitude toward Zionism. In Iraq, Zionism had had an unfortunate recent history. During the struggle for independence the Arab Muslim majority had come to view the country's Zionists as a fifth column, and had tried to suppress Zionist activities wherever possible. The British mandatory authorities, hoping to transfer power to a united Iraqi regime, had also discouraged Zionism. The British adviser to the interior ministry formally requested the Jews of Baghdad to "put an end to all Zionist activities." In the face of this combined opposition, Iraqi Zionists went underground. The Zionist Association of Iraq conducted its work—holding meetings, promoting emigration to Palestine, and instructing members in Hebrew—behind the façade of the "Jewish Literary Society." But the cover failed to provide much protection. When 1936 brought widespread violence to Palestine, many Iraqi Muslims vented their frustrations on the Jews closest at hand. Government officials in Baghdad charged Iraqi Jews with lending support to the Zionists in Palestine; two Jews were murdered in plain sight on a Baghdad street; on Yom Kippur, a bomb—fortunately defective—was thrown into a crowded synagogue. So threatened did the country's Jews feel that Baghdad's chief rabbi publicly dissociated himself and the Jewish community from the Zionist movement, and a leading Jewish scholar published an article in the capital's foremost daily newspaper under the title "We Were Arabs Before We Became Jews." The accession of the pro-German Rashid Ali brought further pressure against Iraq's Jews, who were accused of acting as agents for the British. When approaching British troops drove Ali from power, Arab nationalists blamed the Jews, who indeed cheered Ali's demise, for turning the country over to the imperialists. Before the British secured the capital, the largest attacks to date occurred, resulting in the deaths of more than 150 Jews, in addition to scores of non-Jews, some of whom died attempting to defend their Jewish neighbors.[18]

Henderson knew or soon learned this history, although he got most of his information from non-Jewish sources. After all they had gone through, few Iraqi Jews cared to risk an interview with the representative of a government widely thought to favor the Zionist cause. Exposed principally to the Arab Muslim perspective, Henderson not surprisingly received with skepticism a report from the Jewish Agency in Jerusalem, the shadow government for Palestine's Jews, describing the oppression of Iraq's Jewish community. Henderson considered the portrayal politically motivated and historically overdrawn. He took issue with the report's charge of widespread contemporary violence against Jews, declaring that while occasional incidents occurred their number was "surprisingly small." Much of the violence that did take place, he continued, owed to the "not entirely unfounded" feeling among Arabs that although the Jews publicly proclaimed their loyalty to Iraq, they were "secretly sympathetic to Zionism, which they hope will eventually result in Jewish ascendency in the Near East." He cited as further provocation "the public dishonesty, profiteering and greed of some of the Jewish merchants who play a leading role in the retail trade," and "the fact that the Jews, although relatively few in number and although openly respectful to the Arabs, conduct themselves in a manner which gives the impression that they consider themselves socially and culturally superior to the Arabs." In the latter context Henderson noted the "special status" accorded Jews in Iraqi law, which provided them "an organization which in some respects constitutes a state within a state" and which allowed them "to maintain themselves as a closely knit, well organized group."

Henderson did not hold the Jews entirely at fault. He conceded "the readiness of the Arabs, when angered, to resort to cruelty and violence." But on balance he argued that where Iraqi Jews faced problems, they often had themselves to blame. Describing the anti-Jewish riots attending Rashid Ali's fall in 1941—the extent of which he overestimated, putting the Jewish dead at more than 350—he pointed out certain extenuating circumstances. In the first place, at the time of the killings there existed no responsible government. Second, the Jews had expressed an indiscreet amount of happiness at the arrival of the British. Citing the accounts of Arab eye-witnesses he considered reliable, he asserted that many Baghdad Jews had "openly celebrated the British victory and did not attempt to conceal their scorn and contempt for Arab-Moslems in general, and for the Iraqi Army in particular." Henderson thought it significant that during the same period Christians had largely escaped persecution. He attributed the latter group's relative good fortune to the Christians' "neutral attitude."

Henderson especially contested the conclusion of the Jewish Agency's report, which suggested that Jews had no future in Iraq and implied that they would be safe only in a Jewish state. "This is hardly true," he wrote. Admitting that recent years had witnessed a deterioration in the social and economic position of Iraq's Jews, Henderson described them as still "the most prosperous people in Iraq." As evidence he cited the fact that although Jews constituted less than one-fifth of Baghdad's population, they controlled three-quarters of the city's commercial enterprises—which in general terms was true. He granted that they faced some discrimination but said they managed to offset this by "their amazing adaptability and dexterity in attaining their aims in the face of legal and other obstacles." Henderson believed that such problems as confronted the Jews resulted not from their Jewishness, but from the agitation of the Zionists, which tarred all Jews with the brush of disloyalty. He conceded that in a narrow sense the dire predictions of the Jewish Agency might prove true, but if so they would be self-fulfilling. He considered the report a piece of disinformation designed to widen the rift between Muslim and Jew in Iraq, since, in Henderson's interpretation of the Zionist position, "the broader the gulf, the stronger the feelings of Jewish nationalism are likely to be."

Contributing to Henderson's distrust of the Zionists were reports—accurate, as it turned out—of clandestine paramilitary training among the Jews of Baghdad. Henderson granted that such activities might be intended as a precaution against a repetition of anti-Jewish violence, but he contended that by preparing for violence the Jews might precipitate it. Not only did the creation of an underground army threaten the Arabs, it added to a feeling of "the superiority of the Jews as a nation" and thereby made future reconciliation more difficult. "The Jews under training," Henderson concluded, "naturally begin to regard the Arabs around them as potential enemies."[19]

Henderson's concerns regarding the destabilizing effects of Zionism were reinforced by the emergence of a communist movement in Iraq. To some degree the radical left simply drew on popular discontent regarding governmental corruption, now compounded by war-induced shortages of consumer items. At the same time the British, fighting the fascists in alliance with communists in the Soviet Union, found among local leftists convenient collaborators, who in fact became some of the strongest supporters of the war. Although nominally outlawed, communist activities were tacitly allowed. Party leaders freely visited government offices and attended receptions at the British embassy. Ever impolite, the radicals did not let this toleration deter them from agitating against the perfidy of the imperialists and their stooges. As a result the communists made great inroads among the lower classes in the country. One historian of the period describes the war era in Iraq as "a field day for the left."[20]

This was precisely what worried Henderson. As the end neared for the fascists in the latter part of 1944, Henderson remarked what seemed the opening shots of a new radical propaganda offensive. Commenting on a pamphlet hitting hard at government corruption, he wrote, "If the distribution of this leaflet marks the inauguration of a Communist campaign in Iraq, the campaign has begun in auspicious circumstances." By taking the side of the masses against a venal bureaucracy, the communists were winning "many friends."[21]

But the most potent weapon in the arsenal of the radicals was popular opposition to Zionism. Although Iraq had formally allied with the United Nations after Rashid Ali's fall, the country still harbored significant sympathy toward Germany as being reliably anti-Zionist. In January 1945 Henderson noted this inclination in the Baghdad press and reported that leftist and pro-Soviet groups had achieved "considerable success" turning public antipathy to Zionism into support for their own political agenda. In particular, Iraqi communists had convinced many people that the Soviet Union was a staunch opponent of Zionism and a powerful friend of the Arabs.[22]

Had conservatives in the various Arab states presented a common front, the activities of the radicals would have caused Henderson fewer worries. But with the different royal families competing for primacy, cooperation was a distant prospect. In February 1945 Henderson discussed the dynastic quarrels with Nuri, who despite his political eclipse remained highly influential in Hashemite circles. Nuri had recently returned from several days in the company of Faisal's relative, Emir Abdullah of Transjordan, who was advancing his claim to the throne of Syria. As Nuri explained the situation, Abdullah's ambitions exceeded his grasp, and his attempts to enlist the French on his side were bringing embarrassment upon the Hashemite house. In addition they were undermining Syria's efforts to drive out the French. To make matters worse, Abdullah was encouraging Ibn Saud to meddle in Syrian affairs. Nuri reported to Henderson that he had told Abdullah to give up. The French had hoodwinked Abdullah and were simply using his hopes for a crown as a device to extend their lease on Syria. Nuri added that he had gone on to say that neither the French nor any other external power could place Abdullah on the Syrian throne unless the Syrian people wanted him there. What little chance existed of a popular call would soon be lost unless Abdullah demonstrated he was more interested in Syria's future than his own. As Nuri related the tale, this line of reasoning had succeeded, and Abdullah had assured him he would respect the wishes of the Syrian people.[23]

Listening to this account, Henderson was convinced he was learning more about Nuri than about Abdullah. Nuri heretofore had kept troubles among the Hashemites within the family. Why he chose this moment to unburden himself, Henderson did not know. Further, as Henderson had discovered from evidence he had gathered independently, and as he explained to the State Department, Nuri had "failed to give an accurate description" of the Abdullah matter. Henderson could only guess what Nuri's scheme was this time.[24]

10

The Great Game

Before Henderson had an opportunity to unravel Nuri's latest maneuver, the State Department called him back to Washington. In March 1945 a visitor, a military friend of Henderson who had heard good things about the hunting in the marshes outside Baghdad, arrived from Cairo. Henderson took a rare afternoon off to join him on a shoot. They spent several hours wading through irrigation ditches and crawling over mud fences, but to little avail. Returning to the legation at dusk, frustrated and thoroughly fatigued, Henderson was greeted by the decoding officer, who said he had just received an important message. "Does it need action tonight?" asked Henderson. The officer thought it did. The message was from the acting secretary of state, Joseph Grew, inquiring whether Henderson would accept assignment as director of the office of Near Eastern and African affairs (NEA).

Henderson later said he was "not exhilarated" at the prospect of taking the NEA job, despite the fact that it represented a major career advancement. During his eighteen months in Baghdad he had begun to sort out the complicated politics of Iraq. A 1944 trip to the Persian Gulf and to the mountains of the north had made him appreciate all the more just how complicated things were. (The journey had also prompted London's foreign office to instruct British officials in the area to accord him every hospitality—the better to keep an eye on him.) He hesitated to leave Baghdad just as he was getting a handle on the situation there.[1]

More to point, his sojourn in Iraq had afforded respite from the backbiting of bureaucratic Washington. After fighting the White House for several years—and losing—Henderson had little desire to enter that thankless arena again, at least not yet. The Zionist issue, to name just one of the questions facing an NEA director, would probably make the difference of opinions regarding the Soviet Union look like the mildest of academic debates.

But Henderson had never turned down an assignment, and he was not about to start. He sent off a cable expressing gratitude that the department possessed such confidence in his abilities and accepting the new job.[2]

The office Henderson returned to in the spring of 1945 had been headed for more than a decade and a half by Wallace Murray. In certain respects Murray was to American diplomacy toward the Middle East what Robert Kelley had been for diplomacy toward Eastern Europe. To his supporters Murray was a tower of strength, an individual, as Evan Wilson described him, of "quick mind, a formidable knowledge of Near Eastern affairs, and a lively appreciation of the importance of the area to the United States." Yet even Murray's friends acknowledged his sometimes excessive suspicions of Britain and his impatience toward anyone or anything that intruded on the peace of his bailiwick. Zionists and Zionism fell squarely into Wallace's category of intruders.[3]

Inheriting Murray's division, Henderson inherited the attitudes others felt toward that division. Zionists in particular viewed NEA with distrust, and most found little to choose between Murray's NEA and Henderson's. James McDonald, an outspoken Zionist who became America's first ambassador to Israel, captured a pervasive feeling among supporters of a Jewish state in Palestine when he described Henderson as "intelligent" and "vigorous" but nonetheless "autocratic." McDonald went on to say that Henderson was "prone to consider the Middle East as a personal province and himself as its benign overseer."[4]

Despite Henderson's lingering reservations about Britain's predominance in countries like Iraq, British diplomats deemed the new NEA chief a marked improvement over Murray. Humphrey Trevelyan, who had extensive dealings with each, described Murray as "difficult and openly hostile," while Henderson was "friendly and helpful." After the crises of the early cold war threw Washington and London into close collaboration, Henderson seemed a genuine ally. In 1949 British defense minister A. V. Alexander spoke of Henderson as "a very loyal friend," one with "very sane views from our point of view."[5]

II

The close Anglo-American relationship did not develop at once, however, and when Henderson assumed the reins of NEA in the spring of 1945 he considered the British to be part of America's problem in the Middle East rather than part of the solution. Only the French were worse. At that time the particular difficulties facing the United States in the region centered in the Levant. At the end of World War I the League of Nations had mandated the territories at the eastern end of the Mediterranean to the French and British, with the former receiving Syria and Lebanon, the latter Palestine and Transjordan. During the next twenty years nationalist agitation developed in all four countries. In certain respects the return of war in 1939 had afforded a breathing space, in that the fighting provided the mandatory powers a reasonably persuasive excuse to suppress dissent. At the same time, it increased local pressure to have done with mandates and get on to independence. While both France and Britain had nominally committed themselves to the principles of the United Nations, neither had the slightest intention of abandoning the region simply because international opinion demanded pulling down the Tricolor and the Union Jack. But both Paris and London recognized that where force had formerly sufficed, modern conditions called for ingenuity in maintaining influence once political sovereignty passed to the locals. Henderson had seen the pattern in Iraq. Now the model was being applied in the Levant.

Problems arose from the fact that the British were getting greedy, to the dismay and annoyance of the French. As early as 1916, by the Sykes-Picot agreement, London had accepted a French sphere of influence in Syria and Lebanon, which the League's mandate had confirmed. Following the French collapse in 1940, though, British troops had collaborated with the Free French forces of Charles de Gaulle to occupy the two countries, in the process pledging support for Syrian and Lebanese independence. Yet even as they came to France's aid, the British doubted that the French could play the stabilizing role western interests in the Middle East required. Consequently they laid plans for further entrenchment of Britain's control. "This is what the French suspect we are after," Oliver Harvey, assistant undersecretary at the British foreign office, commented dryly. "How right they are."[6]

Before World War II the majority of Americans would happily have let the two European empires fight it out. But Pearl Harbor had completed the discrediting of the isolationists Munich had begun, and although world-weariness would return after the fighting ended, nothing like the anti-Versailles reaction of a quarter-century earlier would develop. On the contrary, most Americans would accept with minimum persuasion the notion that only American involvement in the peace would prevent later American participation in yet another war. This, in any event, was the argument Henderson began to make as soon as he arrived in the NEA division.

As it applied to the Levant, Henderson's argument required an active American role in forcing France and Britain to live up to their United Nations pledges. With other American officials, Henderson placed most of the blame for problems in Syria and Lebanon on France. Each power had announced its desire to withdraw, but neither would evacuate first. The French particularly procrastinated, causing the American consul in Damascus to write that "the principal point of French policy in the Levant at the moment revolves around a French belief in the necessity for delay." De Gaulle, aware of his junior position in the antifascist alliance, expected his bargaining position to improve with time. Meanwhile, however, the stability of the region and the prestige of the United Nations there declined. According to the American consulate's contact in the Syrian government, if the French did not get out soon, "serious troubles" would develop.[7]

Henderson agreed. Writing to William Phillips, the special assistant to the new secretary of state, Edward Stettinius, Henderson described his fear that the situation in the Levant was becoming "extremely dangerous." Rather than withdrawing, the French were reinforcing their position in the Levant. Henderson suggested that de Gaulle was hoping to provoke an incident that would afford a pretext for a prolonged French presence. "There is the possibility," he told Phillips, "that the disturbances will become so widespread that the French will call in more troops, battleships, and perhaps even airplanes for the avowed purpose of maintaining the peace."[8]

Henderson thought the United States must act quickly to prevent the area from exploding. The British, fearing that French foot-dragging would undermine their own position, had proposed a conference in Beirut of representatives of Britain, France, the United States, Syria, and Lebanon to negotiate the withdrawal of foreign forces. Henderson was not convinced a conference such as London proposed provided quite the right solution, but he did not think the United States could afford to reject the British plan without suggesting something of its own. "If we adopt at this time a negative attitude towards the situation, and if we offer no counter-suggestion, the British might be in a position to state that we, by our failure to take any practical steps, are responsible for what may take place." Henderson recommended a smaller, less formal set of discussions, in London preferably, or Paris if the French insisted on a home-field advantage. Henderson's reluctance regarding a full-blown conference followed from a suspicion that the British and French had already made secret arrangements for the disposal of the Levant. The United States then would face an unpromising choice between accepting a fait accompli and publicly breaking with the two allies. "It is imperative," Henderson wrote, "before we enter into conferences including the British and French, that we know what if any understandings exist between them."[9]

While Britain and France pondered Henderson's counteroffer, the NEA director elaborated his concerns regarding the area. He did not consider the British any more solicitous than the French of the well-being of the locals, but he thought Britain's larger stake elsewhere in the region would incline London toward accommodation in the Levant. He thought the French were defining French interests so narrowly as to throw the Middle East into chaos. Equally damaging to the position of the United States, the French were reverting to gunboat diplomacy at the very moment when the United Nations were meeting in San Francisco to hammer out a framework for the peaceful settlement of disputes. France, Henderson declared, was pursuing a policy

> which certainly does not seem to be consistent with the principles which the United Nations have stated they are upholding, and which apparently ignores the purposes and aims of the International Security Organization which we are trying to form in San Francisco. . . .

While in San Francisco we are talking about world security and are devising methods for combating aggression, France is openly pursuing tactics which are similar to those used by the Japanese in Manchukuo and by the Italians in Ethiopia.

France's policy, Henderson argued, could only erode confidence in collective security. "So far as we are aware, French action with regard to Syria and Lebanon is the first instance since the formation of the United Nations in which a great power has deliberately set about by force and threats of force to work its will upon smaller powers, without provocation and in its own selfish interest." The Russians were employing heavy-handed tactics in the countries they occupied, but Moscow had the plausible excuse of fighting Germans there, and in any event, bullying was precisely what the world expected from the Kremlin. "It is much more serious for a great Western power, possessing democratic traditions, to follow a course similar to that pursued by Russia in Eastern Europe." Henderson contended that the newly emerging countries differentiated between the Soviets and the west. "They are almost certain, however, to judge all the Western powers in the light of policies pursued by any one of them."

Obviously the United States could not break relations with de Gaulle over Syria and Lebanon. Nonetheless, Henderson asserted, Washington must speak out. Describing developments in the region as the first real test of American sincerity in promoting a world organization in which small nations as well as large could feel secure, Henderson said that the peace might disintegrate almost before the fighting stopped. "We are remaining silent and are allowing the exigencies of the moment to prevent us from taking action which might help prevent the world from going back to the practices which, from 1931 to 1939, resulted in the present war." At this point Henderson had no single answer to the problem France posed, but he felt compelled to raise the basic question confronting American leaders.

Are we, at the moment that the International Security Organization is being launched, to tolerate one of our Allies engaging in a policy which partakes of aggression because we do not wish to give offense to that Ally? Or are we to make it clear to that Ally and to the rest of the world that we intend to follow a policy of combating aggression, even though such aggression should be committed by our closest friends and even though the combating of it might be extremely inconvenient to us?[10]

Henderson's argument persuaded William Phillips and, in turn, Undersecretary Grew. "Mr. Henderson's memorandum is worth careful study," Phillips told Grew. Phillips said Henderson had presented an "alarming picture which we must have clearly in mind" while dealing with the French. Grew took Henderson's case to the new president, Harry Truman, who approved a note to the

French foreign ministry explaining America's concerns at the situation in the Levant and requesting the French government to review its policy toward Syria and Lebanon.[11]

But the message had little effect. De Gaulle made clear he intended to continue as before. The general blamed the continuing trouble in the Levant on Syria's and Lebanon's refusal to act in good faith.[12]

Consequently Henderson reiterated his argument. This time he went directly to the top. In a memorandum to Truman the NEA chief described the situation in Syria and Lebanon as "rapidly developing into open warfare." The French had bombed and strafed Damascus and other population centers, killing hundreds. The governments of Iraq and Saudi Arabia were calling for American intervention. The president of Syria had demanded of the American representative there:

> Where now is the Atlantic Charter and the Four Freedoms? What can we think of San Francisco? . . . Your country has encouraged us in our stand to refuse special privilege to France or any other country but you have permitted France to block the adequate arming of our gendarmerie; now the French are bombing us and destroying our cities and towns with Lend-Leased munitions which were given for use against our common enemies.

Henderson commented that despite American efforts to dissuade de Gaulle from sending more soldiers, the French had dispatched not only additional troops but warships as well. Under the guns of these ships they had stepped up their demands for favored treatment. Because the recent American attempt to elicit a review of French policy in the region had yielded no results, Henderson, with Stettinius' approval, suggested a personal presidential appeal to de Gaulle, and he provided a draft telegram.[13]

With minor revisions Truman signed Henderson's cable. The president expressed his "grave concern" at the fighting in Syria and Lebanon and his hope that the French would act at once to restore peace to the area without prejudice to the rights of any nation.[14]

Before de Gaulle responded to this message, a British spanner fell into the works. At the end of May, Churchill, in the belief that neither British imperial security nor British credibility allowed acquiescence in French savaging of the Levant, ordered the British commander in Syria to impose a ceasefire and forcibly restrict French troops to their barracks.[15]

De Gaulle was outraged. Britain, he declared, had "insulted France and betrayed the West." The general added, "This cannot be forgotten." Realizing he was in no position to challenge the British action militarily, de Gaulle chose diplomatic weapons, advocating a conference of the four great powers, including the Soviet Union, to discuss all the problems of the Middle East. It required little imagination on Henderson's part—or anyone else's—to see that de Gaulle was hoping to play the Russians against what he perceived as an incipient Anglo-American coalition in the Levant.[16]

When the Truman administration, with Henderson's hearty concurrence, emphatically rejected the idea of a Soviet role in the Middle East, France shifted its demands slightly. In June the counselor of the French embassy in Washington, Francis Lacoste, explained his government's policy to Henderson. France, Lacoste declared, did not wish differences over the Middle East to undo its entente with the British, but London had insisted on acting as though France were guilty of serious crimes. Such a state of affairs was intolerable. "No country," he said, "will permit itself to be treated as a culprit by its ally." There could be no doubt that the British intended to dispossess France in Syria and Lebanon. Before long France would have nothing left to lose. His government sought peace and did not wish the alliance to rupture over such a matter. To Henderson's query regarding a possible solution to the crisis, Lacoste described a three-step process: first, negotiations between France and Britain to resolve the immediate difficulty; second, negotiations between France and the Levant governments to define France's future relationship with the area; third, a broader conference on the Middle East including all the great powers.

Henderson noted that placing the big-four conference at the bottom of the list marked something of a concession to Washington. On the other hand, the proposal for Anglo-French talks appeared to be an effort to exclude the United States. To this Henderson objected immediately. "We cannot look with favor," he told Lacoste, "on any agreement that would discriminate against the United States, and we would regard it as unfortunate if any agreement contrary to this principle should be entered into." Aside from peace, what the United States wanted in the Middle East was nothing more than "ordinary normal inter-course," an open door "free of artificial restraint." To Lacoste's objection that Britain had for years enjoyed special privileges in the Middle East, Henderson rejoined that the United States had not condoned such favored treatment. In any case, the American government perceived a large distinction between pacts signed decades earlier and expiring shortly—he cited the Anglo-Iraqi agreement of 1930 as an example—and new treaties yet to be concluded. Most fundamentally, Henderson stressed the need for an early settlement of the Anglo-French dispute. Continued quarreling among the allies, he said, would jeopardize the prestige of the west as a whole. Quoting a British diplomat he had known in Iraq, Henderson closed with the comment, "Whenever agitation in the Near East rises against one power, it ends by being agitation against all of us."[17]

The governments of Syria and Lebanon found the French proposals no more satisfactory than did the Americans. Charles Malik, the Lebanese minister in the United States, told Henderson his country did not want its future decided by a group of outside powers. An impending visit by de Gaulle to Washington had prompted Malik's call on Henderson. The Lebanese diplomat feared an attempt by the French general to arrange an imposed solution. A few days later the Syrian minister arrived at Henderson's office with a similar concern. To each, Henderson offered assurances that the United States did not intend to negotiate over the heads of the countries of the region.[18]

When Syria supplemented its warning against de Gaulle with a request for an American military mission, ostensibly to assist in organizing and training Syria's armed forces but with the obvious objective of using the American presence as protection against the French and the British, Henderson quickly endorsed the idea—although for a different reason. Whatever the position of Paris and London at the moment, sooner or later world opinion and domestic politics would force at least a partial French and British withdrawal from the Middle East. Some other great power would fill the void, and if not the United States undoubtedly Russia. Henderson worried about what he called "the emergent role of the Soviet Union" in the area. Further Soviet penetration would pose a serious danger to American interests. The Syrian invitation provided "an excellent opportunity" for the United States to preempt Moscow. "The effect of our entering upon this comparatively small task will undoubtedly extend through the whole region and will serve to strengthen greatly our influence and prestige well beyond the borders of Syria."

Henderson predicted that the British would accept an American military presence in the Levant, if only to balance the French. A few weeks earlier, in fact, Michael Wright of the British embassy had suggested something along the lines of the Syrian request. In addition, Henderson suspected that Britain would welcome American help in Syria as a first step toward American responsibility for solving the more vexing problem of Palestine. Henderson commented that Britain had reason to anticipate support, in Syria at least, since the American government had not objected to Churchill's decision to land troops. As for the French, they would complain but could do little more.

Henderson concluded with a ringing declaration—the first of many—that the choice confronting the United States amounted to a decision between forthright acceptance of America's responsibilities as a great power and abdication of world leadership. If America failed to take up this challenge, he argued, the effect could devastate American credibility in a vital portion of the globe. "Our refusal to meet the present Syrian request would be comparable, in its disillusioning and unsettling effect throughout the Near East, only to the retirement of the United States into isolationism after the First World War."[19]

As Henderson predicted, the British accepted the idea of an American military mission to Syria while France rejected it. The French foreign ministry declared the Syrian request "a manoeuvre intended to offend France" and warned that France would consider a favorable American response "unfriendly." To Lacoste, who delivered de Gaulle's rejection, Henderson commented that the note contained "rather strong language." He assured Lacoste that the American government did not consider the Syrian request antagonistic toward France, and he added, not entirely truthfully, "We, frankly, would prefer not to send a military mission to Syria." At the same time he reminded the French diplomat that Syria was an independent country and that its request and America's response were matters for the Syrians and the Americans alone to consider. He rejected categorically, to Lacoste's obvious displeasure, the notion of a preferred position for France in Syria, which he labeled "outmoded."[20]

What annoyed the French also upset those of Henderson's associates charged with handling de Gaulle. Henderson encountered resistance to an American military mission to Syria among his old office-mates in the State Department's division of European affairs. The Europeanists contended, not illogically, that since the primary importance of the Middle East for the United States lay in its relation to Europe, alienating France for the sake of Syria stood matters on their head.

But Henderson, who in addition to believing in the need to preempt the Soviets was developing a bureaucrat's sense of turf, held his ground. In November 1945 he made his case to H. Freeman Matthews, his counterpart in the European division. "We realize," Henderson said, "that the appointment of such a mission will extremely irritate the French and that they will do their utmost by intrigue in the Levant as well as through diplomatic means to make the mission a failure." Other factors would also conspire to foil the plan: "the confused conditions in the Levant, the backwardness of the Syrian people, and the lack of public honesty on the part of various Syrian leaders." Yet any important initiative involved an element of danger, and to shun all risks would consign America to a "weak and vacillating" foreign policy. For several years the United States had expressed sympathy toward the aspirations for independence of the countries of the Middle East. American leaders had led those countries to believe that the United States would help them achieve independence.

> If we refuse the Syrian request for a military mission, the impression is sure to be created in the Near East that although we are willing to talk glibly regarding our interest in the welfare of the peoples of the Near East, we are unwilling to implement the principle to which we say we adhere by refusing to take any measures which might meet the fierce opposition of any great power. They will feel that when the principles of the United Nations come to close grips with imperialistic ambitions, the proponents of those principles do not have the courage or assurance which characterizes the advocates of Western imperialism.

Under the circumstances, Henderson said, the Syrian request assumed the nature of a test of American credibility. It was "extremely important" that the United States not allow France to veto a favorable response.[21]

Matthews was not persuaded. The director for European affairs told Henderson he appreciated the arguments for acceding to Syria's wishes, but he claimed that the adverse effects of injecting an American military presence in the Levant into U.S.–French relations outweighed the advantages for the American position in the Middle East. Moreover, an American mission to Syria would require congressional approval, which raised problems, perhaps intractable, of another sort. Finally, recent developments hinted that the French and British might manage to work out their problems alone. For the United States to force its way into the region at this time could destroy what chances existed for a peaceful accord between America's two key allies.[22]

Matthews' objections carried the day. Not for the last time Europe outweighed the third world. When Matthews' prediction of an Anglo-French accord proved correct, Henderson had little space left on which to stand. Even so, he defended that space the more. For the American government, he asserted, to accept a solution devised in Paris and London would discredit the United States in the eyes of the international community. Outside observers would see Washington as an accomplice in a cynical subversion of the Atlantic charter and the principle of self-determination. Such collusion might well wreck the United Nations before it had fairly begun its work. The organization "would come to be regarded as an instrument for the extension or perpetuation of imperialism rather than as an instrument for the preservation of world peace on the basis of justice and non-discrimination with respect to all nations, large and small." In addition Henderson pointed out that a provision of the Anglo-French accord, one that delayed evacuation of foreign troops pending a comprehensive peace settlement, established a pernicious precedent. The Soviets, he suggested, would be quick to seize the example as justification for their continued presence in occupied territory, including key countries like Iran.[23]

III

Henderson had good reason to fear a precedent for Iran, for by the latter half of 1945 it was becoming clear that Iran would provide an early test of the possibilities for a continuation of the grand alliance into the postwar period. Henderson, of course, had few hopes. Asked later when the spirit of wartime collaboration had broken down, he responded that there had been "no spirit of wartime collaboration." There certainly was none on Henderson's part, which was why in 1945 he was worrying about the Soviet position in Iran.[24]

Soviet troops remained in Iran under the terms of a 1942 tripartite pact among London, Moscow, and Tehran, which called for Britain and Russia to withdraw their forces six months after the end of the war with Germany and Germany's allies. When the fighting in Europe ended in May 1945, the Iranian government began counting down the six months for the Russians, who had not yet declared war on Japan. British troops posed a different problem, since

Britain continued to fight in Asia and the Pacific. American forces, which had entered the southern part of Iran subsequent to the three-power accord, fell into the same uncertain category. Complicating the matter further was the fact that the Soviets were expected, in fulfillment of a pledge made at the Yalta conference, to declare war against Japan before their six months in Iran ended. Whether this would reset the clock remained a matter of dispute.

By June, the Iranians, having had their fill of foreign troops, were agitating to get the British and the Soviets to honor the principle behind their wartime pledge by withdrawing at once. The Iranian theater meant little in the drive to finish off Japan, Tehran's representatives argued. Iran was suffering from more than the usual postwar confusion and disruption, and with chaos looming the presence of foreign military forces exacerbated the country's problems. As an Iranian diplomat explained to Henderson, his people had no concerns regarding the Americans. Only the British imperialists and the Soviet communists worried them. His superiors had taken some encouragement from London's suggestion that Britain might withdraw before the war with Japan ended, but at the same time London had said British troops would not leave before the Russians did. Conversely, it appeared that the Soviets would not pull out until the British did.

To Henderson's remark that the British might wish to keep troops in Iran to defend the country's oil fields and the large refinery at Abadan, the Iranian diplomat replied that the British had nothing to fear from Iran. In any event, Britain possessed sufficient naval forces in the Persian Gulf to defend the oil supplies. When Henderson asked what effect the Iranian request for immediate evacuation would have on the presence of American troops guarding the airfield at Abadan, Tehran's representative replied that this would create no difficulty and that his government and the Americans could work out a solution "one way or another." He concluded by asking Henderson what position the United States would take on the matter at the Potsdam conference, soon to convene. Henderson replied that he could not say precisely but that the Truman administration would consider the Iranian request for immediate evacuation with "the greatest of sympathy."[25]

When the big three met at Potsdam in July 1945 the British urged the Russians to accept an accelerated timetable for mutual withdrawal. Truman supported the British, adding that he expected American forces to be out within sixty days. Stalin agreed to the evacuation of Tehran, but to no more. Further withdrawals, the Soviet leader indicated, and related questions regarding the future of Iran would have to await the September conference of allied foreign ministers.[26]

The government of Iran received news of the outcome of the Potsdam meeting with "extreme disappointment and regret," as an official from the Iranian embassy told Henderson. Henderson was disappointed too, yet rather than dwell on the setback he immediately began preparing for the foreign ministers' conference. In particular he sought to raise the visibility of Iran in American policy. He wrote a lengthy memorandum to James Byrnes, the recent replacement for Stettinius as secretary of state. Byrnes seemed an improvement over "Brother Ed," who had a reputation among the careerists as a glad-hander who never did his homework, but the improvement was only slight. Although Byrnes had impressive political credentials, having served in both houses of Congress, in the executive branch under Roosevelt, and on the Supreme Court, he had almost no experience in foreign affairs, and he early displayed the politician's penchant for splitting differences in the interest of making a deal. Elbridge Durbrow summarized the bureaucracy's view when he described Byrnes as "the great compromiser."[27]

Deciding that the secretary needed educating on the Iran issue, Henderson assumed the task. He traced the roots of the controversy to the traditional rivalry between Britain and Russia on the frontier of their respective empires. During the nineteenth century each had attempted to secure its borders at the expense of the other. German penetration of the Middle East in the 1920s had complicated matters, setting in motion a train of events leading ultimately to the Anglo-Soviet invasion of Iran in August 1941. The allied occupation had snuffed out pro-German activity but had heightened tensions between competing factions in the country. The authoritarian government of Reza Shah Pahlevi had given way under his young son to what Henderson described as "a weak, constitutional regime for which Iran was ill prepared by tradition or experience." As a consequence the government's administrative machinery had broken down, latent separatism among minorities had flourished, and the economy had collapsed.[28]

At this point, Henderson continued, Iran had turned to Washington for assistance. In 1942 the United States began providing financial, technical, and military aid. Upon specific request from the Iranian government, civilian American advisers took positions in the departments responsible for customs collection, internal revenue, price control, rationing, public health, and municipal police administration. American military officers reorganized and trained the Iranian army and the gendarmerie, or rural police. American assistance, as Henderson described it, had two-and-a-half aims: to demonstrate interest in the needs of a "friendly nation," and to contribute to "the reconstruction of Iran as a sound member of the international body politic, and thereby to remove a future threat to Allied solidarity and international security."

American policy, Henderson said, had achieved limited but not insignificant results. During the war the American presence helped guarantee the uninterrupted flow of supplies to the Soviet Union. Meanwhile American forces, by demonstrating American concern for the security and territorial integrity of Iran, bolstered Iranian morale and perhaps moderated the ambitions of the British and Russians. Furthermore, representatives of American businesses followed the GIs, pursuing trade and investment opportunities for the postwar period. In turn the Iranian government pledged that when it reopened negotiations for oil concessions it would consider American applications along with those from other countries.

Regarding future policy toward Iran, Henderson contended that American leaders would soon have to confront the fact that such subordination of the Anglo-Russian rivalry to wartime cooperation as had occurred would quickly vanish. A conflict between Britain and the Soviets would intensify the "steadily widening politico-social schism between leftist and conservative forces" within Iran, rendering stability and governmental continuity impossible. A vicious circle was closing as internal chaos created a "vacuum," which threatened to make foreign intervention "inevitable." In short, Henderson declared, Iran showed every symptom of becoming "one of the major security problems of the future."

By acting quickly, however, the United States might forestall this discouraging outcome. Henderson told Byrnes that the American government should declare to Britain and the Soviet Union that it would not allow those two countries to divide Iran between them. The ideal solution, he suggested, would be a tripartite—Anglo-American-Soviet—advisory commission, charged with stabilizing and rebuilding Iran. If the British or the Russians rejected such a scheme, the United States should refer the Iranian question to the United Nations. But in any case, Iran must not become a pawn between Britain and Russia.[29]

While Byrnes reflected on Henderson's proposal, the NEA chief detailed his fears to Undersecretary of State Dean Acheson. Augmenting the pressure provided by its continued military occupation, the Kremlin had mounted a propaganda campaign in support of a separatist movement in Iranian Azerbaijan. Henderson remarked to Acheson that it was too early to tell whether the Russians intended to annex Iran's northern provinces to Soviet Azerbaijan—he noted that praise for Moscow's enlightened policies toward the socialist camp's Azerbaijanis accompanied Soviet blasts at Tehran's treatment of their oppressed brethren across the border—or whether they had designs on the country as a whole. In either case, he argued, the Kremlin was demonstrating that it wished Iran no good. Consequently the United States must do "everything possible" to expedite the evacuation of foreign military forces from Iran. As a first step, America should pull its troops out at once, the better to argue for similar action by its allies.[30]

Henderson judged both the British and the Soviets guilty of violating the spirit if not the letter of their 1942 accord with Iran, in that each country was intervening excessively in Iran's internal affairs. But he considered the Soviets the worse offenders. In this opinion he was joined by Wallace Murray, now ambassador in Tehran, who asserted that the British presence, for all its imperialistic implications, was essentially defensive and designed to prevent further

Russian advances, while ultimate Soviet aims included access to the Persian
Gulf and penetration into other parts of the Middle East. Beyond the strategic
dangers inherent in a Soviet position on the gulf, Russian hegemony in Iran
would result in the exclusion of American oil concessionaires and would
threaten American petroleum operations in Saudi Arabia, Bahrain, and
Kuwait.[31]

Through November, conditions in Iran continued to deteriorate. By the middle of the month, Azerbaijan was in armed revolt against the central government of Iran. When Soviet troops in the area refused to allow strengthening of
Iranian security forces—in clear violation of the agreements governing the allied
occupation—Henderson became more convinced than ever that Moscow
aimed to convert at least part of Iran into another Poland. Such an outcome, he
remarked, was "fraught with dangerous possibilities," and Washington must not
stand by and let it happen. As before, he argued to Byrnes that the United
States must press for "the complete withdrawal of Allied forces from Iran at the
earliest possible date."[32]

Acting on this advice, the secretary of state directed the American ambassador in Moscow, Averell Harriman, to remind Molotov of Russia's obligations
under the Tehran declaration of 1943 and to invite the Soviet Union to join the
United States and Britain in evacuating troops and restoring unfettered sovereignty to Iran. Byrnes conceded that the great powers were under no legal com-
pulsion to withdraw before March 2, 1946, but he declared that the American
government, at "considerable inconvenience" to itself, was preparing to evacuate by January 1. He urged the Soviets, as he was urging the British, to do
likewise. While foreign troops remained in Iran, "incidents and misunderstandings" were likely to occur. Rapid withdrawal, on the other hand, would calm
the situation in Iran and reinforce the prestige of the United Nations. "The
Government of the United States is confident," Byrnes said, "that the Soviet
Union and Great Britain are no less anxious than the United States, in dealing
with nations such as Iran, to follow a line of action which will make it clear that
the trust of these nations in the permanent members of the Security Council
has not been misplaced."[33]

The Soviets rejected Byrnes's advice. While Iran sank further into anarchy,
a spokesman for the Kremlin declared that unsettled conditions in Azerbaijan
required the continued presence of Soviet forces and suggested that those
forces, acting under a Soviet-Iranian treaty of 1921, might be compelled to stay
beyond March 2.[34]

As a result the Iran issue went into the hopper for a meeting of the allied
foreign ministers scheduled for December. Henderson, worrying that the politician in Byrnes might be more interested in the symbolism than the substance
of an agreement, wrote a spine-stiffening briefing paper for the secretary. He
contended that the United States was better positioned than either Britain or

the Soviet Union to seize the initiative in enunciating allied policy toward Iran, "because we are freer from suspicion of having selfish interests in that country," and he urged Byrnes to take a firm stand on the principle of Iranian sovereignty. The secretary should emphasize that "we are not concerned with the maintenance of any particular social or economic system in Iran." He should not dispute the right of the Soviets to seek oil concessions in northern Iran, which some observers had suggested as one motive behind the Russian reluctance to evacuate. But he must stress that sovereignty included the right to be free from coercion in deciding for or against such concessions. Soviets troops on Iranian soil, now that the war was well over, could only be viewed as coercive. As something to place on the table at Moscow, Henderson offered Byrnes a proposal comprising three parts: first, that all foreign troops be withdrawn from Iran as rapidly as transportation facilities permitted; second, that the United States, Britain, and the Soviet Union publicly affirm the right of the Iranian government to move its troops without hindrance through the country; third, that the three great powers declare that Iran should be free from external pressure in its decisions to grant or withhold commercial concessions.[35]

Henderson soon discovered that he had wasted his time. At the opening session of the conference, Byrnes, seeking to preserve what remained of allied cooperation, indicated that he did not wish to make an issue of Iran. The secretary did mention Iran in a private meeting with Stalin, but the Soviet leader dismissed Tehran's complaints as attempts to stir trouble among the'allies. Stalin defended the presence of the Red army as in accord with the Soviet-Iranian treaty of 1921 and necessary to the security of the Soviet Union's southern border. When Stalin assured Byrnes he had no designs on Iran, the secretary dropped the matter.[36]

Knowing Byrnes's background and priorities, Henderson had not really expected a tough line at Moscow. The NEA chief was beginning to see that success in the battle for a more active American policy in the Middle East would require a protracted campaign. At the end of December, while Congress and much of the administration was enjoying its first peacetime holiday season in more than half a decade, Henderson prepared a long memorandum entitled "The Present Situation in the Near East—A Danger to World Peace." In important respects Henderson's memo was to American policy for the Middle East what George Kennan's more celebrated "long telegram" of two months later would be to American relations with the Soviet Union.

Henderson described the Middle East as a "breeding ground for international misunderstandings," in that the objectives of two of the great powers, Britain and the Soviet Union, collided "head-on" there. The British, Henderson argued, were engaged in a holding action, attempting principally to use their influence in the area to build a "great dam" against the Russians. Moscow, by contrast, had more-aggressive designs. "The Soviet Union seems to be determined to break down the structure which Great Britain has maintained so that Russian power and influence can sweep unimpeded across Turkey and through the Dardenelles into the Mediterranean, and across Iran and through the Persian Gulf to the Indian Ocean." During the past five years the Kremlin had witnessed and helped achieve the destruction of two formidable barriers to its expansion: Germany in the west and Japan in the east. "Russia now appears to be concentrating upon the removal of a third barrier in the south."

Because the United States had not made its presence felt in the Middle East, the region by default had become a diplomatic battleground between the British and Russians. The British, severely weakened by the war, were leaning toward compromise with Moscow, in the hope "that the Soviet Union may be satisfied by obtaining the control of certain territory now belonging to third powers and of achieving strategic defensive positions at the expense of other members of the United Nations." The United States could not sit back and allow the British to continue on this course without consigning to oblivion all chance that the principle of collective security against aggression would become a meaningful reality. If the British had their way, the United Nations would "either disappear as a force in world affairs or would tend to become merely an instrument for the use of the Great Powers in carving up the world." More tellingly, an appeasing approach would fail with Stalin as it had with Hitler. The Russians, "once in possession of the new positions conceded to them by the British, would undoubtedly begin preparations for further attacks upon such barriers to their emergence into the Indian Ocean and the Mediterranean as might remain."

To keep the British from emasculating the United Nations, to prevent the Soviets from expanding south, and to avert the strategic erosion that might "eventually lead to a third World War," the United States must adopt a more forceful policy toward the Middle East. In particular, Henderson argued, the American government should call for a conference of the four great powers with interests in the region. The purpose of such a conference would not be to impose a four-power condominium on the area, which of course would itself violate the principles of the United Nations, but would be to determine whether the big four could agree among themselves on a common framework for solution of the region's problems. Should they succeed in this endeavor, they would then consult the countries of the Middle East to work out arrangements acceptable to all. Henderson recognized the long odds against the success of such an endeavor. London would look askance at the idea as a means for stripping Britain of its empire. The Soviets, already succeeding in enlarging their influence in the region, would have little incentive to agree to anything that would put roadblocks in their way. The French, being French, would find some cause for taking offense. Certain groups in the United States would attack the plan as unnecessarily involving America in the affairs of ungrateful foreigners. The countries of the Middle East would not look kindly on what could be construed as another round of great-power dictation. But the need was great enough to make the risks worth hazarding. "The situation in the Near East is fraught with so much danger that nothing which might offer some hope of alleviation should be left undone." Besides, even if the conference failed to achieve the comprehensive solution it sought, it would serve the useful purpose of bringing the issues involved into the "court of world opinion."[37]

A more direct route to the court of world opinion was the United Nations, and judging from Henderson's solicitude for the prerogatives of the UN, one might have expected him again to advocate referring Iran's troubles to the international body. In fact Henderson now sought to keep Iran out of the UN. As troubles with the Soviets mounted, he preferred to leave the UN on the sidelines, lest the organization be destroyed. Henderson feared that the UN might be placed in the position of the League of Nations when the latter had found itself having to deal with aggression by one of its members. "It is important to spare UNO this supreme test at the very outset of its existence, for it might not survive such a test."[38]

But the UN was not spared. Despite Henderson's best efforts to convince the Iranian ambassador in Washington, Hussein Ala, that referral to the UN would bring more trouble than relief, the Iranians took their complaints against the Russians to the security council.

While the council deliberated, the cold war took two turns for the worse. The first came at the beginning of February 1946, when Stalin proclaimed the fundamental incompatibility of capitalism and communism and declared that the continued existence of the former rendered war inevitable. Although a few self-confident types like Henry Wallace considered Stalin's speech a healthy test, most commentators found it ominous. *Time* characterized Stalin's address as "the most warlike pronouncement uttered by any top-rank statesman since V-J Day." Even Supreme Court justice William O. Douglas, a confirmed liberal, considered it "the declaration of World War III." To George Kennan, the speech provided the inspiration for the eight-thousand-word cable that would make the chargé d'affaires in Moscow the administration's most influential voice on the Soviet Union.[39]

The second ratcheting of tension occurred early in March, when Winston Churchill, now out of office but still hawking an alliance of the "English-speaking peoples," announced that "from Stettin in the Baltic to Trieste in the Adriatic, an iron curtain has descended across the continent." The Kremlin's agents, Churchill said, at work throughout the world, constituted "a growing challenge and peril to Christian civilization." The Russians did not want war, but they wanted "the fruits of war and the indefinite expansion of their power and doctrines." When Stalin responded, in a rare public interview, that Churchill's words constituted a "call to war with the Soviet Union," the lines of ideological and political battle became more firmly drawn than ever.[40]

Amid the provocative phrasemongering, the deadline for the evacuation of Soviet troops came and went. The troops remained. If anything, it looked as though the Soviets intended to entrench themselves for an extended stay. From Tabriz, American vice consul Robert Rossow wired reports of heavy Soviet reinforcements from the north, and he declared that he could "not overstress the seriousness and magnitude" of Soviet troop movements. An intelligence veteran who spent days in the consulate in Tabriz and nights counting Soviet tanks in the moonlit mountains outside the city, and who on at least one occasion was arrested by Soviet troops for precisely such activities, Rossow warned, "This is no ordinary reshuffling of troops but a full scale combat deployment."[41]

Upon receiving this news Henderson directed his special assistant, Edwin Wright, to prepare a map depicting the movements of the Soviet units. On the afternoon of March 7 Henderson and Wright took their presentation to Byrnes's office, where they explained what the Russians were up to. Superimposed on the map were large arrows depicting the main lines of advance: one aimed at Turkey and the Turkish straits, a second heading toward Iraq, a third targeting Tehran and the oil fields of the Persian Gulf. The show was a smashing success. Byrnes agreed that the Kremlin was adding military invasion to its repertoire of political subversion. Beating one fist into the other hand for emphasis, he declared, "Now we'll give it to them with both barrels."[42]

The next morning Henderson and Wright repeated their performance for Dean Acheson, Charles Bohlen, and Alger Hiss. "All agreed," Wright recalled later, "that these Soviet moves were clear violations" of numerous commitments governing the presence of foreign troops in Iran. "Only one conclusion could be drawn—the USSR seemed to be determined to face Iran and the rest of the world with a fait accompli."[43]

To put Moscow on notice, as Acheson remarked, that American officials were "aware of its moves," Hiss and Henderson collaborated on a message for Kennan to deliver to the Kremlin. Asserting that the American government was receiving reports of "considerable movements of Soviet combat forces and materials of war" from the Soviet frontier toward Tabriz, the letter requested confirmation or denial. "In case Soviet forces in Iran are being increased," the note concluded, "this Government would welcome information at once regarding the purposes therefor."[44]

This cable was not quite the ultimatum Truman later said it was, but it did demonstrate the depth of American feeling regarding Soviet activities in Iran. Even so, the Red army continued to go forward for a time before eventually pulling back. Precisely why Stalin changed plans remains something of a mystery. Probably he was surprised that the Americans would make such an issue

over Iran, which at that time lay beyond the sphere of interestedness of all but
a few specialists like Henderson. The American economic stake in the country
was only prospective, and distantly at that. Domestic American political concern over Iran bordered on the nonexistent. Whatever the reasons, Stalin chose
not to challenge the Americans' resolve, and at the end of March the Soviet
delegate at the UN, Andrei Gromyko, pledged withdrawal of Russian troops
within six weeks.[45]

IV

Henderson was greatly pleased with the outcome of the affair. The United
States had held its ground and demonstrated that it took seriously the principles
it espoused. The Soviets had backed down. But Henderson also realized that
the danger to Iran, and by extension to the entire Middle East, had not disappeared. Rather it had changed form. If the Soviets could not overawe Tehran,
they might yet achieve their objectives through subversion. The need for American action remained as great as ever.

For a few months during the spring and summer of 1946 other problems occupied most of Henderson's time, but an October meeting with Ambassador Ala
again pushed Iran to the top of NEA's stack of troubles. Ala explained that the
Iranian left, including the Soviet-backed Tudeh party, was exploiting the country's chronic economic problems to make trouble for the government of Prime
Minister Ahmad Qavam. In particular the radicals were demanding cooperation
with the Russians as a lever to loosen the British grip on the country's economy,
especially the oil sector. The Qavam government could not long withstand the
pressure unaided. The situation, Ala asserted, was "very critical." If the government gave in to the leftists' demands the Russians would "consolidate their
position in Northern Iran and eventually gain control of the entire country."
Only American assistance could stand between Iran's freedom and an "exclu-
sive orientation toward Russia."[46]

After checking with George Allen, who had replaced Wallace Murray in
Tehran, to verify Ala's assessment of Iran's predicament, Henderson forwarded
the ambassador's request for help to Dean Acheson with a strong recommendation for approval. "The situation is critical," Henderson wrote. "We should
do everything within our power to prevent Iran from slipping into the Soviet
orbit." Such a development would damage American interests, immediately by
limiting access to Iran's resources but more permanently and widely by diminishing American credibility among countries disposed to look to Washington
for protection. Although the Soviets had withdrawn their troops, an American
failure to follow up diplomatic support with economic sustenance might negate
that victory. The Iranian economy was disintegrating and with it Qavam's political base. The prime minister had been forced to make "concession after con-

cession" to the Russians. Without prompt measures from Washington the situation might be irretrievably lost. "Unless we can consistently show Qavam by action that he can count on the support of this country, in and out of the United Nations, Iran will, in our opinion, inevitably give way to Russian pressure, with all that such yielding entails for the interests of this country."[47]

During the next week Henderson and his assistants fleshed out these thoughts. On October 18 the NEA chief handed Acheson a set of specific proposals. Asserting that Qavam had become "virtually a prisoner of his own policy of retreating before Soviet pressure" and that Iran was "daily losing what remains of its independence," Henderson recommended four courses of action. First, Ambassador Allen should be directed to tell the shah and the prime minister that the United States was prepared to support the independence of Iran "not only by words but by appropriate acts." Second, the White House or the State Department should announce that the administration would consider favorably an Iranian request for a loan from the Export-Import Bank, with the object of funding development projects designed to boost the Iranian standard of living, thereby undercutting the radicals' popular appeal. Third, the United States should furnish Tehran with weapons and ammunition sufficient to maintain "internal security," along with the advisers necessary to ensure these weapons' competent use. Fourth, such programs of aid should be backed by appropriate legislation in Congress.

Henderson understood that he was asking a lot for a country not intrinsically vital to American security. But he continued to believe that Iran was a test of American intentions, and not only toward the Middle East. Characteristically, Henderson viewed the struggle for control of Iran as an aspect of a larger game. It constituted, he said, "an integral part of the broader question of United States relations with the Soviet Union."[48]

11

The Battle Joined

At the end of 1946 Henderson was ahead of the rest of the Truman administration on the Iran question, and despite his best efforts he failed to generate sufficient support to make Iran the site for a demonstration of American resolve. Before the next year was two months old, however, he did succeed—with help, to be sure—in prodding the president to make just such a stand in a different arena.

Almost from the moment he took charge of NEA, Henderson had worried that Greece would prove to be the Achilles' heel of the west in the Mediterranean. His fears were well founded. In the summer of 1945 Greece was staggering from the triple blows of war, occupation, and civil conflict. Between 1939 and 1945 some 8 percent of the population had been killed (a death rate ten times that of Britain), and destruction to property totaled more than $1000 per capita, in a country dismally poor to start. The defeat of the Axis powers brought not recovery but a continuation of the misery as monarchists and communists turned from fighting outsiders to battling each other. When Henderson inherited the Greek problem a fragile ceasefire prevailed, but sporadic violence still plagued the country. Both sides recognized that the truce was no more than that, and each was simply catching its breath before the next round.[1]

Henderson's worries were more than matched in London. In October 1944 Churchill and Stalin had agreed that Britain should take the predominant role in Greece, in exchange for a Soviet sphere of influence in Yugoslavia and elsewhere in the Balkans. Roosevelt, on learning of the arrangement, acquiesced. During the next six months top British officials demonstrated their proprietary interest personally, with Churchill making two visits to Greece, Anthony Eden three, and Harold Macmillan several more. The British established a military liaison with the Greek government, coordinated relief and reconstruction aid, trained police, and generally attempted to put Greek life back on a normal, peaceful, and noncommunist footing. But their efforts failed, and in the late winter of 1946 the communists, with the encouragement of co-ideologists across the border in Yugoslavia, again took up arms.[2]

Even while the uneasy truce continued, Henderson had come to the conclusion that the United States would be sucked into Greece sooner or later. British officers in the field were suggesting that London ask Washington for help. Whether the British would lodge a formal request remained to be seen. Henderson recognized the drawbacks of a forward American role in Greece, in terms of domestic politics and diplomatic precedent, yet he believed that the situation was growing critical enough to warrant a forthright response. "A weak and chaotic Greece," he told Byrnes in November 1945, "is a constant invitation to its already unfriendly neighbors on the north to take aggressive action and constitutes a menace to international peace and security."[3]

At the moment, the Greek government most needed financial support. Athens' halting efforts had failed to stabilize the currency, to get a grip on government revenues and expenditures, or to revive agricultural and industrial production. "It appears, therefore," Henderson continued, "that some measure of responsibility and firmness must come from outside Greece." Several months earlier the United States, through the United Nations Relief and Rehabilitation Administration, had begun furnishing economic aid. Henderson believed this aid might provide a lever to encourage the government to put its house in order. More precisely, he suggested that American pressure might be just the "added bulwark" to persuade the government to impose needed but unpopular decisions. Henderson drafted a three-point program for Byrnes to recommend to the president and, if the president approved, to the Greeks. First, the United States should urge the government in Athens to undertake a stringent program of economic stabilization. Second, Washington should offer technical assistance in designing and implementing such a program. Third, the administration should promise further monies and services in the future as an incentive to keep the Greeks on the straight and narrow.[4]

Byrnes found Henderson's arguments sufficiently persuasive to send them to Truman, and the president directed the secretary of state to move forward along the suggested lines. Byrnes did so, after checking with the British, who thought the scheme sound.[5]

The Greek government had a different opinion. American ambassador Lincoln MacVeagh in Athens would prove a staunch ally for Henderson in the fight for aid to Greece, but at this point he sympathized with the Greek government for rejecting the American plan. As MacVeagh pointed out, the Greeks were living literally hand-to-mouth, and under the circumstances they resented suggestions from well-fed foreigners for further belt tightening. The financial situation in the country was already "very serious." Reconstructing the Greek economy in peacetime would have been a herculean task; amid a civil war people at the starvation line naturally preferred bread today to factories tomorrow. "It is scarcely to be wondered at," MacVeagh wrote, "if they are hesitant to take a long-term view in utilizing what liquid capital remains to them in efforts to expand production." Besides, they could not know how long they would remain the object of foreign interest. "What assurances have they that British Labor Party preoccupation with Socialist dogma, or an American lapse of interest in Balkan affairs, may not deliver Greece to Communism in much the same way as Yugoslavia?" Lately the Greek government had discovered that bowing to foreign demands could backfire. "As a reward for maintaining freedom of its press, at the same time permitting foreign correspondents to circulate about the country and report as they see fit, Greece has been misrepresented and maligned in the American and British press in a most undeserved fashion." MacVeagh did not deny that the Greek government might have made better use of the foreign aid already provided, but he asserted that to "the quick Greek mind, impatient of detail often to the point of being superficial," the arguments for drastic and unpopular economic reforms simply did not outweigh those against.[6]

A more attractive package of aid might have changed minds in Athens, but through the end of 1945 Henderson failed to find support in Washington for sweetening the deal. In January 1946 the administration agreed to send a representative to sit on a committee appointed to help stabilize the drachma—but only, as Acheson put it, "if the Greek Government clearly understands that the U.S. Government assumes no responsibility for the operations of the currency committee or for the determination of exchange rates." He added, unnecessarily, that the United States would be "in no way committed to additional financial support" for Greece.[7]

MacVeagh considered this half measure almost worse than nothing, and he warned that unless Washington came up with something better, Greece would face grave danger. The situation in the country, he said, was "highly flammable." "If we fail to deal with the Greek problem with imagination and understanding at this moment, it is our view that the present democratic government will certainly fall and probably be succeeded by a regime of the extreme right which in turn could scarcely fail to produce in due course a Communist dictatorship." MacVeagh urged Washington to adjust its thinking about Greece to a "higher plane." "Time and stability are of the essence," he said. "What Greece needs is a plan (1) which gives her the reassurance of continued economic existence after the present year; and (2) which prevents the Greek vices of extravagance and incompetence from wrecking the plan."[8]

Henderson would gladly have supplied such a plan, but at the beginning of 1946 neither the department nor the rest of the administration would go along. During the next several months, therefore, he focused his attention on preparing for the day when attitudes in Washington might change. Following a successful—from the Greek government's perspective—plebiscite in the summer of 1946, Henderson told the Greek chargé d'affaires in Washington he hoped Athens would not interpret the vote as carte blanche to crack down on political opposition. He urged a course of "prudent moderation," one aimed at "consolidating the varied political views of the great majority of the Greek people." He added that although the regime could justly adopt measures to ensure law and order, "the U.S. Government would find it very difficult, in the face of adverse opinion from the American public, as well as the rest of the world, to look with favor on a Greek government which would follow the plebiscite with terroristic or unnecessarily repressive steps to get rid of all Greek political elements unfriendly to the government."[9]

Simultaneously, Henderson worked to raise the consciousness of the Truman administration regarding the overall Soviet threat to the Middle East and eastern Mediterranean. Fortunately, following the confrontation over Iran, and while the Russians were giving every indication of making their stay in Eastern Europe permanent, administration opinion was moving in his direction. In September, Acting Secretary of State William Clayton remarked that the United States had come face to face with the problem of "whether in view of the policy which the Soviet Union appears to be pursuing of endeavoring to undermine the stability and to obtain control of the countries in the Near and Middle East such as Greece, Turkey and Iran, we should make certain changes in our general policies." The changes Clayton referred to would "enable us to strengthen the will and ability of the various Near and Middle Eastern countries under Soviet pressure to resist that pressure."[10]

Clayton advocated making such changes, as did Henderson in forwarding an NEA memorandum describing circumstances in Greece. "It is evident," Henderson wrote, "that the critical Greek situation will require very close attention and active U.S. interest." The NEA paper, which laid the groundwork for what became the Truman doctrine, described Greece as a "focal point in strained international relations" and declared, "Its fate during the next few months may be a deciding factor in the future orientation of the Near and Middle East." By "unceasing and virulent" propaganda attacks the Soviet Union and its agents on the "Extreme Left" in Greece were endeavoring to destroy the legitimacy of the Greek government. By massing troops in Albania, Yugoslavia, and Bulgaria and by provoking border incidents the communist countries were waging psychological war against the Greeks. By a policy of harassment in the United Nations they were sapping Greece's international credibility. By charges of illegal western "interference" in Greek affairs the Russians were trying to deter the United States and Britain from coming to the aid of the government. If present trends continued, Greece would find its ability to deal with the crises it faced mortally impaired. "Its economy is still shattered as a result of enemy occupation, and public order is at a low ebb as a result of hatreds engendered by partisan cruelties and strife." The country had polarized: the "Extreme Left," an "apparently well-organized and armed Communist-dominated minority supported by the U.S.S.R. and Soviet satellites," confronted the "Extreme

Right," some of whose members had entrenched themselves in the government and who were "not averse to playing on the fears of the Greek public in order to brand all opposition as Communistic and foreign-inspired, with the hope of justifying strong measures to stamp out Left factions and to render impotent any real Center republicanism."

A glance at a map demonstrated the strategic significance of Greece to the United States. "It is the only country of the Balkans which has not yet fallen under Soviet hegemony. Greece and Turkey form the sole obstacle to Soviet domination of the Eastern Mediterranean." Should Greece collapse, the Russians could apply "irresistible pressure" upon the Turks. From a military perspective alone the United States had no choice. "We cannot afford to stand idly by in the face of maneuvers and machinations which evidence an intention on the part of the Soviet Union to expand its power by subjecting Greece to its will, and then using Greece as an important stepping-stone for a further expansion of Soviet power."

But Greece's importance transcended military strategy, for the country served as a proving ground of America's seriousness in upholding the purposes of the United Nations. Of course, success in the diplomatic realm would contribute to success in the military realm. "The moral strength imparted by high principles and the conviction that the U.S. is defending not only its cause but that of all free nations is a tremendous factor in world affairs and would contribute greatly to our strength should matters ever come to a military test." The present assault on the Greek government followed less from a desire on the Kremlin's part to communize Greece directly than from "the clear intention of making it impossible for any country in the geographic position of Greece to remain friendly to the Western Allies and to Western ideals of democracy." If Greece succumbed, "there could not fail to be most unfavorable repercussions in all of those areas where political sympathies are balanced precariously in favor of the West and against Soviet communism."

The NEA paper turned from a description of Greece's importance to recommendations for American action. Such action must be prompt and vigorous. Should America procrastinate, it might find itself confronting a treacherous choice between intervening in a civil war and risking the loss of Greece to the communists—whose own intervention would be "cleverly disguised to conceal its outside character." Past and present experience demonstrated the inability of the American political system to deal adeptly with such a choice. "Decisions of this nature have been difficult in China, and may become more so; they were impossible in the Spanish civil war." Consequently the United States must move "*before the fact*" to reinforce the Greek government.

To this end American officials "should make it clear to the world that we are determined that Greece remain independent and in charge of her own affairs and that we are prepared to take suitable measures to support the territorial and political integrity of Greece as important to U.S. security." The American government should enunciate this determination by public statements and by confidential messages to interested parties. It should give active support to Greece in the United Nations whenever possible. It should facilitate lending to Greece by the World Bank and the Export-Import Bank. It should assist Greek representatives in creating markets for Greek products in the United States and in third countries, and it should take measures to ease current shortages in the Greek shipping industry, an economic sector essential to Greek prosperity. It should provide expert financial and technical advice in the form of an American economic mission to Greece.

Finally and most importantly, the United States should be ready to pick up where the British were beginning to let down in the area of military assistance. London was showing signs of weakness; the weakness could be expected to spread. Without American help the government in Athens would find itself in a dangerous lurch. "Withdrawal of British forces from Greece, as it progresses, will leave the Greek government with the complex problem of maintaining internal order and protecting her borders under conditions where none of her Soviet-inspired neighbors wish her to succeed. The stability of the Greek Government must be regarded as questionable unless given vigorous external support, including support by the U.S." For the moment the Truman administration need not commit itself explicitly to providing military aid. The British retained primary responsibility for arming the Greek government. Nonetheless the United States "should be prepared, in case of British inability, to sell to Greece sufficient arms for the maintenance of internal order and for the defense of Greek territorial integrity until such time as military forces of the UN are prepared to undertake guarantees against aggression."[11]

II

For Britain, the crux of the Greek issue was money, of which London had little. The destruction of the war had greatly eroded Britain's productive capacity, and now the country faced a balance-of-payments crisis and an external debt of daunting proportions. Determined to cut costs wherever possible, the Labour government decided to retrench in Greece.

Facilitating this decision was the perception that the Americans were ready, or nearly so, to take up slack in British support. By the end of 1946 other Truman administration officials were joining Henderson in calling for a larger American role in Greece. Byrnes and Acheson agreed that while it would be preferable for Britain to continue to serve as the principal arms supplier to the Greek government, the United States must provide the weapons if the British fell short. London may not have known the details of American thinking, but the British possessed sufficient knowledge of Washington's thought processes to realize that the administration was moving toward a more active approach to the problems of the eastern Mediterranean.[12]

As 1947 began, backing for a higher American profile in the region continued to grow. From Moscow, American ambassador Walter Bedell Smith stressed the Soviet threat to Turkey. For centuries the Turkish straits had been to Russia what Gibraltar was to the British, except that Moscow never succeeded in acquiring possession. The British, and increasingly the Americans, were intent on seeing the Kremlin continue to fail. American leaders had begun to worry about Turkey in the last months of the war, as Moscow's propaganda machine mounted an offensive against the Turks, who had sat out nearly all of the fight against the fascists. In negotiations with Turkey as well as in conferences with the British and Americans, Stalin pressed for Soviet participation in the control of the straits. At the same time the Kremlin revived claims to portions of eastern Turkey that Russia once had ruled.

By the summer of 1946 the Turkish problem had reached the stage where Henderson felt obliged to warn the president of an impending disaster. Henderson and his assistants in NEA drafted a memo stating forthrightly that "the primary objective of the Soviet Union is to obtain control of Turkey." Should the Russians succeed in this endeavor the United States would find it "extremely difficult, if not impossible, to prevent the Soviet Union from obtaining control over Greece and over the whole of the Near and Middle East." The danger permitted no delay. "The time has come when we must decide that we shall resist with all means at our disposal any Soviet aggression and in particular, because the case of Turkey would be so clear, any Soviet aggression against Turkey."[13]

The warning had the desired effect. Henderson accompanied Acheson to the White House, where they presented the NEA case to Truman, Undersecretary of War Kenneth Royall, Navy Secretary James Forrestal, and the military chiefs of staff. After weighing NEA's argument, Truman approved a heartening note to Ankara and the dispatch of an American naval force to the eastern Mediterranean. Shortly after this the State Department sent a message to Moscow reiterating the American position that the question of the straits was "a matter of concern not only to the Black Sea powers but also to other powers, including the United States."[14]

The Russians backed away slightly, dropping a demand for a base in the Dardenelles, but they continued the psychological campaign against the Turks. In the process they forced Ankara to devote increasing sums to military preparedness. By January 1947 Smith in Moscow considered Turkey in imminent peril of its national life. The ambassador described Moscow's ambition of "subjugating (or in Soviet jargon 'liberating')" the country as being "grounded in Czarist history and reinforced by Communist conviction." Confronted by the "chill menace" of the Red army, Turkey had little hope of survival as an independent nation unless assured of "solid long term American and British support." Likewise, Edwin Wilson, the American ambassador in Ankara, wrote to Henderson predicting that if the United States did not demonstrate that American support was more than a "mere matter of words," the Turks would conclude that Washington could not be counted on. For America to fail to demonstrate in tangible form its concern for Turkey's future would be "most unwise."[15]

Meanwhile the news from Athens afforded no relief. An American economic mission sent to survey the situation in Greece declared the fiscal outlook for the government "dismal." The political prospects were no better. "There is really no State here in the Western concept. Rather we have a loose hierarchy of individualistic politicians, some worse than others, who are so preoccupied with their own struggle for power that they have no time, even assuming capacity, to develop economic policy." The civil service, which included the individuals who would have to implement any reforms, was a "depressing farce." The only solution was for the United States to take prompt and forceful measures. With American "guidance" and "substantial financial aid," Greece might yet be saved.[16]

III

By February 1947 Henderson had been arguing for a higher American profile in Greece for a year and a half. His concern regarding the situation in Turkey was of slightly more recent vintage but no less seriousness. Although he could hardly claim sole credit, he must have felt a certain satisfaction in noting that the administration was swinging toward agreement with him. All that remained to set the new attitude in motion was a signal from the British.

This came in dramatic form on February 21. At three o'clock on that Friday afternoon, Henderson received a message that the first secretary of the British embassy, Herbert Sichel, had urgently requested an interview. Henderson replied that he would receive the first secretary at once. The latter arrived with two notes addressed to the secretary of state. George Marshall, recently confirmed by the Senate as Byrnes's successor, had left the office early to go to Princeton, where he would receive an honorary degree and give his first public address in his new capacity. In Marshall's absence Henderson greeted Sichel. Henderson had been acquainted with Sichel for some time, and they spoke cordially when the British diplomat entered Henderson's office. But Sichel knew, and Henderson guessed, the gravity of Sichel's errand, and the small talk soon gave way to Henderson's reading of the notes.[17]

"They were shockers," Dean Acheson later said. The notes announced that in six weeks Britain would pull the plug on aid to Greece and Turkey. The British government did not do this from any lack of appreciation of the importance of those two countries to western security. On the contrary, it considered their defense vital to the security of Britain, the United States, and the other democracies. Therefore Britain hoped the American government would cover the deficit, and in light of the "extreme urgency" of the matter it hoped Washington would do so quickly.[18]

Henderson, who just the day before had written another jeremiad on the "crisis and imminent possibility of collapse" of the Greek government, immediately took the notes to Acheson. The acting secretary looked them over and told Henderson to gather his staff and "work like hell" through the weekend in order to have a response ready for Marshall on the secretary's return Monday morning.[19]

In Moscow during the 1930s, Henderson had demonstrated his ability to put the talents of assistants and associates to good use. His map performance with Edwin Wright at the time of the Iran crisis of 1946 had shown this ability again. But never was he more successful in melding the efforts of those working below and alongside—and above—him than during the last days of February 1947. After leaving Acheson's office, Henderson summoned the top officials from NEA and the department's European division to a late Friday meeting. Henderson also invited George Kennan, whom Marshall had appointed to head up State's policy planning staff and who could be counted on to advocate an affirmative response to the British request. Convening far into the evening, Henderson's group agreed that the administration had no choice: without quick American action, Greece would fall, Turkey might go as well, and the eastern Mediterranean would disintegrate.

Henderson and the others could not help feeling excited. As Joseph Jones, an assistant who became the principal drafter of the speech enunciating the Truman doctrine, recalled,

> The members of the Office of Near Eastern and African Affairs were quite openly elated over the possibility that the United States might now take action on a broad enough scale to prevent the Soviet Union from breaking through the Greece-Turkey-Iran barrier into the Middle East, South Asia, and North Africa. They had long felt themselves virtually unarmed in trying to deal with this problem, which was to them as real as the walls about them and held frightful potentialities for the security of the United States and the future of the world.

The next morning Henderson and John Hickerson, the deputy director for European affairs, met with Vice Admiral Forrest Sherman, the deputy chief of naval operations, and Major General Lauris Norstad, the planning and operations director of the War Department's general staff. The four discussed the details of military aid to Greece and Turkey: how much it would cost, what form it ought to take, how it should be supervised.

Simultaneously the members of Henderson's staff in NEA were furiously composing, debating, revising, and polishing the memo Acheson had requested. They worked through Saturday until nearly midnight, then went at it again Sunday morning. By noon they had a paper for Henderson to deliver to the undersecretary's Georgetown home. Acheson read it, made some minor changes, and gave it his approval.[20]

Closely following the argument Henderson had made just days before, this memo reiterated the issues in the balance. Three sentences summarized its views.

> Unless urgent and immediate support is given to Greece, it seems probable that the Greek Government will be overthrown and a totalitarian regime of the extreme left will come to power. The capitulation of Greece to Soviet domination through lack of adequate suppport from the U.S. and Great Britain might eventually result in the loss of the whole Near and Middle East and northern Africa. It would consolidate the position of Communist minorities in many other countries where their aggressive tactics are seriously hampering the development of middle-of-the-road governments.

As for recommendations, the NEA paper cited the need for encouraging the Greeks to straighten out their financial and political affairs and for expediting American aid already allocated. Following London's change in policy, the United States must move at once to fill the void Britain was leaving in terms of economic and military assistance. Most important, President Truman must bring the challenge of aid to Greece to the immediate attention of the American people. "We recommend presenting a special bill to Congress on an urgent basis for a direct loan to Greece, stressing the fact that if inflation and chaos are not prevented within the next few months, the gravest consequences will ensue and the country will be beyond our help."[21]

On Monday morning, February 24, Henderson briefed Marshall regarding the notes the British ambassador would formally present in a few hours. Henderson suggested, in view of the major importance of the issues involved, that the secretary refrain from immediate comment. Marshall concurred, and when Lord Inverchapel arrived, Marshall limited himself to a few questions about why, in the opinion of the British government, the Soviets had chosen this moment to heighten the pressure on the eastern Mediterranean. Inverchapel could not say. "No foreigner," he replied, "knows why Russia takes or fails to take certain actions." After Marshall commented that the United States would give its urgent attention to the situation described in the British notes, the interview ended.[22]

At three o'clock that afternoon Henderson held a meeting of sixteen State Department officials drawn from the Near Eastern and European divisions and from various offices connected with arms, intelligence, and foreign aid. He summarized the content of the British notes, adding that Britain's withdrawal from

Greece and Turkey should be viewed in the context of Britain's prospective pull-outs from India and Palestine. The British government, he remarked, had concluded that it was "unable to maintain its imperial structure on the same scale as in the past." He then had an assistant read a paper of the previous summer delineating NEA's evaluation of Soviet designs on the region from the eastern Mediterranean to India and urging a forceful American response. "The time has come," the report asserted, "when we must decide that we shall resist with all means at our disposal any Soviet aggression." Henderson commented that if the time for decision had come in August 1946, the date of the paper, by now it was long overdue.

Henderson next presented the memo NEA had prepared for Acheson and Marshall over the weekend. This prompted lively discussion, with most of those in attendance taking the position that the United States should move quickly to accept the responsibility thrust upon America by the British. The question of how to finance this responsibility could wait until later. Arms specialist General James Crain dissented on this issue, asserting that it was precisely a failure to count costs in advance that had led the British into the fix they were fleeing. Crain advocated instead that the United States "conserve its resources for the final trial of strength" with the Russians. If Greece and Turkey mattered as much as the majority at the meeting evidently thought, the American government should inform Moscow of its intention to defend those countries with its own troops.

Several discussants wondered whether the United States should prepare to go to the rescue of other regions abandoned by the British. What sort of precedents would aid to Greece and Turkey establish? All concurred that regardless of the answer, administration officials should expect the question from skeptical senators and representatives. Hubert Havlik of the office for economic development advocated a global approach to the problem currently confronting Greece and Turkey. Whatever the department chose to recommend, Havlik said, its recommendations should form part of a "worldwide program." On this point, Hickerson of the European division agreed. The administration should avoid a nickel-and-dime policy, instead presenting a comprehensive package that would "electrify the American people." Henderson expressed no opinion on the matter at this meeting, and the committee declined to make a recommendation.[23]

The next day, Henderson, in his capacity as chairman of the department's special task force on Greece and Turkey, explained to Acheson the thinking of the members regarding the British démarche. Some persons in the group had raised the possibility that the British were exaggerating the severity of their financial troubles and that strategic necessity would require them to continue assistance to Greece and Turkey if the United States did not accept responsibility—that the notes were something of a bluff. Others had suggested that the

British announcement signaled a reconsideration by London of Britain's role in the cold war. According to this hypothesis, Britain had already decided to change its basic policies toward the Soviet Union and was now planning to accommodate rather than confront the Kremlin. The British expected an American refusal of their request for aid to Greece and Turkey; this refusal would justify their rapprochement with the Russians.

Henderson did not buy either interpretation, although he considered both plausible. He explained his and the majority's views.

> After examining carefully the notes in the light of the present international situation and of the economic conditions in Great Britain, we are inclined to believe that the British Government is really convinced that it is unable any longer to expend funds, supplies and manpower in the Near East in the future as it has in the past; that it hopes the United States, realizing how important it is that the independence of Turkey and Greece be maintained, will undertake to relieve Great Britain of these financial responsibilities and that the two Governments will be able in cooperation to resist Soviet pressure in the Near East.

Henderson conceded that at least part of the appeasement scenario might materialize should the administration fail to respond favorably to the British request. If the United States did not summon the wherewithal to counter Soviet pressure in the eastern Mediterranean, "the British Government may well find that it will be compelled to approach the Soviet Government in an effort to work out some arrangement which would have the effect of slowing up the Russian advance in the Middle East and elsewhere." To the degree the British felt obliged to seek a "breathing spell," they might be inclined to make "widespread concessions" to the Soviets.

Regarding the adverse effect on the potential recipients of a decision against aid, Henderson had little to add to what he had said before, although with each telling the apocalypse broadened. In this latest version "the resulting chaos would be accompanied by an immediate weakening of the strategic and economic position of the whole western world, particularly of Great Britain, and the very security of the United States would be threatened." Should Greece and Turkey fall, or should Britain be forced to an accommodation of the Soviets, the United States would be "much closer to a third world war in which we would find ourselves in a much more disadvantageous position than that in which we are at present."

Henderson granted that great difficulties might attend a decision in favor of aid. "Certain responsible officials of the Administration and members of Congress, as well as large sections of the general public, are not as yet fully cognizant of the seriousness of the situation and would not like for the United States to expend large sums of money in the Near East or for it to undertake to play a leading role in that part of the world." But such a failure of comprehension could be remedied by educating the public, so that the American people would come to appreciate that "the future of the Near East is no less important to the security and welfare of the United States than is that of the Far East and Europe."

As for specific recommendations, Henderson advocated "top secret conversations at a high level" with British civilian officials, joint Anglo-American military talks, the creation of an American economic planning unit with "wide powers over Greek economic life," the immediate transfer of available funds to Greece on an emergency basis, consultation with the leaders of Congress, and the drafting of appropriate legislation.[24]

When Henderson repeated his arguments at a meeting with Marshall, War Secretary Robert Patterson, and Navy Secretary Forrestal, the three secretaries agreed to the recommendations with slight changes. One change involved Henderson's admonition that a rescue program for Greece and Turkey be an all-or-nothing affair. "Halfway measures," Henderson said, "will not suffice and should not be attempted." Putting a price tag on his proposal, he asserted that saving Greece alone would require "several hundred millions." The three secretaries worried that legislators would balk in the face of such a massive aid request, and they insisted on scaling it down. Yet all three agreed that the Greek and Turkish problems constituted but one facet of a critical world situation, and they concurred in Henderson's contention that the United States must deal with the problem conceptually, if not tactically, as a whole.[25]

Later that day the State Department forwarded Henderson's program to the White House, and Truman approved it as amended by Marshall, Patterson, and Forrestal. While the secretary of state began lobbying congressional leaders, Henderson set to work making the Truman doctrine a reality. He wrote a reply to the British. He informed Ambassador MacVeagh in Athens and the various interested agencies in Washington of the president's decision. He directed preparation of a message for Truman to deliver to Congress and the American people. He coordinated a committee assigned to draw up enabling legislation. To ensure accountability, he suggested to Acheson that all departments and divisions involved in the matter indicate which of their personnel would play substantial roles in implementing the program, and that these individuals be summoned to what would amount to a pep rally. He invited Acheson to speak, saying the meeting "would be most effective if you could make the opening remarks and set the tone for any discussion which might follow."[26]

In composing the reply to the British, Henderson stated that while the administration intended to do what it could to provide aid, Britain might help matters by stretching out its timetable for withdrawal.

In view of the need for Congressional action and of the obvious difficulty of organizing any program of assistance in so short a time, this Government doubts that despite its best efforts it will be prepared within the next few weeks to undertake substantial financial responsibility for Greece. It trusts, therefore, that the British Government will continue on an emergency basis such financial advances to Greece as may be necessary to prevent the collapse of the situation there.

Following this request, London agreed to extend aid for three months beyond the original March 31 deadline.[27]

On the afternoon of March 4 Henderson met with the committee appointed to frame the legislation authorizing American assistance. The group, recognizing the basic problem in the case as one of providing adequate congressional control without sacrificing administrative flexibility, opted for a single comprehensive resolution rather than several separate bills. In the face of uncertainty as to what Congress might be persuaded to accept, the committee presented three alternatives for the omnibus measure. Each recognized that Greece and Turkey would set a precedent for American aid to foreign governments, but they differed in the guidelines they offered regarding such future aid. The first declared that American help should be used to sustain only countries with democratic forms of government. The second applied a much less stringent test, giving the president almost complete freedom to furnish aid whenever he deemed such assistance in the American national interest. The third course—the Goldilocks option—came down between the two, although it landed considerably closer to the second. It asserted that the president should be empowered to provide assistance to any country striving to defend its independence against outside influence when that independence served American interests. The committee went on to describe the nature of the aid the president should request: money, military equipment, industrial and agricultural supplies, and military and civilian advisers. On the vital—especially to congressional overseers—issue of accountability, Henderson's committee recommended requiring recipient countries to permit full disclosure of the kind and amount of American assistance they received and to allow American inspectors to supervise distribution of the aid. As to repayment, that would be left to the president's discretion.[28]

By the time the proposed legislation went to the White House for approval, Henderson and his colleagues estimated the cost at $400 million for Greece and Turkey, to cover the period through June 1948. On March 12 Truman delivered the resolution to Congress. Defending the bill, in a speech based in part on a draft by Henderson, the president declared that "it must be the policy of the United States to support free peoples who are resisting attempted subjugation by armed minorities or by outside pressures."[29]

The open-ended nature of this commitment occasioned some concern at the time. George Kennan thought the administration should focus attention on Greece, which faced a far more serious challenge than Turkey did. Henderson wanted to include Iran as well as Greece and Turkey, but perhaps because of the boundaries of his portfolio he did not argue for extending the American sphere beyond the eastern Mediterranean and the Middle East. Truman, however, largely on the basis of soundings of the congressional leadership, decided that wrapping the issue in high principle would make it more acceptable to the legislators. On this point he had the backing of Arthur Vandenberg, the Republican senator who would become the symbol of cold war bipartisanship. Vandenberg asserted, "The problem in Greece cannot be isolated by itself. On the contrary, it is probably symbolic of the world-wide ideological clash between Eastern communism and Western democracy; and it may easily be the thing which requires us to make some very fateful and far-reaching decisions." Following Vandenberg's advice, when Truman received the draft message from the State Department requesting aid for countries of the Middle East and eastern Mediterranean resisting communist subjugation, the president personally decided that the idea of American support ought not to be geographically circumscribed.

IV

With the delivery of the resolution and the president's speech, the contest for the Truman doctrine moved out of the bureaucracy and into the political arena. Acheson handled most of the testimony before congressional committees; Henderson receded to the background.

In this case the background encompassed oversight of the aid program Congress subsequently approved. As would happen in other countries receiving American assistance, once the dollars started flowing to Greece the government there began to consider itself indispensable to the United States. Athens grew disinclined to follow American advice, which included admonitions to broaden the regime's political base and reach out to other noncommunist elements in the country. In 1947 the focus of right-wing activity was the Populist party, headed by Constantine Tsaldaris. The Liberals of Themistocles Sophoulis controlled the center. During the summer of 1947 Tsaldaris and the Populists held the balance of power in Athens, with the apparent aim of excluding Sophoulis and the Liberals. On grounds of simple expediency the State Department would have advocated a common front among Populists and Liberals, the better to withstand the military and political pressures from the communists and other elements on the left. More to the immediate point was the fact that the White House had come under siege by Greek-Americans who feared that the Populists were using American money to entrench themselves as much against the Liberals as against the communists. Consequently, Henderson was not surprised at the end of August when the president asked him to travel to Greece and remind Tsaldaris of the purposes of American aid.[31]

Henderson had met Tsaldaris in Washington several weeks earlier. At that time the NEA chief had warned the prime minister to diversify his cabinet. The American government, Henderson said, was "extremely anxious that there should be a maximum degree of unity among all national-minded Greeks in the present emergency." He added that it was "extremely important that all Greek leaders regardless of party or personal rivalries begin to cooperate for the purpose of saving Greece." Tsaldaris had rejected the advice. He told Henderson he had had enough experience with coalition governments to know that if he constructed a cabinet embracing the Liberals "it would fritter months away in long-winded debates and would accomplish little." At present, he declared, Greece required not consensus but efficiency. The country needed and he intended to provide a "compact, determined government" that would "understand what the situation demands and would be sufficiently flexible to act quickly and decisively."[32]

Tsaldaris evidently believed that the Americans would not go beyond verbal remonstrances in urging him to bring the Liberals into the government, and Henderson's arrival in Athens at the beginning of September caught him off guard. "Nervous and uncertain" was how Lincoln MacVeagh described the prime minister at the latter's first meeting with Henderson. Tsaldaris began the discussion with an emotional speech about the crisis his government faced and about his vain efforts to obtain the cooperation of his opponents. He complained that the United States was attempting to split his party and had placed him in an intolerable position. He said he had always tried to adopt policies in line with those of the United States, implying that he expected more gratitude and less pressure from Washington.[33]

Henderson himself had reservations about his mission. He respected Tsaldaris as a man who had consistently opposed the communists throughout a period when, as the NEA director put it, "most Greek political leaders, not knowing whether Greece would fall to the Communists like its Balkan neighbors, sat on the fence." Henderson recognized the greater stability a broader base in Athens would provide, but he recoiled at the fact that this particular trip was largely the result of political pressures in the United States. Initially, Henderson had hoped the department would send someone else, but when he realized that anyone else would probably be less sympathetic to Tsaldaris—and when the president requested that he, specifically, go—he yielded.[34]

For all his uneasiness Henderson handled his task with professional aplomb. He told Tsaldaris that although the United States did not propose to dictate Greece's internal policies, the prime minister should know that it would be extremely difficult to maintain public support in America for the Greek government if it appeared that American aid was benefiting a single group or party rather than the country as a whole. Should the government not demonstrate that it was making the best possible use of American assistance, Congress might curtail that assistance. Tsaldaris, evidently suspecting that this time the Americans meant business, replied that he appreciated Washington's concern and would make "every effort" to achieve political unity.[35]

A few hours later Henderson paid a visit to King Paul, to whom the Liberals had been directing their demands for participation in the government. The Liberals' latest proposal called for the appointment of a nonparliamentary prime minister who would select a cabinet of national unity. Henderson thought this a reasonable solution. He suggested that the king bring the leaders of the contending parties together and lock them in a room until they reached an agreement. The king replied that he would give the suggestion a try.[36]

Henderson then spoke with the opposition leaders. He made the same argument he had made to Tsaldaris and the king, tailored in this case to the prejudices of those on the outside. Eventually, and after no little haggling, the opposing parties cut a deal, with the Liberals gaining the premiership for Sophoulis, who then chose a coalition cabinet.[37]

Although he had achieved his objective, Henderson doubted he had done much good for Greece or for the United States. He worried that the new prime minister lacked the will to stand up to the communists. "I was unhappy at the result of my mission," he commented later. "Tsaldaris, in my opinion, was the kind of strong man that Greece needed at the time, whereas the aged Sophoulis was inclined to waver when it came to making tough decisions."[38]

Whether or not Greece needed Tsaldaris, the country certainly needed something. The summer of 1947 brought scant respite from Greece's problems; by the beginning of autumn Henderson feared another full-blown crisis. In October he remarked that recent developments had "greatly reduced the hopes held last spring that Greece could be substantially helped within the limits of time and funds provided by the Greek-Turkish Aid Bill." The developments in question included a severe drought, which had driven up the cost of food; delays in getting the aid program off the ground, which had left the Greek government more strapped than ever; price rises in the United States and elsewhere, which had diminished the purchasing power of appropriated monies; and competition for Export-Import Bank loans, which reduced credits to Greece and intensified the country's needs. Most ominous was what Henderson called the "increasingly grave security situation," which not only diverted funds from production to defense but threatened to undo at a single blow the entire experiment.[39]

Events of the next several weeks convinced Henderson that the military threat constituted the largest danger to Greece and to American interests in the eastern Mediterranean. By December he had concluded that Washington had no choice but to consider direct American military intervention. "The argument has been made," he wrote, "that it would be better to lose Greece than to send armed forces there since Greece would be strategically difficult to defend and since, if serious fighting should develop, American forces might be defeated." Such an argument, he said, missed the point. "The despatch of

American forces to Greece would not be for the purpose of turning Greece into the chief battlefield between the forces of the West and those of international communism. The despatch of these forces would be a political gesture made for the purpose of showing that we are so determined that we will, if necessary, resort to force to meet aggression." Beyond serving as a test of American credibility, Greece would measure Soviet intentions. "In case international communism responds by sending even stronger forces to combat us in Greece, then we shall know that the Soviet Union prefers war to the abandonment of its aggressive policies, and we can take the appropriate measures on a world-wide scale."

With luck, Henderson said, the UN would back the United States in support of Greece. The American government must spare no efforts in working through the international organization (in whose stability Henderson now had greater confidence than at the time of the Iran crisis). Should Greece's communist neighbors recognize a rebel regime in Greece, or recognize any government other than that currently represented in the UN, or should the Russians or their allies step up military assistance to the rebels, American officials ought to denounce such action as aggression against Greece. In addition the United States ought to call on the member nations of the UN to render all requisite assistance to Greece, up to and including military force.

But if the UN, through Soviet veto or otherwise, failed to answer the challenge to peace, the United States must go it alone. "If it should become clear that Greece will be lost unless troops are sent, we should, in accordance with procedures outlined in the Charter of the United Nations, send troops."[40]

12

In the Palestine Labyrinth

The United States did not send troops to Greece. The Greek government managed to hold on until events within the communist movement, notably Tito's 1948 defection from the Cominform, conspired to weaken and finally end the insurgency. As the situation stabilized, Henderson and others in the Truman administration came to look on Greece as the scene of one of the great triumphs of the postwar era, and the formulation of the Truman doctrine as a high point in their careers.

For Henderson it certainly was a high point, since the Truman doctrine embodied the principles he had been espousing for more than a decade. The Soviets, he believed, were out to conquer the world, and if they did not start each insurgency and foment every instance of anti-western agitation, they would surely attempt to benefit from them. When the Russians pushed in Iran, the United States must push back. When they pressured Greece, America must come to Athens' aid. Henderson—and the rest of the Truman administration—overestimated the degree of Russian responsibility for the Greek troubles, which owed more to the kind of local and regional rivalries that have vexed the Balkans since Thucydides than to Stalinist instigation. Yet Henderson's prescription would not have differed materially had he known. "Objectively"—to use a Marxist term Henderson avoided but appreciated—unrest in Greece, of whatever origin, opened the area to Soviet adventurism.

The measure of Henderson's success in the Greek case was the degree to which his view of the Soviet Union, definitely a minority position five short years before, had achieved the status of orthodoxy. Henderson was not so eloquent as George Kennan, nor did he have the patrons in high places who amplified Kennan's voice and made Kennan the spokesman of containment. But Henderson worked the channels of the bureaucracy, and through persistence and determination—and with the help of Stalin, who reverted to pre-war form—he played a major role in bringing the official mind of American foreign policy around to his way of thinking.

The logic of Henderson's convictions regarding the Soviet Union led, in another area, to his advocacy of policies that proved considerably less congenial to governing American wisdom. For more than three years Henderson wrestled with the Palestine issue, convinced that American support of a Jewish state would be a terrible mistake. When the Truman White House decided in favor of such support, Henderson perceived the decision as a gross and despicable example of political expediency overriding diplomatic wisdom. Henderson never jumped the chain of command, and he never sought solace in leaks to the press. But the issue proved so divisive that even his in-house dissent became intolerable to the president and the president's men. *They* leaked all manner of nasty news about Henderson, until he became an object of vilification in Congress and the press. The anti-Henderson campaign succeeded, and once more he found himself banished. Not surprisingly, the experience reinforced the persecution complex he had begun to develop on being exiled from the European division to Iraq in 1943. Then he had spoken what he judged to be true and been punished; subsequent events had vindicated him. Once more he was suffering for what duty compelled him to do; he fully expected the future to bear out his warnings again.

II

"When I arrived in the department in April 1945 ... ," Henderson remembered afterward, "I realized that the problem of Palestine would be one of the crosses which I would have to bear." At the time American policy toward Palestine was in flux, if not disarray. The State Department, as it had for a generation, frowned on Zionism as a troublesome new element in the Middle East. The White House, in closer touch with American politics, looked more favorably on the Zionist dream. Franklin Roosevelt had held the balance until his death, but he generally followed an ad-lib approach, partly because the war and relations among the great powers absorbed his dwindling energy, partly out of confidence in his personal diplomatic touch, and partly from ignorance of the essential irreconcilability of Arab and Jewish aims. In a conversation with Zionist leader Chaim Weizmann in 1940, Roosevelt had expressed the view that Arab objections to Jewish immigration might be overcome by "a little baksheesh." As late as 1945 the president predicted that "he could do anything that needed to be done with Ibn Saud with a few million dollars." Shortly before he died, however, Roosevelt met Saud and discovered first-hand that things would not go as easily as he hoped. "There was nothing I could do with him," the president said after the meeting. "We talked for three hours and I argued with the old fellow up hill and down dale, but he stuck to his guns."[1]

The news of Roosevelt's death in April 1945 reached Henderson on the way from Baghdad to Washington. From what little he knew of Truman, Henderson expected that the new president's Palestine policy would generally parallel

Roosevelt's—although what this meant was unclear, since Roosevelt's position was so inchoate. Henderson's first brush with Truman provided some insight. Shortly after taking up his post as NEA director, Henderson orchestrated an official visit to Washington by his old friend Nuri, now back in power, and the Iraqi regent Abdul Illah. Just before the Iraqis' arrival, Truman summoned Henderson to the White House for a briefing on his guests. "I spent nearly an hour with the President," Henderson remembered, "and was impressed with the quickness with which he grasped the situation and the subjects to be discussed." When Truman met Nuri and the regent, Henderson was impressed still more. "He conducted the discussion with an easy grace, and I marveled at his memory of the details that I had given him during our previous conversation."[2]

Henderson had hardly settled behind his desk at NEA before he found himself caught up in the fight for Palestine. Nahum Goldmann, chairman of the administrative committee of the World Jewish Congress, quickly came calling. Goldmann spoke in somber tones of a "grave crisis" confronting the Zionist movement. For half a decade, he said, moderates among the Zionist leadership, including himself, Weizmann, and Stephen Wise, had urged patience upon their followers, telling them that a solution to the Palestine question must await the end of the war in Europe. In doing so they had relied on the promises and assurances of President Roosevelt and Prime Minister Churchill that once the allies won the war the Zionists might expect support. The moderates had managed to impose restraint upon the Jews of the world. Some "extremists," Goldmann admitted, especially in Palestine, had refused to cooperate, but "on the whole the Jews had shown great moderation." Now, however, the mood of the Jews was changing to "desperation." They had seen millions of their people murdered, their homes destroyed, and their culture all but annihilated in large parts of Europe. The time had come for the redemption of promises. Patience was wearing thin. Jews everywhere were asking: "How long?" In such an atmosphere, momentum was shifting to persons "not averse to violence."[3]

Henderson said almost nothing at this meeting, only thanking Goldmann at the end for his opinions and assuring him he would pass them along to the appropriate officials. Five days later, Goldmann returned, this time bringing David Ben Gurion, chairman of the executive branch of the Jewish Agency in Jerusalem. Ben Gurion outlined the development of the Zionist agenda, from the time of the Balfour declaration through the 1939 British white paper and the recent world war. He denounced current British policy, asserting that if "this intolerable regime" were not modified, there would be trouble. The Jews of Palestine wanted only the right "to set their own house in order without interference from outside elements." But the British were letting their Palestine policy be determined by every "Egyptian pasha" or "Bedouin shaikh" or "Iraqi bey." The idea that such foreigners possessed a legitimate interest in the affairs of Palestine was "preposterous." Ben Gurion reiterated Goldmann's message that the time had arrived for redemption of pledges. Zionists did not want trouble with Britain, and they knew that if the worst came they could not hold out against the full strength of the British empire. But they would fight anyway. They were determined to have their country, and if violence resulted, responsibility would lie with Britain.

Henderson asked whether the Arabs might not object to an effort by the British to impose a pro-Zionist solution. Ben Gurion asserted that he and his associates could deal with the Arabs. He knew the Arabs. They would not put up much of a fight. The Bedouins he respected, but they had no interest in Palestine. The others presented few problems.

To Henderson's query about the short-term goals of the Zionists, Ben Gurion and Goldmann replied that liberalization of immigration quotas was imperative, and at the earliest possible date. They warned, however, that the United States must not think "piecemeal methods" such as immigration reform would solve the Palestine problem. The moment for such measures had passed. Zionists could no longer accept anything less than the granting of all their demands, including the immediate establishment of a Jewish state.[4]

In contrast to the Zionists, Arab representatives in Washington were slow off the mark in making the acquaintance of the new NEA director. Part of the problem resulted from the fact that Palestinian Arabs lacked any organization comparable in reach and influence to the Jewish Agency and the various Zionist support groups. Nor could they claim the votes or the political facility of Jewish Americans. Finally, those who did speak on their behalf—diplomats and leaders of the Arab states—had other fish besides Palestine to fry.

For whatever reasons, Henderson's first conversation as NEA chief with a prominent spokesman for Arab opinion did not take place until the middle of August, nearly four months after he assumed his new job. Mahmoud Fawzi, the counselor of the Egyptian embassy, told Henderson of his government's concern that recent remarks by President Truman indicated a shift in America's Palestine policy. On his return from the Potsdam conference the president had answered a question whether the issue of a Jewish state had arisen in the discussions in Germany by saying he had spoken with British leaders about it. He had added that "the American view on Palestine is that we want to let as many of the Jews into Palestine as it is possible to let into that country."[5]

Henderson's years in the State Department under Roosevelt had accustomed him to the idea that career diplomats were sometimes the last to know of shifts in American foreign policy. He probably expected the less-experienced Truman to be more amenable to advice from professionals. On certain topics—aid to Greece and Turkey, for example—the president indeed did defer to the diplomats. But Truman, like many politicians, especially those from small towns in the heartland, harbored an instinctive distrust of bureaucrats. "The difficulty with many career officials in the government," Truman wrote later, "is that they regard themselves as the men who really make policy and run the government. They look upon the elected officials as just temporary occupants." Truman considered the formulation of policy a power struggle, and he determined that he would not be bested. "I wanted to make it plain that the President of the United States, and not the second or third echelon in the State Department, is responsible for making foreign policy, and, furthermore, that no one in any department can sabotage the President's policy." Truman's suspicions surfaced most clearly on matters relating to Palestine, where opposition to his Zionist preferences could plausibly be ascribed to a variety of motives. "There were some men in the State Department who held the view that the Balfour Declaration could not be carried out without offense to the Arabs. Like most of the British diplomats, some of our diplomats also thought that the Arabs, on account of their numbers and because of the fact that they controlled such immense oil resources, should be appeased. I am sorry to say that there were some among them who were also inclined to be anti-Semitic."[6]

Given Truman's feelings, it was hardly remarkable that the president did not always consult the State Department before announcing policy decisions. As one consequence, Henderson had to tell Mahmoud Fawzi that he had no information from which to speak about the significance of the president's comments. But if he learned anything he thought the Egyptian government should know, he would call the embassy at once.

Fawzi took the opportunity to declare that Cairo was "extremely anxious" to have close and friendly relations with the United States. At the same time, his country had obligations to its Arab neighbors and to the peace and stability of the Middle East. He hoped the United States appreciated the delicate position of his government and would do nothing to jeopardize it. He trusted that American officials would not underestimate the significance of the Palestine question, for the quiet surface of the Arab world masked intense feeling on the matter. A sudden move by one of the great powers might "set the Arab world in motion and result in violence on a wide scale."[7]

With the Zionists in Palestine promising violence if they did not get their way, and the Arabs pledging trouble if the Zionists did, Henderson sat down to figure out what America's policy ought to be. The most pressing issue at the moment involved the possible immigration to Palestine of some 100,000 Jews displaced during the war in Europe. Both Zionists and Arabs took the question of the displaced persons as a test of intentions regarding the future of Palestine. The Zionists demanded the immediate admission of all the refugees, on humanitarian as well as historical and ideological grounds. The Arabs insisted that Palestine not be burdened with Hitler's guilt and declared that immigration, in accord with the spirit of the 1939 British white paper, should remain strictly limited or, preferably, cease altogether.

At the end of August, Henderson sent Secretary Byrnes NEA's recommen-

dation on the subject. Prefacing the paper with the comment that any fair decision would annoy both sides, Henderson advocated a hands-off policy. He said
the administration should not directly oppose letting the 100,000 into Palestine,
but neither should it support entry. Until the British, Jews, and Arabs reached
an accord among themselves, Washington might favor a moderate loosening of
immigration quotas, but it ought to leave the imposition of any such plan to the
British.

The NEA report identified several drawbacks associated with large-scale
immigration. First, Palestine did not possess adequate housing. Indeed it lacked
accommodations for nearly 200,000 people already there. Second, although
Jews owned 90 percent of Palestinian industry, they could never provide jobs
for all the immigrants. As matters stood, Palestine faced serious problems of
reconversion to peacetime production. Third, the Arab inhabitants of Palestine
would oppose immigration, probably by armed force. Finally, for the United
States to support immigration would place on American shoulders a moral
responsibility it could do without. "No government should advocate a policy of
mass immigration unless it is prepared to assist in making available the necessary
security forces, shipping, housing, unemployment guarantees, etc."[8]

Henderson found himself speaking into a void. On the same day that NEA
forwarded these recommendations to Byrnes, Truman wrote British prime minister Clement Attlee urging the admission of the 100,000. The president later
claimed that the State Department was "fully alert" to the change in policy; in
fact, Henderson and NEA did not learn of Truman's letter until they read about
it in the *Washington Post* ten days later.[9]

Henderson's initial reaction to the news was irritation. He protested to Acheson that this was no way to conduct diplomacy. He and his associates had to
instruct American officials in the field. "We feel that we should be in a position
to tell our representatives what the facts are."[10]

His second reaction was a redoubling of his efforts to prevent the administration from going down what he considered the wrong road. The president's letter
to Attlee, Henderson told Acheson, involved the American government in
breaking its word. In recent years American leaders had repeatedly assured the
Arabs that any decision Washington arrived at regarding Palestine would not
be made without consulting both Arabs and Jews. Now the government had
done precisely what it had vowed not to do. Already the damage was visible.
The Iraqi prime minister was describing the president's letter as "contrary to all
the promises and undertakings, oral and written," that the United States had
given the Arabs. The Iraqi leader added that his country could not imagine that
America, regarded by the world as "a citadel of liberty and international justice," would cast down the principles of the Atlantic charter. American representatives in other Arab capitals reported similar responses. Henderson argued
that these responses constituted grounds for shunning what Truman seemed to

be planning. Should the United States side with the Zionists, Henderson warned, "much of the work done in the Near East in recent years in building up respect for, and confidence in, the United States and in increasing American prestige, will be undone."[11]

More Arab complaints arrived in short order. On October 3 Acheson and Henderson received a visit from a combined Egyptian-Iraqi-Syrian-Lebanese delegation. Besides charging Truman with breaking America's promises of consultation, the group asserted that allowing the 100,000 into Palestine would violate international law. A mandatory power, they said, had no right to effect a basic change in the administration of a mandate without the consent of the inhabitants. They averred friendly feelings for Jews and asserted that Jews in Arab countries had generally received respectful treatment. They hoped the great powers would not take any steps that might change the situation.[12]

Ever vigilant, the Zionists countered each Arab initiative with one of their own. In this case the Zionist response took the form of a letter from Weizmann calling for the immediate abrogation of the 1939 white paper and its limits on immigration. Humanitarianism alone, Weizmann declared, demanded such a change, since in light of the physical condition of the refugees, "time means lives." Weizmann included a memorandum from the Jewish Agency arguing that the white paper had no standing in international law. According to this memo the white paper represented nothing more than an edict of the British government, acting on its own. The United States had no responsibility for upholding the British policy.[13]

Regardless of the white paper's international legitimacy, Henderson continued to believe that the American government could not help overturn it without incurring severe liabilities. He worked the bureaucracy to bolster his case. He asked the War Department for an estimate of the force levels required to keep order in Palestine in the face of large-scale Jewish immigration. The American military had no desire to fight for a Jewish state, and the strategists delivered a reassuringly dismal estimate of 400,000 troops, of which as many as 300,000 would have to come from the United States.[14]

Henderson then solicited the opinions of American representatives in the Arab countries as to how support for Zionist policies would affect American relations with the Arab world. The minister to Syria and Lebanon responded that American relations with Damascus and Beirut would be "distinctly affected," since there was "no more burning issue" in Arab politics than Palestine. The minister to Saudi Arabia declared that if the United States went ahead with support for the immigration of the 100,000, "we shall be accused of bad faith, and our prestige with the Saudi Arabian Government will be liquidated." The United States government would find it difficult to continue operations at the military airfield at Dhahran, and American petroleum firms would risk losing access to the oilfields and pipelines of Saudi Arabia. The minister to Egypt predicted that American relations with Cairo would be "immediately and adversely affected."[15]

To strengthen his case further, Henderson sought to arrange publication of letters between Roosevelt and Ibn Saud committing the American government to consultation with the Arabs. The Saudis had raised the issue of the letters, and they were pressing for the publication of transcripts of conversations between the two heads of state as well. Henderson thought release of the transcripts, which touched sensitive issues among the Arab countries in addition to the Palestine question, might have unfortunate consequences. He advised against it. But to placate the Saudis—and to undercut the case of administration Zionists—he recommended publishing the letters.[16]

This proposal produced an uproar at the White House. The strong Zionist faction there wanted Truman to distance himself from any promises to the Arabs. Aide Samuel Rosenman argued—incorrectly if not dishonestly, since Truman in May 1945 had assured the future king of Jordan that "no decision should be taken respecting the basic situation in Palestine without full consultation with both Arabs and Jews"—that Truman had never endorsed Roosevelt's pledges. Rosenman said that Truman should not start doing so now. In any case, he argued—with no greater basis in fact—that admission of the 100,000 would not constitute a fundamental change in the Palestine situation, and thus did not require consultation. Finally—effectively conceding the weakness of his previous points—Rosenman pointed out that the president could fulfill any commitments he was said to have made by a strictly pro forma consultation. "Then," he told Truman, "you can take whatever action you wish."[17]

In this case Truman chose to follow Henderson's advice, if only because he realized that the Saudis had the letters and could publish them on their own—a circumstance Henderson had taken care to point out. The president agreed to release Roosevelt's side of the correspondence, leaving Ibn Saud's to the Saudis. The decision signified no more than a minor victory for Henderson, to be sure. But it encouraged the NEA director to believe that the administration's Zionists had not won the war just yet.

Others in the State Department were less sanguine. Henderson's deputy, George Allen, complained, "It seems apparent to me that the President (and perhaps Mr. Byrnes as well) have decided to have a go at Palestine negotiations without bringing NEA into the picture for the time being. The question we must answer is: Should we nevertheless inject ourselves actively into the negotiations with further recommendations at this stage, or should we wait to be called?" Allen admitted that as public servants they had a responsibility to offer their best judgments on important questions, regardless of the political unacceptability of such judgments. But if NEA persisted in pushing unpopular policies, it ran the risk of being isolated permanently. Allen advocated lying low. "I see nothing further we can appropriately do for the moment except carry on in our current work, answering letters and telegrams, receiving callers, etc., as best we can, pending the time (which will come soon) when the whole thing will be dumped back in our laps."[18]

Allen possessed more political savvy than Henderson, but he lacked some of the latter's stubborn devotion to duty. If the tide was flowing against a particular policy, Henderson interpreted this as greater reason to offer counterarguments. He understood the advantages of swimming with the current, yet he believed he owed it to calling and country to give his honest opinions on vital matters. If he was right—and Henderson was doubting his own opinions with decreasing frequency—time would prove him so, as it had done with respect to the Russians. In any event he was not ready to surrender to pessimism.

III

Meanwhile the British were deciding to try to convert the Americans from irresponsible commentators on Palestine to responsible participants. Attlee chose to interpret Truman's suggestion for opening Palestine to mass Jewish immigration as an offer to help seek a comprehensive solution. The prime minister proposed what took form as the Anglo-American Committee of Inquiry, in the hope that the two countries together might have better luck finding and perhaps imposing a settlement on Palestine than Britain was having alone. To some degree as well, the British proposal reflected annoyance at the sideline quarterbacking of the Americans. Ernest Bevin's friend and biographer, Francis Williams, wrote that Truman's pushing for immigration of the 100,000 drove the foreign secretary into the "blackest rages." Bevin complained that of course the Americans wanted to see the Jewish refugees let into Palestine, because "they did not want too many Jews in New York."[19]

Shortly before Attlee offered his proposal, Henderson developed an idea more ambitious than the Anglo-American committee scheme. At the end of August 1945 he recommended to Byrnes an effort to hammer out a common Palestine policy among the United States, Britain, the Soviet Union, and perhaps France. When Henderson made this suggestion the grand alliance had not yet disintegrated, and such coordination appeared not completely out of the question. Still, it was a measure of Henderson's fear for the future of the Middle East that he even considered inviting the Russians in. He believed that if the great powers agreed on a settlement, neither Arabs nor Jews could hold out for long against it. On the other hand, should the powers disagree, "both Arabs and Jews might have grounds to hope that with a sufficient amount of agitation on their part the decision could be revised. Such a situation would almost inevitably lead to years of political instability in Palestine and in the Near East."

To provide a basis for negotiations among the powers, Henderson offered an analysis of the most likely solutions to the Palestine question. The first involved the establishment of a "Jewish commonwealth" in Palestine. This obviously would appeal to the Zionists, and it would garner the support of both political parties in the United States. But it would probably lead to civil war in Palestine, it might provoke reprisals against American oil concessions in Saudi Arabia, and it would seriously damage America's moral standing throughout the Middle East. Henderson's second option, an "independent Arab state," also promised to provoke civil war, started this time by the Zionists. Henderson thought such a solution could only be imposed militarily, and he added that American participation in an endeavor of this sort would have "very serious political repercussions" in the United States. Plan three envisaged partition of Palestine, on the premise that "irrreconcilable antagonisms" between Arabs and Jews precluded the creation of a single administrative entity. Hardliners among Arabs and Zionists would object, as would nearly all the Palestinians, but moderates on both sides might—just might—accept partition as the only way to avert a war. Yet even this scheme would have to be imposed from without, and it would only succeed with "the unanimous backing of the three great Powers." Henderson's last alternative, the one he favored, involved a variation on the mandate theme. Palestine would become an "international territory under the trusteeship system with Great Britain as the administering authority." This arrangement would suit neither Arabs nor Zionists, but it possessed the merit, in Henderson's estimation, of being insufficiently repugnant to the Arabs to trigger armed resistance. The Zionists would resist; how seriously he did not know.[20]

Henderson was far from certain that even this last option would work. He was beginning to doubt that any plan would. Yet a trusteeship arrangement possessed the cardinal advantage that it allowed the United States to share the burden of a solution with the Russians. If it succeeded, good. If it failed, the Arabs would be as angry at Moscow as at Washington.

Byrnes refused to act on Henderson's advice. The secretary decided to excuse himself from the battle for Palestine. "He rather pointedly washed his hands of the problem, and repeatedly stated that he was leaving it to the President to handle," Evan Wilson wrote. After Truman grew disenchanted with Byrnes, for reasons that had nothing to do with Palestine, George Marshall likewise found less controversial matters—European recovery, for example—to attend to. Obviously Marshall could not avoid the Palestine dispute entirely, but he refused to stake his personal prestige or much of his energy on the outcome. Acheson was another Atlanticist, who, besides, resigned as undersecretary in June 1947 just as the Palestine issue was really heating up. Acheson remained out of office until the beginning of Truman's second term. Acheson's replacement, Robert Lovett, also focused on European and east-west relations, and on seconding the ailing Marshall, who by this time was, in Acheson's words, "a four-engine bomber going on one engine." As a consequence of all this, Henderson became the State Department's point man and lightning rod on the Palestine question.[21]

Through the end of 1945 and the beginning of 1946, the Anglo-American committee commanded the attention of Palestine watchers. The committee began its work with hearings in Washington in January 1946. Witnesses included noted Jewish Zionists and some Jewish anti-Zionists. The biggest name was Albert Einstein, who denounced British imperialism, the 1939 white paper, and, more surprisingly, the idea of a Jewish state. ("The audience nearly jumped out of their seats," British committee member Richard Crossman wrote in his diary. But all was later forgiven, and Israel's government offered Einstein the country's presidency.) The State Department's Middle East experts were not invited to testify. From Washington the committee flew to Britain, Germany, Austria, and Poland, gathering testimony along the way. The members then journeyed to Cairo and Jerusalem, where they learned, if they had not known already, that the views of Zionists and nearly all Arabs remained as irreconcilable as ever. After an exhausting three months the committee retired to Switzerland to produce its report. Like Moses, the members came down from the mountain with a prescription in ten parts. Like the Mosaic report—as reinterpreted by Jesus, at any rate—the committee's ten recommendations boiled down to two: the immediate entry of the 100,000 and the creation of a binational state in which Arabs and Jews would receive equal representation.[22]

Both recommendations could have been expected to outrage the Arabs, the first since it amounted to abrogation of the white paper, the second because it contradicted the majoritarian principle implicit in the idea of self-determination. The expected occurred, and within days Henderson began receiving storm warnings from all parts of the Middle East. The American consul-general at Jerusalem wrote that Palestinian Arabs "were not surprised at the contents of the report but have nevertheless reacted violently." Worse, they were blaming the Americans, whom they considered the prime movers behind the committee's report. The minister at Cairo characterized the Egyptian reaction as "bitterly critical" and likewise anti-American. American officials in Baghdad and Jidda described similar responses.[23]

On May 10 the ranking diplomats in Washington of five Arab governments delivered their personal protest to the State Department. Reiterating the Arabs' demands for consultation in advance of any basic change in the Palestine situation, the Egyptian minister declared that the report had made a "painful impression" on the Arab world. He reminded the Americans that the Anglo-American committee possessed only advisory status and that its recommendations bound neither Britain nor the United States. The Arab League would be meeting within a week to consider its response to the report. He hoped the American government would make use of the time to distance itself from the committee's proposals and to reaffirm its commitment to consultation.[24]

Fortunately for Henderson—or at least it seemed fortunate at the time—the Anglo-American committee's recommendations went nowhere. The Zionists liked the lifting of immigration restrictions, but they rejected a binational state. Though Truman shared the Zionist view, Attlee insisted that the recommendations be taken as a package. The Arabs continued to denounce the report root and branch. With few defenders and plenty of enemies, the report died in the summer of 1946.

Upon the death, Washington and London tried again. In June the American and British governments appointed representatives to a new commission. Henry Grady, formerly assistant secretary of state, headed the American delegation. Herbert Morrison, Lord President of the Council, chaired the British group. Morrison proposed a scheme for the cantonization of Palestine, with Arabs and Jews occupying separate and largely autonomous provinces. Grady accepted the idea. Although the Anglo-American committee had considered and rejected such a plan, Grady became convinced that cantonization afforded the only hope of getting the British to agree to letting in the 100,000, and he threw his support behind it. Dissenters on Grady's staff, however, leaked the package to the press, touching off a Zionist campaign to defeat it. Zionists and allies attacked the Morrison-Grady plan as a sellout to the British and an effort to create a Jewish ghetto in Palestine. Truman, while still very sympathetic to the Zionist cause, was so put out with the Jews at this point that he declared, "Jesus Christ couldn't please them when he was here on earth, so how could anyone expect that I would have any luck?"[25]

Truman underestimated himself. The Morrison-Grady plan passed into oblivion, but a few weeks later the president made a statement that pleased the Zionists greatly. On the eve of Yom Kippur—and one month before the 1946 congressional elections—Truman reaffirmed his backing for immigration of the 100,000. He went on to advocate efforts to reconcile a fresh proposal by the Jewish Agency for a Zionist state with certain features of the Morrison-Grady plan. "I cannot believe," he said, "that the gap between the proposals which have been put forward is too great to be bridged by men of reason and good will." Truman predicted that a Jewish state "would command the support of public opinion in the United States."[26]

As before, Henderson and the State Department were left to puzzle out what the president meant. The problem was a pressing one, since a strong possibility existed that the United Nations would take up the Palestine question, and the department needed to instruct the American delegation. Henderson recommended using the president's Yom Kippur statement as a point of departure for a policy of "bridging the gap" between provincial autonomy and partition. Speaking for NEA, he wrote, "We believe that the President had such a solution in mind when he issued his statement of October 4."[27]

In fact, Henderson probably believed no such thing. Like most members of

the State Department, Henderson gave Truman credit for well-developed political instincts but not for an understanding of the intricacies of the Palestine
question. Henderson likely suspected that the president had not thought the
matter through much beyond the November elections. Yet if the department
pushed hard for a compromise settlement of the sort Henderson recommended,
it might convince the president that such was what he had in mind.

Henderson had little expectation of persuading Zionist radicals like Ben Gurion to accept less than immediate partition. But he thought he might have better luck with moderates like Weizmann and Goldmann, and he rested in them
such hopes as he retained for a peaceful settlement. At the beginning of
November he met with Goldmann, who reminded him that the World Zionist
Congress would soon meet in Basel. Goldmann desired to be able to report
motion toward a solution on Palestine. Otherwise, he feared, "extremists" like
Ben Gurion and Abba Hillel Silver might "do a great deal of damage." By contrast, should the Americans and British produce a realistic plan, the moderates
could "keep matters in line."[28]

But no such plan emerged, and the confrontation Goldmann predicted
between moderates and radicals took place. At the end of December, John
Lehrs, the American vice-consul at Basel and Henderson's informant on the
proceedings of the Zionist congress, reported the bad news that the Ben Gurion–Silver group had overthrown Weizmann and the moderates, although the
extremists had not managed to produce an agenda acceptable to a majority of
the delegates.[29]

Henderson was beginning to find the whole situation extremely distressing.
The Truman administration, he wrote, had been "practically been forced by
political pressure and sentiment in the U.S." to support a Zionist state. Not only
was this entirely the wrong way to make policy, it corrupted the fundamental
standards of self-determination and equality the United States stood for. What
was Palestine compared to America's good name?

Henderson complained that American policy toward Palestine was a sham-
bles, tending to produce just the results it aimed to prevent. The Palestine question was "an open sore, the infection from which tends to spread rather than to
become localized." American actions, reflecting the pressures of "a cosmic Jewish urge with respect to Palestine," had in fact accomplished almost nothing to
ease those pressures or satisfy that urge. The crux of the problem was that
American policy was "one of expediency, not one of principle." Or if a principle
did exist somewhere beneath the politics, it consisted of going "as far as we can
to please the Zionists and other Jews without making the Arabs and the British
too angry." This approach must not continue; it was a "dangerous business
which we ought to get out of with all speed possible."

What, positively, should be done? Henderson did not have a solution that
would satisfy both sides in Palestine. No such thing existed. But he did rec-

ommend a swift transition to independence—whatever form that took—since the present situation was rapidly tending toward explosion. More important, the United States must regain its moral standing in the matter. Pandering to special interests must cease; the American government must eschew any further actions based simply on a desire "to thread a way between the specific projects and plans of the contending pressure groups." The administration should announce its intention not to support any solution not agreed to by both sides in Palestine and approved by the United Nations. It should try to pursue "a policy of principle and general procedure which will be approved as fair and reasonable by the general public in this and other countries."[30]

IV

But, as 1947 began, fairness and reason were fast vanishing from the Palestine debate. Whether intransigence among the Zionists had begotten intransigence among the Arabs, or vice versa, the extremists were winning. With the British retreating from positions throughout their empire, a pull-out from Palestine could not be far off. Just what would happen then, no one could say. Certainly not Henderson. Ever more he realized the irrelevance of NEA and the State Department to the basic decisions on Palestine. If not for his success on the Greek question, the first months of 1947 would have been discouraging indeed.

In January, Ernest Bevin led negotiations in London between representatives of Arabs and Jews, seeking a solution based on the provincial autonomy scheme. Henderson approved of the talks, primarily because such a plan would "reduce Palestine from a world issue to a local issue," thereby limiting the damage the problem could do to American interests in the Middle East. But he held no hopes that the talks would progress far, and events proved him right.[31]

The British next began hinting that they would turn the matter over to the United Nations. Henderson prepared contingency plans. He recommend caution. "We should move slowly in committing ourselves in any direction," he wrote. No matter what course the UN followed, someone's oxen would be gored, and the United States ought to stay out of the skewering. Eventually, of course, the American government would have to settle on a position; in the meantime Washington should get its affairs in order. "We should decide only after discussions with Congress, with the White House and with American Jews and other interested groups what our policy is to be. We cannot afford in the forum of the United Nations to retreat from a position once taken as a result of pressure brought upon us from highly organized groups in the United States."[32]

Of pressure from organized groups Henderson felt plenty. Moshe Shertok, the Jewish Agency representative in Washington, came calling to outline the policy his organization wished the United States to adopt in the UN. Shertok argued against a recently floated proposal to appoint a committee of neutral nations to consider the Palestine problem. Palestine, he said, was "primarily a great power question." Although the Jewish Agency preferred to keep the Soviet Union off the committee, the Jews would accept a Soviet role if such was the price of American participation.

Henderson and Acheson listened politely to Shertok, but they offered no encouragement. Now that the grand alliance was thoroughly shattered, they rejected the idea of a great-power committee, favoring the neutral-nation approach. Acheson said the latter would bring an objectivity and credibility to the Palestine issue the great powers could not command.[33]

Undiscouraged, Shertok and his associates persisted. Henderson received a Jewish Agency proposal for American funding of development projects in Palestine. According to the proposal, American support, to the amount of $100 million, would make "a great constructive contribution" to the resolution of "an unnecessarily tangled and embittered situation."[34]

Henderson thought such an aid program was the last thing the United States ought to get involved in. For Washington even to begin talking about financial aid, he said, risked "the most serious repercussions." No matter how closely the administration tried to hold news of such talks, the Zionists would surely leak the information, and much of the world would be convinced that America had thrown its weight irrevocably behind the Zionist cause. The Arabs would react violently, destroying what little chance remained for a peaceful settlement.[35]

Acheson agreed, and when Shertok arrived for a follow-up visit the undersecretary put him off. He said he recognized the importance of economic development. The subject merited examination. But not now.[36]

Reinforcing Acheson's reluctance to commit the United States to economic development in Palestine was the fact that the administration at precisely this moment was preparing a major program of aid for Europe. Within the week George Marshall declared that the United States would look favorably upon proposals to assist in the reconstruction of the continent. Applied to countries whose economies remained in chaos, and whose politics seemed open to communist encroachment, the Marshall plan would serve the dual ends of humanitarianism and American security—an appealing combination. But Marshall's idea also required convincing Congress to appropriate more money than the legislature had ever dreamed of spending on foreigners. Under the circumstances, an aid program for Palestine, even had it been diplomatically advisable, was politically improbable.[37]

By this time the British had referred the Palestine problem to the UN, which appointed a committee of neutral countries to consider the matter. Through the summer of 1947, while the UN committee heard testimony, the Zionists stepped up pressure on the administration. In June, Rabbi Silver complained to Marshall and Henderson that the Zionists were "completely in the dark" as to

American policy. Silver said he had the impression that the United States favored partition, but partition could mean many things. One partition plan might elicit the support of nearly all Jews; another would provoke almost unanimous opposition. He hoped that whatever the administration did, American officials would meet privately with Zionist leaders before making any public statement. He wished to avoid misunderstanding, and he did not want to see American Zionists placed in a position where they would feel "compelled to oppose the policies of their own Government."

Marshall gave Silver no grounds for optimism. The secretary, a master of taciturnity, replied that he had nothing to add to the previous statements of the administration regarding Palestine. Unhelpfully, or so it must have seemed to Silver, he said that if the Zionists wished to convey to the department any information or opinions they had not delivered already, they should speak to Mr. Henderson.[38]

During the next several weeks Henderson counseled keeping America's distance from the UN deliberations. "We must not permit ourselves to be maneuvered into such a position that the plan finally adopted by the General Assembly should be considered as primarily an American plan." If this perception developed, "we should probably be held primarily responsible for the administration and enforcement of such a plan."[39]

Henderson guessed that the committee would deadlock and produce a divided report. It did. The majority favored partition of Palestine into three parts: an Arab state, a Jewish state, and the city of Jerusalem, the last under trusteeship of the UN. The minority proposed a single federal state with Jerusalem as its capital.

At the time the committee unveiled its proposals, Henderson was in Greece persuading Tsaldaris to broaden his regime. When he returned to America, Marshall summoned him to New York, where the secretary was preparing to address the UN. On entering the secretary's suite Henderson discovered that Marshall had also invited members of the American UN delegation, including Eleanor Roosevelt and John Foster Dulles. Although unprepared for a formal review of the UN committee's report, Henderson did his best to explain the drawbacks of the majority's partition plan. In the first place, he argued, the plan rested on no discernible principles. It was "full of sophistry," largely because those who devised it had no responsibility for its implementation. He was certain that partition could be enforced only by arms, and since the British would not risk alienating the Arabs for the sake of the Jews—as he had confirmed in London on his way back from Greece—the burden of enforcement would fall on the United States. Henderson did not shrink from committing American troops when vital American interests were involved. But partitioning Palestine did not qualify. On the contrary, American support for partition would work against American interests. It would alienate the Arabs and open the Middle East to the Soviet Union.

Eleanor Roosevelt was not convinced. "I think you are exaggerating the dangers," she said. "You are too pessimistic. A few years ago Ireland was considered to be a permanent problem that could not be solved. Then the Irish Republic was established and the problem vanished. I'm confident that when a Jewish state is once set up, the Arabs will see the light; they will quiet down; and Palestine will no longer be a problem."[40]

Henderson had handled the State Department's correspondence on Palestine with Mrs. Roosevelt, and he had suspected that she was as ignorant on this subject as he thought she was on Soviet-American relations. Her comments removed his last doubts. That she knew nothing about the Irish problem— which, as Henderson could testify from personal experience, had not exactly "vanished" after Ireland's partition—seemed entirely in character.[41]

On leaving the meeting, Henderson believed he had not done an adequate job explaining the dangers involved in backing the majority report. As soon as he returned to Washington he prepared a more comprehensive brief. Accurately describing his comments as representing the views not only of himself and NEA but of "nearly every member of the Foreign Service or of the Department who has worked to any appreciable extent on Near Eastern problems," Henderson proceeded to deliver a scathing attack on a policy of American support for the creation of a Jewish state in Palestine. He listed the policy's drawbacks. First, it would seriously damage American interests in the Middle East "at a time when the Western World needs the friendship and cooperation of the Arabs and other Moslems." Without this cooperation, the United States might find that its efforts to stabilize Greece and Turkey were wasted. American communication facilities throughout the Middle East might be lost. The region's oil, on which the reconstruction of Europe depended, would be at risk.

Second, partition was unworkable. Of all the previous committees and commissions charged with finding a solution to the Palestine problem, only one, the Peel commission of 1937, had recommended partition, and the Peel plan had fallen apart on closer look. The UN committee's majority plan called for an economic union between the Arab and Jewish states. Irrigation ditches, railways, roads, and telegraph and telephone lines would cross the frontiers. In the face of fierce Arab hostility to partition, these ties would not last long. The problems attending the birth of the two states would only increase with age. The proposed economic union would break down, if it ever functioned at all, exacerbating the political problems. Under the most favorable conditions, the Arab state would be marginally viable. Under conditions actually likely to obtain, it would fare far worse.

Third, the majority plan left a number of refractory problems unaddressed. An international supervisory board would referee economic disputes between the Arab and Jewish states. Who would sit on this board? Americans? Russians? Would American representatives come under the same political pressures the Truman administration currently felt? Henderson predicted that the Palestine problem, instead of achieving solution with the majority plan, would return to the United Nations and fester there for years, poisoning relations among members of the organization.

Fourth and finally, the majority plan flew in the face of the UN charter and well-established American ideals, particularly self-determination and majority rule.[42]

The White House found it impossible to ignore Henderson's brief. A few days later he received a call inviting him to the president's office for a chat. When he got there he discovered the real reason for the invitation. Henderson commented afterward, referring to David Niles, Clark Clifford, and additional Zionist sympathizers at the session, that "the group was trying to break me down and humiliate me in the presence of the President." But Henderson held fast, reiterating the arguments he had made in his memo. Niles and the others impugned his judgment and expertise, harassing him to the point where Truman finally stood up and said, "Oh, hell, I'm leaving."

Henderson won no converts, but he did gain a certain appreciation for Truman's position.

> From his bearing and facial expression that evening I was not at all convinced that even at that late date the President had made the final and definite decision to go all out for the establishment of the Jewish State.... I had the impression that he realized that the Congress, the press, the Democratic Party, and aroused American public opinion in general, would turn against him if he should withdraw his support for the Zionist cause. On the other hand, it seemed to me, he was worried about what the long-term effect would be on the United States if he should continue to support the policies advocated by the Zionists. He was almost desperately hoping, I thought, that the Department of State would tell him that the setting up in Palestine of Arab and Jewish States as proposed by the U.N. Commission would be in the interest of the United States. This, however, the State Department thus far had not been able to do.[43]

The possibility that the department might discover such capacity diminished at the beginning of October, when the Saudi representative at the UN and Nuri of Iraq delivered a message to the American delegation in New York, saying that the Arab countries had met the day before to consider asking Soviet help to prevent partition. The Russians earlier had indicated a desire to trade the votes of the Soviet bloc in the UN for the votes of the Arabs on matters of interest to each group. The point of the visit by Nuri and the Saudi diplomat was to let Washington know that unless the United States could offer assurances against partition, the Arabs would accept the Soviet offer.[44]

This was blackmail, of course, and if Henderson had not agreed with Nuri's conclusion that partition was a disaster about to happen he would have considered such a maneuver intolerable. But under the circumstances he found the Arabs' position entirely understandable. "The question of Palestine is to them the most important question in their international life," he said. "They consider it their duty to use all means available in order to block the setting up of a Jewish state." Henderson commented that the Arabs suspected the Soviets of making similar advances to the Zionists. During the past several months the Russians had been hedging their bets by hinting that they might back partition. "The Arabs now undoubtedly feel that the Soviet Union will turn against them if they continue to refuse to bargain with it. If the Soviet Union should turn against them and if the United States should also be against them, their hope of preventing the establishment of a Jewish state would indeed be small." Henderson suggested that despair informed the Arab threat. "What they are really trying to do is to persuade us to take an attitude which will enable them to reject Soviet overtures."[45]

The Arabs were out of luck. On October 11 the American ambassador at the UN, Herschel Johnson, with Truman's approval announced that the United States would support the majority plan, subject to slight modifications.

Henderson's initial reaction to this development was concern that the Arabs would make good their threat to walk into the Soviet camp. But two days later Moscow's representative also announced for partition. Betrayed, as it appeared, by both sides, the Arabs called a plague on both houses. An Iraqi diplomat told a member of the American delegation that the Arabs would side with neither east nor west in the UN. "Since the United States is apparently no longer an ally and Russia does not seem to desire to be so," the Iraqi said, "we shall now vote on each matter in accordance with our own interests."[46]

If Henderson's concern about an Arab turn to the Soviet Union proved overdrawn at this time—the turn would come later—his worries that the United States, by supporting the majority plan, would be identified with partition, did not. Delegates at the UN spoke of the "American plan" for the partition of Palestine. Henderson rejected the label and argued against associating the United States with the plan. Such association would yield evil consequences. "We shall inescapably be saddled with the major if not sole responsibility for administration and enforcement," he predicted. Partition would produce violence for which the United States would be blamed. The Arab countries would intervene. Eventually the Soviets would work their way in.

Though the die seemed nearly cast, Henderson still refused to concede that partition was inevitable. He told Undersecretary Lovett, "Our people in New York feel that time is working against partition as more and more of the delegates come to appreciate the difficulties." The United States had announced for partition, but if the American delegation did not lobby actively in support, the majority plan might fail to gain approval.[47]

Henderson considered himself a faithful public servant, a professional who could support a decided policy over his own personal misgivings. In forwarding his point-by-point dissent on the majority plan, he had assured Marshall that "in spite of the views expressed in this memorandum, the staff of my Office is endeavoring loyally to carry out the decision which you made." Yet Henderson's hope that the American-backed majority plan would fail indicated, if not active sabotage, at least latent subversive tendencies. It could hardly be said he was doing his best to ensure the passage of partition.[48]

To the degree he tried to justify this discrepancy to himself, Henderson doubtless believed he was attempting to protect the president from the bad advice Truman was receiving from his pro-Zionist aides. Henderson thought Niles and Clifford were using political pressure from American Zionists to push the president into a dangerous policy, one threatening serious repercussions upon American security.

Unfortunately for Henderson, Niles and Clifford had more clout than he did. As the UN general assembly took up the majority plan the White House organized a forceful campaign to convince wavering members that their interests dictated approval. Henderson did not know the details of the campaign, but he feared the worst and prepared for it. On November 10 he recommended to Marshall that the increasing likelihood of conflict made it advisable for the United States to embargo weapons to Palestine and the neighboring countries. "Otherwise," he argued, "the Arabs might use arms of U.S. origin against Jews, or Jews might use them against Arabs. In either case, we would be subject to bitter recrimination."[49]

Marshall approved this suggestion, and the arms shipments—the legal ones, at least—stopped. Yet Henderson was hardly reassured. With the general assembly's decision on the majority plan just days away, his forebodings intensified. On November 24 he wrote to Lovett that once again he felt it his duty to sound the warning that "the policy which we are following in New York at the present time is contrary to the interests of the United States and will eventually involve us in international difficulties of so grave a character that the reaction throughout the world, as well as in this country, will be very strong." The United States was committed to the idea that the Middle East was vital to American security. Good relations with the Arabs were a prerequisite to American influence in the region. But the administration's Palestine policy was guaranteed to alienate the Arabs.

That policy also seemed likely to alienate Britain. Ambassador Johnson had publicly denounced Britain for not agreeing to recent UN recommendations on Palestine. Had Britain accepted the recommendations in question, Henderson said, they "would certainly have ruined British relations with the Arab world and would probably have resulted in the British being forced to withdraw from the whole Middle East." Henderson accounted Johnson's attack on Britain "extremely unfortunate" and liable to have damaging repercussions. "I am afraid that the reactions in London to criticisms of this kind will not help us in our efforts to prevail upon Mr. Bevin not to withdraw the remaining British troops from Greece."

Unwilling to indict the president, Henderson suggested that Truman had not been sufficiently informed regarding crucial details of the policy the United States had committed to. Niles and Clifford did a good job of filtering out unwelcome intelligence. "I wonder if the President realizes that the plan which we are supporting leaves no force other than local law enforcement organizations for preserving order in Palestine." Evidently Ambassador Johnson had intimated to the UN that the United States was prepared to join a peacekeeping force, presumably in coordination with the Soviet Union. Was the president aware of this? Henderson warned, "We ought to think twice before we support any plan which would result in American troops going to Palestine. The fact that Soviet troops under our plan would be introduced into the heart of the Middle East is even more serious."[50]

V

On November 29 the general assembly approved the partition plan, and the possibility that a peacekeeping operation would be necessary appeared to increase. Early in December, Henderson received a message from Cairo relaying King Farouk's vow to resist partition in Palestine by force of arms. The king added that if the United States took part in a peacekeeping operation the world might witness the "tragic spectacle of Arabs fighting Americans." A few days later the Iraqi foreign minister made the same point to Henderson personally. Fadhil Jamali said the Arab people would never accept the UN partition plan. "They would die in defense of their rights rather than accede to it." Jamali described his "two principal fears": that Arabs would fight Americans and that the Soviets would gain a foothold in the Middle East, from which they might never be dislodged. Was there any way to avert this outcome?, he asked.

Henderson, by now convinced that the administration's Zionists had won, replied sadly that he would be misleading the foreign minister if he offered any hope. The United States government had chosen to support partition after long and careful consideration. The decision was final. Henderson said the United States valued a close relationship with Iraq and other Arab countries. But regarding Palestine, he could see "no other solution than the acceptance by the Arabs of the UN solution."[51]

The Zionists were also preparing their response to the assembly's decision.

On December 8 Moshe Shertok of the Jewish Agency came to the State Department to try to obtain American weapons and military advisers. He told Henderson that as the British departed the Jews would have to assume the task of keeping order. He pointed out that his people were giving Washington first chance to provide such help, and he hoped the Americans would avail themselves of this opportunity. If the United States could not accommodate Zionist needs, he would be obliged "to turn elsewhere."[52]

Henderson found the brazenness of this request stunning. To think that the United States might send military advisers to support the Zionists boggled the mind. At least it boggled his.

Whether or not it boggled Truman's mind, the president rejected the idea. He backed the creation of a Jewish state, but not to this extent. The arms embargo stayed on and American military advisers stayed home.

Through the early spring of 1948 Henderson attempted to soften the impact on American interests of Washington's support for partition. By the end of March, fighting in Palestine had escalated to the degree where the Truman administration was considering a temporary trusteeship following the British evacuation. Henderson warned that such a path had no end, that it would lead to "the presence of American troops in Palestine and even the shedding of American blood." To prevent such an outcome Henderson advocated some sort of final try at collaboration with Britain. Admittedly, collaboration might prove hard to elicit, since the British were "extremely bitter" at the administration's deference to the Zionists and Washington's unreliability on the Palestine issue. The administration might retrieve its reliability, but only by deciding "once for all that it will not permit itself to be influenced by Zionist pressures."[53]

On this last point Henderson was whistling in the dark, as he realized. The White House had already sided with the Zionists, and it would not change its course at this hour. In a private letter, notable for its prescience, Henderson spelled out his worries regarding Palestine. He forecast that the Zionists would win "the first few rounds" but would be unable to establish anything like lasting peace or stability. The American people, unwilling to countenance the destruction of the Jewish state, would find themselves increasingly drawn to the Zionists' defense. Anti-western elements would batten on the chaos; present regimes, "too Western in spirit for the aroused fanatic masses," would be overthrown. The region would experience "the rise of fanatic Mohammedanism" of an intensity "not experienced for hundreds of years."[54]

When Henderson during this period advocated increased efforts on behalf of Anglo-American cooperation, he was motivated at least partly by the knowledge that although few persons of influence in Washington listened to what he was saying, the British would give him a sympathetic hearing. Critics of the State Department on the Palestine issue hit the mark squarely when they asserted that America's professional diplomats held views more in line with Britain's than with those of their own president. Henderson certainly did, and during the last few weeks leading to the creation of Israel he confided his fears to his British counterparts. "Upset and bewildered" was how R. N. Hadlow described Henderson after a conversation at the beginning of May. Harold Beeley remarked that "Henderson seemed extremely discouraged." Henderson predicted to Beeley that the declaration of the Jewish state, anticipated within days, would be followed by an invasion by the Arab nations. Should this likely event come to pass he saw no way for the United States to remain above the fray. At the least Washington would have to lift its arms embargo, for once Arab armies crossed the borders of Israel there would be "no holding" American feelings in favor of aiding the Jews. Israel, he told Beeley, would be "a source of grave trouble in the Middle East for an indefinite period."[55]

Through the first two weeks of May, violence in Palestine intensified. Attacks on Arabs by Jews and on Jews by Arabs set in motion massive transfers of population, especially of Arabs from the sector of Palestine assigned to the Jews. Washington offered to mediate, but each side preferred its prospects on the ground to its chances at the negotiating table. As before, Henderson laid most of the blame on the Jews, who, he said, were "intoxicated" by their success.[56]

On May 15 at one minute past midnight, Zionists in Palestine proclaimed the existence of Israel. Ten minutes later, which in Washington was shortly after six o'clock on the evening of May 14, the White House announced de facto recognition.

To the last, Henderson and the State Department dissented. At a crucial meeting on May 12 Lovett cautioned against "premature recognition." The United States did not know what kind of government the Zionists would establish, and to recognize before acquiring this information would be "buying a pig in a poke." Marshall criticized the arguments of Clifford, the principal advocate of early recognition, as "a transparent dodge to win a few votes." The secretary added that precipitate action would "seriously diminish" the presidency. Marshall, the epitome of the nonpolitical soldier, made it a practice not to vote in presidential elections, but he told Truman bluntly that if he were a voter and the president followed Clifford's advice, he would vote against him.[57]

To Henderson fell the task, following the president's decision in favor of Clifford and immediate recognition, of passing the news to the Israeli government. Beyond backdating a letter from Marshall to Eliahu Epstein conveying the message of recognition, to reflect for the historical record the priority the administration placed on Israel, Henderson hewed strictly to protocol. He addressed Epstein, who had been the representative of the Jewish Agency in Washington until six in the evening on May 14, simply as a private individual. Explaining to Lovett, Henderson commented, "We do not yet know officially what his capacity and title are after 6 p.m., and will not know until we have been informed." Of greater importance, Henderson made certain that the administration's message did not commit the United States to accept the frontiers of Israel, either as they were then being fought for or as specified in the UN decision of the previous November. He could not resist observing that although the Israelis took the UN vote as a basis for the legitimacy of their state, this was a selective use of the general assembly's resolution. The majority plan had called for an economic union of the Jewish and Arab states and a special international regime for Jerusalem, neither of which the new government of Israel seemed inclined to accept.[58]

VI

"I am unfortunate," Henderson wrote in March 1948, "in that I always seem to be doing a job which calls for the making of enemies. I imagine that the campaign against me is only getting under way and that the future holds much more than the past."[59]

Henderson was right, although he was imagining a lot. By the time the struggle for Palestine reached its final months the NEA director was already one of the most criticized individuals in the history of the American foreign service. For legislators, editorialists, and other keepers of the public conscience to attack political appointees raises few eyebrows, and those persons benefiting from what remains of the spoils system know that criticism, fair and foul, comes with the territory. Acheson may not have liked being labeled "Red Dean," but he could not have expected to get out of office without being tagged with some such title. Nor is it unusual for unimaginative populists to lash out at the nameless bureaucracy that allegedly perverts the will of the people and prevents America from achieving its democratic destiny. But castigating a particular bureaucrat by name is relatively rare. For all the hundreds of persons on McCarthy's wish list of disloyalists in the State Department, the American people were treated to the public execution of only a few. Viewed charitably, this diffidence reflects a feeling that such assaults are low blows, since the bureaucrats usually lack a forum to defend themselves. More realistically it demonstrates an inability to pin particular policies on specific individuals. The buck always stops in the Oval Office, but where it starts is harder to determine.

In Henderson's case, however, the Zionists had an easy target. The NEA chief made no effort to adjust his views to the needs of domestic politics, and when Byrnes and Marshall demonstrated little desire to take on the White House over Palestine, Henderson assumed most of the burden of making the case against supporting the Zionists. He already was suspect in the eyes of liberals who during the war had correctly considered him an impediment to friendly Soviet-American relations. The fact that the ranks of the Zionists included many of these New Deal-types did not help Henderson's chances for a pleasant time at NEA. Nor did the fact that Marshall could still claim most of the war hero's immunity from criticism contribute to Henderson's peace of mind. In May 1947, Emanuel Celler, a Democratic congressman from New York and an outspoken Zionist, demonstrated a willingness to absolve Marshall of the sins of State. "Perhaps Palestine is a new subject for Mr. Marshall," Celler said. "Perhaps he is being briefed by Mr. Loy Henderson, the Arabphile. I do not wonder that he is possibly confused." Celler went on to brand Henderson a "mischief maker" and a "striped-trousered underling saboteur" who was "more Arab than the Moslems."[60]

Celler's blasts displayed energy but little imagination. He was simply following the tracks of those who had been lashing Henderson for many months. Nearly a year earlier, Bartley Crum, a member of the Anglo-American committee, had demanded Henderson's resignation. To his self-described shock, Crum had discovered during his service on the committee that the State Department in May 1943 had sent King Saud a "highly confidential note" assuring the monarch that the United States would make no changes in its policy toward Palestine without consulting both Arabs and Jews. Describing this and other similarly outrageous items in a "secret file," Crum intoned, "It was a sorry and bitter record for an American to read." Crum laid the blame for this history of duplicity on the "middle level" diplomats in the State Department who persisted in keeping the president in the dark regarding their actions—and who were "captives of the British social lobby" in Washington to boot. Asked to name names, Crum replied, "It would be a salutary thing if Mr. Loy W. Henderson's resignation were requested."[61]

As the going got really rough on the Palestine issue, Henderson became a veritable celebrity. His name surfaced in a criminal libel suit involving the grand mufti of Jerusalem and the operation of a submachine-gun factory in Pakistan. The *New York Post* ran a four-part series on him under the head "Man in the Saddle: Arabian Style." According to the *Post*, Henderson, whose office sported "a brilliantly colored tiger-skin rug and an 8-foot oil painting of a nineteenth century Bey of Tunis" was "the man who, more than any other individual, influences day-to-day American foreign policy in the critical hot spots of Greece, Palestine, and the whole Near East." Arthur Klein, a New York colleague of Celler's, did the House of Representatives the favor of reading the *Post* series into the *Congressional Record*. Klein congratulated the newspaper for revealing "the true danger involved in permitting the direction of the enormously powerful and strategic Near Eastern Section to rest in the hands of Loy Henderson." Adding his own gloss, the Democratic congressman charged Henderson with being "a virtual propagandist for feudalism and imperialism in the Middle East" and "a fellow traveler of the British Colonial Office as regards Palestine." Klein also lambasted Henderson for attempting to subsidize "barbarism, polygamy, savagery, and banditry"; for trying to yoke the United States to "every wazir, satrap, and pasha of the Near East"; and for "willfully and wantonly distorting the foreign policy of the United States as enunciated by the Congress and the President." Catching his breath, Klein proceeded to pronounce Henderson "completely unfit" for his position.[62]

The criticism did not take Henderson by surprise. In the late spring of 1945 an acquaintance, who happened to be a Jew, offered some friendly counsel. "Loy," he said, "I want to tell you that you will be in deep trouble if you continue to adhere to what is believed to be your present attitude with regard to the establishment of the Jewish State. Can't you at least modify your stand and just let affairs take their natural course? If you continue to advise our Government against supporting the setting up of a Jewish State, your career can be ruined, and I don't know what else might happen to you."[63]

Henderson decided he could not in good conscience temper his advice, and he resigned himself to suffer the slings and arrows of outraged Zionism. While he considered the attacks unfair, he took it as a point of professional pride to bear his burden in silence. Or at least when he complained he did so privately. In the summer of 1946, following charges that he had already made up his mind on the Palestine issue and thus could not be objective, he wrote an old associate,

> I do not think that I need to convince you or my friends in the Department who know me that I have not "taken a position" with regard to Palestine. I have been merely trying to do my job, which is to apply the various policies of the President and of the Secretary of State to an extremely complicated situation. Without endeavoring to defend myself, I may state that I am convinced that no conscientious American official could act in this situation much differently from the way in which I am acting. I am a prisoner of circumstances which are not of my creation and my freedom of action is therefore rigidly circumscribed. I am not anti-Zionist nor pro-Arab. I am merely trying to apply the policies of my Government in as humanitarian a manner as possible. I venture to assert that I have had, and continue to have, a keen sense of our responsibility for the welfare of the displaced Jews in Europe and I feel that I am doing everything that I can in the circumstances to alleviate their situation. I sometimes feel that some of us in the Department who are being criticized by some of the more extreme Zionists for failure to produce at once the results which they demand are really more interested in helping these victims of the Nazi regime than some of those who criticize us.
>
> When I took over the Directorship of this Office a little more than a year ago, I realized that I would be sure to come in for considerable criticism on the Palestine question since it is manifestly impossible for the Department to satisfy the more radical demands with regard to Palestine and since, for various reasons, it might seem desirable in certain quarters to place responsibility for failure to receive satisfaction upon the permanent personnel of the Department.[64]

The predictable subtext of much of the criticism of Henderson was that he took the position he did on Palestine because he was antisemitic. This was a bum rap, as Allen Podet, writing in the *American Jewish Archives*, has clearly demonstrated. Podet examines the matter at length and finds "no evidence of a predisposition towards anti-Semitism in any of Henderson's official or unofficial communications, the files he generated or those kept on him." Neither does Podet's search in other relevant areas turn up evidence of antisemitism. Podet concludes, "It is time that the record be set straight." Henderson occasionally lapsed into stereotypes regarding Jews, as his remarks from Baghdad about Iraq's Jews showed, but ignorance rather than mean-spiritedness inspired the lapses—something that could not be said of more than a few American diplomats of his day, who really *were* antisemitic.[65]

Nor, narrowly speaking, was Henderson anti-Zionist. Having no personal interest in the matter, he cared little whether the Jews got a state of their own. What he objected to was the idea that the American government ought to assist in the project, against the wishes of the majority of the inhabitants of the region. If the Zionists on their own accomplished their goals, good for them. If not, too bad. But the United States should stay out of the matter. He could sympathize with the sufferings of the Jews, but he did not agree that outrage and guilt over the holocaust ought to be allowed to disrupt American relations with a significant portion of the world's population—especially the portion in his bailiwick.

Unfair criticism confirmed Henderson in his conviction that he was doing the right thing. "I don't see how any man can hold the position which I occupy at the present time," he remarked in March 1948, "without doing almost precisely what I am doing. If a successor should follow a different line, he would, I am afraid, become so deeply involved in adventures contrary to our basic interests that there would soon be real explosions." Several weeks later he declared, "I do not see how I can advocate full satisfaction of the Zionists' demands and at the same time live up to my responsibilities." Occasional letters of support, as from William Phillips, his former boss and Anglo-American committee member, provided encouragement. But he did not really need it. Just as during his troubles in the European division regarding the Soviet Union, Henderson thrived on the adversity. It made him feel more principled.[66]

Henderson read the criticism of his position on Palestine in light of the criticism of his stance regarding the Soviets. In February 1948 he commented, "I sometimes have feelings of discouragement similar to those which I used to experience during the years 1939 to 1943 when I was doing my best to prevent Eastern Europe from going behind the iron curtain." A short while after this he wrote a friend,

> As you probably know, during the years 1940 to 1943, when I was assisting in the handling of our Russian affairs, I was under particularly vicious attack by communists in the United States, by fellow-travellers, and by some undoubtedly patriotic liberal Americans for "sabotaging President Roosevelt's policies" with regard to the

Soviet Union. I was charged with being anti-Soviet, reactionary, etc. It is interesting
to note that among those who have led the campaign against me on the Zionist
issue are a number of writers who led the campaign against me on the Russian
issue.

Henderson did not claim that all who criticized his position on Palestine were
motivated by distaste for his views on the Soviet Union. But he thought many
were. "The move to push me out in the front as the arch-opponent of Zionism
was planned by persons whose real grievance against me was that they considered that I was advising the American Government to adopt a firm stand with
regard to the Soviet Union."[67]

Henderson exaggerated this last point. The fact that he did so is significant.
It indicates the degree to which he perceived an essential unity in the various
attacks he was subjected to, and it reveals his continuing preoccupation with
matters relating to the Soviet Union. Five years after being relieved of responsibility for policy toward the Kremlin, he was still fighting the old battles. He
even thought the enemies were the same. It is hard to avoid the impression that
whatever the merits of Henderson's arguments on the Palestine issue—or on
Greece, or later India or Iran—he was motivated at least in part by a desire to
again prove wrong those who had refused to listen to his warnings about the
Soviet Union in the late 1930s and early 1940s. For Henderson, all roads led
from Moscow, and back.

Whatever the source of their complaints, the Zionists in the Truman administration determined that Henderson had to go. For some time Clifford and
Niles had been trying to get him out of the way. In June 1948 Clifford alleged
that Henderson had tried to blackmail Israel into giving up territory to the
Arabs in exchange for de jure recognition by the United States. The conversation in which Henderson's attempt was supposed to have taken place went
unrecorded in the State Department. Henderson claimed it never occurred.
Considering Henderson's conscientiousness and his conception of duty, Clifford's tale is hard to credit. Henderson just might have suggested such a trade
to Marshall, but he would never have freelanced with foreign officials to the
degree Clifford claimed.[68]

In any event, Truman hardly needed Clifford's advice to realize that with a
reelection campaign in progress Henderson was a liability. The president
decided to reassign Henderson to the field, out of Washington and far from the
Arab-Israeli conflict. The State Department suggested Turkey, but the Zionists
considered even Ankara too close to Israel for comfort. Early in July Henderson
learned he would be heading to India.[69]

IV

THE WAGES
OF NEUTRALISM

13

The Most Charming Man
He Ever Despised

The circumstances of his removal from the NEA directorship disappointed Henderson, who sometimes wondered whether a democracy might ever manage a rational foreign policy. But on the whole he deemed his most recent assignment to Washington a success. At the time he returned to America from Baghdad in 1945, a tolerant view of the Soviet Union had still generally informed American foreign policy; by the summer of 1948 a far harder line prevailed. Henderson would not have claimed credit for the transformation. He would have argued that it was all Moscow's doing, in that the Russians were simply reverting to Stalinist form now that Hitler and the Japanese no longer enforced good behavior. Equally significant and encouraging, from Henderson's perspective, was the fact that the United States government and the American people were recognizing the possibilities for, and need of, a vigorous assertion of American principles in the world arena. Here Henderson would have acknowledged a somewhat greater role, since in the formulation and implementation of the clearest statement of American assertiveness—the Truman doctrine—he had played a signal part.

Henderson would not have admitted that the enlarged American sphere that was resulting constituted an empire. Nor did it, in the traditional sense. The United States did not exercise direct political control over the countries and peoples within its sphere. Yet to an ever increasing degree, Americans were coming to recognize the preeminent position the recent world war had conferred upon them, as it augmented their country's strength and undermined if not destroyed altogether that of potential rivals. Ten years earlier a statement like Truman's of March 1947 would have been inconceivable. Then the American people were running away from responsibility for international order, and five major military powers were in a position to contest an extension of the American reach. But the war had leveled Germany and Japan, enfeebled France, and bankrupted Britain. The war had knocked the Soviets to their

knees, costing Russia fifteen to twenty million lives and ravaging its economy—
although even on their knees the Russians appeared formidable. In the wake of
the war, American leaders and the American people found themselves in a position to remake much of the world in their image. They conceded the eastern
half of Europe to the Kremlin, but in the name of halting communism they
eventually endeavored to retain or establish—by economic and political means
if possible, by force if necessary—a predominant western influence nearly everywhere else. Call it what they might, in practice they produced an empire.

II

For strife and bloodshed, the Middle East in the late 1940s had nothing on the
Indian subcontinent. If anything, the communal mayhem between Hindus and
Muslims made the fighting between Jews and Arabs look like a parliamentary
exercise, and while the partition of Palestine cost thousands of casualties, the
sundering of South Asia produced a hundred-fold more. Just as the termination
of the British mandate in Palestine left the locals to fight over boundaries and
other details of sovereignty, so the collapse of the raj sparked a war between
the successor regimes in India and Pakistan. When Henderson arrived in New
Delhi after a long-deferred vacation and a stopover in London, the two governments remained at bayonets' points in the contested province of Kashmir, and
a resumption of fighting appeared likely at any time.

But if these parallels made Henderson feel at home, in one crucial respect
India differed from Palestine: almost nobody in America in 1948 cared a hoot
for India. There was no Hindu lobby—or Muslim lobby either, but the Muslims' absence had been evident in the Palestine fight. India was not the holy
land to any significant portion of the American people. Hitler's gruesome grasp
had not extended to the peoples of the subcontinent. India and Pakistan lacked
oil. They did not guard the path to any place particularly important. Communism posed no immediate threat to the region. In sum, neither from the heights
of Capitol Hill nor from the plains of the heartland could Americans see over
the Himalayas and the Hindu Kush.

The view from Foggy Bottom—to which the State Department moved in
1947—was hardly clearer. For a brief moment during World War II, American
diplomats had worried about the subcontinent. In the first months of 1942, as
the Japanese swept through Indochina and sacked Singapore, a near-revolution
broke out against British rule, with nationalists demanding that the colonialists
"quit India" at once. Aiming to bring peace to the British empire and, more
importantly, Indian troops to the battle against the fascists, Roosevelt urged
Churchill to compromise. But the "former naval person" told his esteemed ally
to mind his own business and proceeded to slam 100,000 of the nationalists into
jail. The mass jailing calmed the situation, which was what Washington really

wanted at the moment, and the Americans turned their energies to more-pressing matters.

After the war American sympathies lay with the Indian independence movement, but Churchill's fall from power in July 1945 deprived American anti-imperialists of their most likely villain. Clement Attlee could not compare, committed as he was to retiring the raj. Mahatma Gandhi's unorthodox style attracted some attention, but the parlous condition of Europe and China elicited considerably more. Millions of Americans had fought in Europe and the Pacific. With those regions they could identify. Who had fought in India?

At the policy level, American officials adopted a hands-off approach to the transfer of power and the issue of Indian partition. In the ensuing conflict between India and Pakistan, the Truman administration immediately opted for neutrality. Some, including Henry Grady, the American ambassador in New Delhi, thought the administration pushed neutrality to illogical limits when it refused to lend the Indians ten transport planes to evacuate Hindu refugees, on grounds that Karachi failed to concur in the request. But Washington held out against Grady's counsel, and it avoided being drawn into efforts to mediate between the two sides. Even as they picked up Britain's slack in the Mediterranean and Middle East, administration leaders believed South Asia ought to remain a British sphere of influence. When the British secretary for Commonwealth relations, Philip Noel-Baker, came to Washington at the beginning of 1948 to recommend an American initiative in the Kashmir dispute, and warned of a possible "holocaust" on the subcontinent, Robert Lovett replied that entangling the United States in Kashmir would distract the American Congress from rebuilding Europe and would draw "undesirable Russian attention."[1]

What Lovett only implied, Henderson stated explicitly in a memo to the undersecretary at just this time. Referring to consideration of the Kashmir issue at the UN, Henderson wrote,

The problem is one in which British initiative is clearly indicated not only in view of the strength of the British Delegation and the familiarity which the British have with the problems of the area but also because in essence the present situation is a further development in the evolution of the political problems connected with the British withdrawal from India.

A pronounced American initiative, on the other hand, would not only carry with it the danger of extending our already heavy commitments in various parts of the world but might also involve an American formula which would require making a choice between giving support to the interests of India or of Pakistan, a course which we have thus far assiduously avoided.

In other words, if someone had to crack eggs for a Kashmir omelette, better Britain than the United States.[2]

III

En route to New Delhi, Henderson stopped in London, where he attended a dinner hosted by the British foreign secretary. Ernest Bevin had just come from a Commonwealth conference at which the Kashmir question formed a principal topic of discussion. He was encouraged. Quoting the proverbial Irishman who had been on strike for six months, he said there existed "a danger of a settlement." Prime Minister Nehru had particularly impressed the gathering. Nehru had asserted that nationalism, not communism, was the great force in Asia, and that thanks to the enlightened policies of the British government, nationalism in India and Pakistan was oriented toward the west—in contrast to the situation in the Dutch East Indies, for example, where suppressed nationalism was turning toward communism. Bevin repeated Nehru's remarks to the effect that "when people are fighting for their independence they are inclined to make common cause with any ally—even the Communists. When they achieve independence they realize the falsity of Communism and fight against it."

Bevin dominated conversation at the dinner table, regaling the group with stories of his days in the trade-union movement. After dessert, however, in response to a question from Michael Wright of the foreign office regarding the prospects for continued Anglo-American cooperation, Henderson managed to slip in some observations. Henderson said he had every reason to believe Washington and London would be able to coordinate their actions in the future as they had in the past. Of course the transatlantic relationship would have its ups and downs, but this was to be expected in any alliance of equals. Bevin, sitting next to Henderson, patted him on the arm and said, "I like to hear that, Loy." When Wright queried how the Palestine question would affect the relationship, Henderson said he hoped that after affairs in Palestine calmed, matters between Washington and London would proceed more smoothly than in recent months.

Returning to the question of India, Bevin declared that he was delighted that Henderson would be the American ambassador in New Delhi. "There is a country where we must keep together," the foreign secretary asserted, "although you must let us be in the shop window." Henderson answered, "The United States fully realizes the importance of keeping the United Kingdom in the shop window, but you must keep us informed." Henderson remarked that the American people, and to some degree the American government, were easily distracted from the affairs of India. If London allowed coordination to slip, there would be a temptation in America to leave India to the Brits.

Henderson added that for all the importance of cooperation, a common front would only be effective if he was "100 per cent the American Ambassador." He must deal with the government of India in such a way as to avoid the least impression that he was a mouthpiece for Britain, or that Washington and London were colluding against India. Bevin leaned back and laughed. "I know exactly what you mean," he said. Referring to the British high commissioner in New Delhi, he stated, "If it is ever convenient for you to have a public row with Archy Nye, I'll be happy to play that game with you."[3]

On the same layover, Henderson made his first contact with an individual who would puzzle Americans and bedevil U.S.-Indian relations for years, V. K. Krishna Menon. Few people who observed Menon in action doubted his ability, but most found him extremely abrasive. An Indian colleague, C. S. Jha, perhaps best captured Menon by describing him as "an outstanding world statesman but the world's worst diplomat."[4]

Henderson would not have argued with the latter part of Jha's comment. When Henderson visited the Indian high commissioner's office in London to pay his respects, Menon rose from his desk without coming forward. "This is interesting," he said, accepting Henderson's outstretched hand only slowly. "You are the first American ambassador who has ever darkened my threshold." Speaking of Lewis Douglas, lately appointed by Truman to London, Menon continued, "Yes, Ambassador Douglas has not seen fit to call on me." Henderson apologized for Douglas, explaining that the press of various world crises had prevented the ambassador from making his courtesy calls. As for his own visit, he assured the high commissioner that the honor was entirely his. The interview otherwise included little of note, beyond Menon's parting shot that the time would come when American officials really would feel honored to be received by an Indian high commissioner.[5]

IV

Henderson's arrival in New Delhi coincided with a period of transformation in the cold war. In certain respects the lines between east and west were hardening. On the left side of the Elbe, the Marshall plan was beginning to revive and stabilize the economies of Western Europe. Across the river, on the leftist bank, the Russians were entrenching themselves further. In February 1948, Stalin's allies in the Czech Communist party, fortified by the presence of Red army tanks massed on the Czech-Soviet border, snuffed out the opposition led by the liberal anti-Stalinist, Jan Masaryk—and snuffed out Masaryk as well. In June the Kremlin responded to a decision by the United States, Britain, and France to fuse their three zones of Germany into a single entity and to commence economic reconstruction, by cutting off land access to the isolated outpost of West Berlin. Although Truman chose to answer the blockade with an airlift, rather than the armored column his more-belligerent assistants advocated, many observers believed general violence imminent. The commander of American forces in Germany, Lucius Clay, described "a new tenseness" in U.S.–Soviet relations and said war might come "with dramatic suddenness."[6]

But the diplomatic deck contained two jokers. The first was Yugoslavia. In June 1948 Tito challenged Stalin's claim to dictate to the world socialist community; for his temerity the Yugoslav boss was read out of the Cominform. American officials were quick to realize the significance of Tito's defection, and although political considerations kept them from openly embracing the tough, old, and still thoroughly communist dictator, they prepared to exploit this breach in the eastern front.

The second source of puzzlement was China. The official American line on Mao Zedong and his communist cadres was that they were in deep cahoots with Moscow, but a significant sector of opinion among the China-watchers in the State Department held that Mao might eventually turn out to be as fractious an ally as Tito was proving to be. In fact, one reason for the care the Truman administration displayed in cultivating Tito was that a judicious policy toward Yugoslavia might stimulate a similar split between China's communists and the Kremlin. As matters developed, the administration adopted a policy almost guaranteed to produce the opposite result—a consolidation of the Moscow-Beijing axis. But the hard line toward China owed principally to the outbreak of McCarthyism and the onset of the Korean war. In 1948 the future was more open-ended.

This future provided the central topic of one of Henderson's first conversations with Indian officials in New Delhi. Henderson visited the foreign ministry, where he spoke to the secretary-general for foreign affairs, Girja Bajpai. With a communist victory in China all but assured, Henderson asked Bajpai what he thought it would mean. Bajpai responded that as far as Sino-Indian relations went, he foresaw little immediate impact. Communism in China posed no threat to India. Problems might arise if communists elsewhere in Asia—in Indochina, for example—took heart from the establishment of a communist regime in China and succeeded in seizing power. But this did not seem likely in the near future. As for communism in India itself, his government had the situation well in hand. Barring a marked deterioration of the Indian economy, the communists would not generate sufficient support to mount a serious challenge.

In subsequent conversations Henderson discussed the same issue with Nehru and Governor-General Chakravarti Rajagopalachari. These two seemed even less concerned regarding a communist victory in China than Bajpai. Nehru made the point—which State Department officials knew but had scared out of them by the McCarthyites—that a communist China need not be a China subservient to the Soviets. If the noncommunist countries handled China correctly, Nehru averred, they might prevent an alliance from forming. Rajagopalachari went further, positively welcoming Mao as a successor to Chiang Kai-shek. According to the governor-general, a China under communist control would take a more Asian line than Nationalist China had, and therefore would make a more suitable partner for India. Henderson thought Nehru's and Rajagopala-chari's comments revealed "a certain smugness." The ambassador wrote, "There is apparently a feeling that China is destined to disappear for some time as a world force leaving India as the foremost Asiatic power, courted on one side by the capitalistic powers of the West and on the other by the Communist powers of Eastern Europe and Asia."[7]

China would form a staple topic of discussion between Henderson and Indian officials throughout his tenure in New Delhi. So would Kashmir, although this was not apparent at the beginning of 1949. At the moment, India and Pakistan had tired of fighting and were indicating interest in a ceasefire. A year earlier, Henderson had advocated leaving this Commonwealth problem to Britain; but then he had had Palestine and the rest of the Middle East to worry about. Now, with nothing but India—and the Soviet Union, as always—on his mind, he rec-ommended a more active role for the United States. In particular he suggested an initiative by which Washington would tell New Delhi and Karachi that it would lift the American arms embargo—imposed at the outset of fighting in 1947—as a demonstration of its confidence in the good faith and peaceful inten-tions of the two disputants. Without being so blunt as to link repeal to any specific development in negotiations, the United States should drop a broad hint that if fighting resumed, the embargo would too.[8]

Acheson, recently appointed secretary of state, agreed, as did Truman. But the leverage provided by the ending of the arms embargo proved insufficient to produce a settlement, and the ceasefire talks droned into the summer, with no resolution in sight.

In July, Bajpai invited Henderson to discuss what the secretary-general called a "serious matter." He said he was disturbed by recent reports that the Amer-ican government was blaming India for the lack of progress on Kashmir. Amer-ican officials at the UN seemed especially distrustful, although Bajpai thought skepticism in New York must have percolated down from Washington. Describing frankness as essential to good relations between their two countries, Bajpai asked Henderson whether the United States government did indeed deem India responsible for the present failure of the talks, and if so, to indicate what his country might have done differently.

Henderson denied that any feeling such as Bajpai described had crystallized in the State Department, though he conceded that it was only natural that doubts should arise regarding India's intentions. After all, Indian officials were not exactly going out of their way to be conciliatory. It was no secret that cer-tain influential circles within the Indian government preferred stalemate to a plebiscite, which was the most likely means of settling the problem. Anyone could see that delay worked in India's favor, since Indian troops controlled the most productive parts of Kashmir. Henderson went on to say that the absence of a settlement made it difficult for American officials to provide the support both Washington and New Delhi considered essential to India's economic development. It was, he said, "almost hopeless" to think about American aid so long as "the running sore" of Kashmir undermined India's financial position and threatened the stability of South Asia. Furthermore, despite the lifting of the arms embargo, the American government hesitated to make special efforts to expedite shipments of weapons lest they be used against Pakistan.

Bajpai thanked Henderson for speaking so directly. Only "complete frankness" would conduce to profitable relations, the secretary-general said. Henderson agreed, adding that he would inquire of the State Department regarding further details Bajpai would find useful. Bajpai said he would appreciate it if Henderson ensured that the two of them handle the matter themselves. He did not want the State Department to take up the issue with the Indian ambassador in Washington, Nehru's sister Vijaya Lakshmi Pandit, since he did not believe she would be as open to forthright discussion.[9]

Henderson did not know Pandit well, but he suspected that if she were anything like her brother, Bajpai was right. Henderson's relationship with Nehru formed one of the central, and one of the least fortunate, aspects of the ambassador's stay in New Delhi. The two men did not get along at all. Journalist C. L. Sulzberger, whose work entailed close contact with both, put the matter succinctly: "Henderson detested Nehru and Nehru knew it." Henderson had learned enough about the prime minister before arriving in New Delhi to have doubts regarding his character, but their first meeting went surprisingly well. "My initial impression of Nehru was favorable," he wrote. "I found his personality more sympathetic than I had anticipated." Before long, however, Henderson's views began to change. In July 1949 he wrote to Acheson recommending that either the president or the secretary send a letter to Nehru urging the prime minister to compromise on Kashmir. Henderson added a caution, though, that such a letter would have to be drafted with the utmost circumspection. Plain language any other politician would read without batting an eye would "mortally offend" Nehru, who was "morbidly sensitive to criticism which might reflect on his motives or good faith." The White House or the department must not avoid telling the truth—that would be worse than saying nothing at all—but Washington should exercise care "not to present the kind of statement which would invite an argumentative reply and might lead us into profitless wrangling."[10]

Two weeks later Henderson wished he had put his warning in stronger terms. He had delivered to Bajpai a note elaborating American thinking, and although the secretary-general had taken the message well, Nehru had not. The prime minister summoned the ambassador for a talk. As Henderson described the scene afterward, Nehru greeted him affably, but "his whole attitude changed when he began discussing Kashmir." The prime minister's "accumulated pique found expression in a characteristic emotional tirade."

He said he was tired of receiving moralistic advice from the U.S. India did not need advice from the U.S. or any other country as to its foreign or internal policies. His own record and that of Indian foreign relations was one of integrity and honesty, which did not warrant admonitions. So far as Kashmir was concerned he would not give an inch. He would hold his ground even if Kashmir, India and the whole world would go to pieces.

The Kashmir issue, Nehru said, went straight to the heart of the difference between India and Pakistan. His country was a secular, progressive democracy; Pakistan was a theocracy. India considered Kashmir primarily a political matter; the Pakistanis wanted to make it a religious crusade—which was why his government could not support a plebiscite. At the present time and under the conditions Pakistan proposed, a plebiscite would only arouse communal hatreds in both countries. (Nehru neglected to mention what Henderson knew: that most Kashmiris were Muslims and presumably would favor Pakistan over India if given a choice.)

"I was not entirely unprepared for the Prime Minister's tirade," Henderson commented. "This was but one of the irritating incidents which are inevitable on serious issues with a person of Nehru's temperament and character." Henderson continued, "I said nothing, remained calm and looked him straight in the face." After about ten minutes Nehru settled down and "turned on his well-known charm." He remarked that perhaps he had been over-forceful in his comments. Henderson replied that if the prime minister's statement reflected his true feelings, it was better that he had spoken directly. When Henderson offered to consider the discussion off the record, Nehru replied that he would let the ambassador decide for himself. In that case, Henderson said, he would convey the substance of the conversation to Washington, although he doubted he could be "as eloquent an advocate" of India's views as the prime minister. Henderson closed the interview by saying he hoped the prime minister would find the occasion on his upcoming visit to the United States to make India's case personally.[11]

The possibility of a visit by Nehru to the United States had surfaced some time earlier, but various crises and elections in both countries had combined to postpone the event until the autumn of 1949. When the idea first arose, India had been merely one, albeit the largest, of the countries rising from the wreckage of the colonial empires, and Nehru was simply the best-known spokesman for a middle path between east and west. By the time he arrived in Washington, however, events had conspired to confer heightened significance upon his appearance. The Soviets had just tested an atomic bomb, and the communists had gained their final victory on the Chinese mainland. Although the first development weighed more in the balance of strategic power, the latter carried heavy symbolic freight, especially as it related to India. At the moment when the American love affair with China was crashing to an end, India afforded an alternative; after their jilting by Mao, Nehru caught Americans on the rebound. American legislators hailed the arrival of the great man in the most effusive terms; American pundits praised the pandit as Asia's new savior.

Henderson warned against the administration's getting swept up in the enthusiasm. Nehru, he pointed out, did not reciprocate America's warm feelings, but deemed the United States a land of crass commercialism and undeveloped intellect. Sketching a psychological portrait for Washington's benefit, Henderson depicted Nehru as "a vain, sensitive, emotional and complicated person." In part, the prime minister's vanity reflected an enlarged ego, but it also indicated, at least to Henderson, "certain furtive doubts" about his ability. For many years Nehru had been the outsider, the critic, and the agitator. Now that he was on the inside and responsible for constructive work, he was not sure he could do it. Perhaps as a result of the Indian caste system, perhaps from his educational background, the prime minister was fully at ease only with those he considered his social and intellectual inferiors. "It seems that during his schooldays in England he consorted with and cultivated a group of rather supercilious upper-middle-class young men who fancied themselves rather precious. . . . He acquired some of their manners and ways of thinking." Britain had also imparted a scorn for and distrust of things American. The companions of his youth had taken pains to demonstrate the deficiencies of the American way of life. More recently the Mountbattens, the last viceroy and his spouse, had conveyed the notion that Americans were "a vulgar, pushy lot, lacking in fine feeling," and that American culture was "dominated by the dollar." Nehru appeared to accept the opinion that Washington possessed "an ulterior design for United States economic domination of the world." Henderson noted that the British campaign continued to the present. "When Nehru visits the U.K., he is worked on by a varied collection of notables ranging from Churchill to Laski and from Bernard Shaw to Ernie Bevin."

For all Nehru's faults, Henderson had to acknowledge the prime minister's better qualities. "In spite of his vanity and of his petty snobberies, he is a man of warm heart, of genuine idealism, of shrewd discernment, and of considerable intellectual capacity. He is also an expert politician and a natural leader." In light of these latter traits, Henderson believed Nehru's American tour might yield benefits to both countries—if the United States could "capture his imagination instead of getting on his English-strung nerves or of stirring his jealousy."[12]

Unfortunately, things did not work out as Henderson hoped, and the trip proved an undiplomatic disaster. To Assistant Secretary of State George McGhee, it seemed that Nehru arrived with a "chip on his shoulder toward American high officials, who he appeared to believe could not possibly understand someone with his background." McGhee also suspected that Nehru took offense at an overheard discussion between Truman and Vice-President Alben

Barkley on the virtues of bourbon. Acheson came away convinced that he and the prime minister were "not destined to have a pleasant personal relationship." The secretary later added that Nehru was "one of the most difficult men" he had ever encountered. Henderson commented that while the prime minister's American tour attracted "pacifists, friends of Asia and well-meaning cranks," to those of sounder judgment the prime minister came across as "wooly and evasive" and as making "no contribution" to solving world problems.[13]

Nehru found the visit equally unprofitable, and he left America with his prejudices reinforced. According to a CIA informant, evidently a person close to Nehru, the prime minister brought away an impression of Truman as "a mediocre man who, as a result of unexpected circumstances, had been placed in a role far superior to his capacities." He considered Acheson "equally mediocre" and the career officers in the State Department "uncertain, confused, superficial, too much inclined to improvisations and at the same time pretentious and arrogant." As for the American people generally, they were "elementary and material," wanting only "to eat and drink and live comfortably." Indian illiterates, Nehru believed, possessed "a greater propensity for culture."[14]

V

Returning from the United States in a nasty mood, Nehru was in no frame of mind to give ground on the issue that most separated India and the United States at the moment. While the prime minister was in America, the People's Republic of China proclaimed its existence. During the weeks and months that followed, the rest of the world faced the problem of deciding whether to recognize the revolutionary regime.

This question, of course, caught neither Washington nor New Delhi unawares. As early as May 1949 Henderson had discussed recognition with Bajpai. Henderson argued that the great noncommunist powers—here he tactfully included India—should not make the Chinese communists' victory easier by premature recognition. If and when it became clear that Mao's forces had actually consolidated their hold on China, there would be plenty of time to decide about opening relations. In the meantime there was no point adding to the Nationalists' troubles. In any event, the democratic nations should consult each other before making decisions. For various countries to compete with each other to be first to recognize, and to allow the Chinese communists to manipulate them in this matter, would damage the interests of all concerned. Bajpai agreed.[15]

During the summer of 1949 India held to the position that it would not act abruptly. At the end of June, when reports filtering back from China cast doubt on India's intentions, Bajpai responded to Henderson's request for clarification by reaffirming that his government would not move unilaterally on recognition or other issues relating to China. "It intends to keep in touch with other governments," the secretary-general said, "and see what they decide to do."[16]

By the beginning of September, however, Henderson sensed a weakening of this commitment to democratic solidarity. Nehru, it now appeared, would recognize as soon as the communists eliminated organized resistance. New Delhi would certainly consult with the British and other members of the Commonwealth, and would probably raise the issue with Washington as well. Depending on the circumstances at the time, the Indians might delay recognition momentarily at the Commonwealth's behest. But Henderson did not think they would tarry long, since Nehru would want to demonstrate India's friendly attitude toward its Asian neighbor. Supporting the prime minister in this matter was the Indian ambassador in Beijing, K. M. Panikkar, who had substantial influence on Nehru's China policy and who, Henderson believed, fancied himself a mover and shaker in the diplomatic community in Beijing.

As to how India viewed recognition of China by third countries, including the United States, Henderson remarked a spectrum of opinion within Indian elites. On the political right, a minority of individuals considered communists in China, as elsewhere, implacable enemies of the noncommunist world. This group looked to the United States and Britain to hold the ramparts against communism and therefore against recognition. Unfortunately, as Henderson assessed things, their number was small and their voice no larger. The political near-left, which Henderson deemed dreamily naïve, held an opposing view. "Indian Socialists, most of whom have not yet learned what European Socialists now know from bitter experience, will probably be more critical at withholding than at granting of recognition." As for the far left, the communists and fellow travelers, who though limited in number had "considerable influence particularly among half-baked and maladjusted 'intellectuals,'" they would object no matter what the United States and its allies did. At the center of Indian politics, in the Congress party and the mainstream press, an American decision against recognition would provoke more criticism than praise. Henderson added that should the United States find itself alone in non-recognition—if Britain and France recognized, while America did not—criticism would be "wider and more sharp."[17]

After the official birth of the PRC, the consultation Bajpai had promised turned out to be what Henderson had guessed: simply an exercise in informing the Americans and others of India's views. At the beginning of December, Bajpai said he was polling the members of the Commonwealth to determine their positions. A few days later he indicated that recognition would come soon. In telling Henderson this, Bajpai assured the ambassador that his government fully appreciated the danger communism posed to India and to Asia. He and his government were under no illusions regarding Beijing's and Moscow's intentions. But India would find itself in an embarrassing position if Burma and other Asian powers recognized, while India held back. Moreover, London was planning to recognize momentarily, and his government could not afford domestically to be seen as following Britain's lead. Finally, the secretary-general reiterated that recognition of China did not presage a turn toward communism. If anything, although his government formally maintained a policy of neutrality between the blocs, in practice it was increasingly making decisions more in harmony with American policies than with those of the Soviet Union.[18]

While Henderson understood that little he or the United States could do would prevent India from granting recognition, he felt obliged to advance the negative case. In a conversation with Bajpai's deputy, K. P. S. Menon, Henderson said his own experience with communists in Moscow and elsewhere had convinced him that they paid no heed to gracious gestures. They responded only to "displays of firmness." He asserted that the communists would be delighted at a break in the ranks of the free nations. They could be counted on to exploit this division "to their advantage and the disadvantage of the democratic world." Menon replied that his government did not underestimate the risks involved in its action, but it had to make its own decisions.[19]

With recognition by India nearly a certainty, Henderson opposed recognition by the United States. In some measure his opposition followed the same line of thought that had led him and the others of the Kelley school to reject recognition of Moscow. In particular he wanted to see proof that China would act as a responsible member of the world community. At the same time, his opposition reflected his new perspective as American ambassador in India. He contended that swift American recognition of China would do more harm than good to American relations with India—this in contrast to his earlier argument that American recognition would go down well with the Indians. At year's end, just as Nehru announced in favor of recognition, Henderson made his argument against. With more experience of India under his belt, he commented that although elements of the Indian press and many Indian officials would criticize the United States for withholding recognition, such persons would find fault with the United States on other counts, if not on recognition. And recognition of Beijing would not satisfy them for long. On the other hand, to recognize China when that country continued to display a "contemptuous attitude" toward the United States would simply make America appear "weak and vacillating." Furthermore, to reverse course and follow India's example on the China question might momentarily gratify Nehru, but it would "undoubtedly cause the Government of India to give less weight in the future to our views on international problems, to have less respect for our judgment of the international situation, and to create doubts about our tenacity of purpose." Finally, if China by its actions demonstrated a desire for normal relations, the United

States could recognize at a later date without the sacrifice of principles immediate recognition would entail. On this subject, haste implied uncertainty.
Deliberateness connoted strength.[20]

Henderson's arguments hardly decided the recognition issue for Washington;
the current was already flowing strongly against Beijing. But his comments
added to the force of those counseling caution. In February 1950 Senator Joseph
McCarthy launched his assault on the loyalty and judgment of the State Department, which combined with previous allegations of incompetence and appeasement in allowing the "fall" of China to incline the administration toward even
greater circumspection. When the Korean war broke out in June the possibility
of recognition vanished entirely.

While congressional anticommunists were making big noises in Washington,
Henderson set off a few firecrackers in New Delhi. As a rule Henderson tried
to stay out of the public eye. After his experiences on the Moscow desk and in
NEA, he recognized that career officers who find themselves at the center of
attention usually find themselves at the center of trouble. As a result, he only
rarely took it upon himself to explicate American policy in open forums. But an
ambassador in a country as relatively off-the-beaten-track as India could not
decline all invitations to speak and in March 1950 Henderson addressed the
Indian Council on World Affairs.

He began by conceding the Eurocentric emphasis of American foreign policy. It was the result, he said, of the European origins of most American citizens
and of the easier accessibility, especially until recently, of Europe as compared
to Asia. For many years only Europe figured significantly in American diplomacy. Of late, however, and "by force of circumstances," the United States had
assumed a global role. Asia now mattered greatly to America. Yet Asia
remained just one part of a world-wide network of interests, and American
officials had to weigh their actions and designs in Asia against those in other
places.

Henderson acknowledged the existence of inadequacies in the American
approach to Asia. These he attributed partly to a deficiency of knowledge and
understanding in Americans regarding Asian affairs. But he also asserted that
American efforts to assist countries like India were hampered by uninformed
and unfair criticism of American motives by Asians.

> There are, for instance, still many people in Asia who are sincerely convinced that
> efforts on the part of the United States to extend technical or financial aid are
> prompted not by a desire for a peaceful, orderly progressive world but by some kind
> of economic imperialism. There are other people who really believe that action
> taken by the United States through the United Nations or through other channels
> for the purpose of effecting peaceful settlements of disputes and dissipating hatreds
> and rivalries are motivated by great power politics and selfish considerations.

To the ranks of these honestly misguided souls, Henderson added others:

influential groups who apparently do understand and appreciate the objectives of the United States with regard to Asia, but who shrink from close cooperation with the United States lest such cooperation create hostility toward them on the part of powerful forces of the world which feed on human poverty and suffering, which rely on force and terror to achieve their ends and which look with disfavor upon any association of free nations that might be effective in overcoming poverty, liquidating strife, or discouraging aggression.

Having insulted most of his audience as either ignorant or cowardly, Henderson proceeded to imply that India's position on Kashmir was at least partly irrational. Without denying the political, economic, and strategic aspects of the Kashmir question, Henderson argued that the issue might already have yielded to a solution but for the existence of national and religious rivalries, some that had a "historical basis"—and therefore were more or less legitimate—but others that were "of a comparatively recent artificial creation." Under the influence of such rivalries, persons tended to "lose all perspective and fail to act as rational human beings."[21]

Henderson must have anticipated that his address would cause a ruckus. It did. From left and right, Indian commentators assailed the American ambassador for arrogance, insensitivity, and most of the other bad traits Indians have often expected to see in Americans. But Henderson had heard worse; by this time he was nearly immune to criticism of his judgment on the communists—the "powerful forces of the world which feed on human poverty," and so on. He believed that just as Americans until about 1947 had needed to have their eyes opened to the dangers the Kremlin and its agents posed to world order and peace, so now the Indians did. Having served as a scapegoat before, he was willing to take the brunt of Indians' displeasure. He thought it would serve a useful purpose to let them blow off steam. "Although the cutting criticism and the type of jeers to which Indians, particularly those with a thin veneer of British culture, are addicted are not pleasant, it is probably preferable that they frankly and publicly give vent to their pent-up resentment rather than remain silent while seething inwardly."

Henderson noted that the charges against him were really charges against America. Among them he listed

our treatment of American negroes, our tendency to support colonialism and to strive for the continued world supremacy of white peoples, our economic imperialism, the superficiality of our culture, our lack of emotional balance as evidenced by our present hysteria in combatting Communism and our cynical use of "witch-hunting methods" in promoting domestic political ends, our practice of giving economic and other assistance to foreign peoples only when we believe such assistance will aid in our struggle against Communism, our assumption of superiority merely because we have higher standards of living, our hypocrisy, etc.

Some of these complaints, Henderson believed, should be ignored, representing nothing more than trivia inflated for domestic Indian consumption. Others he took seriously, as reflecting either genuine deficiencies in American society or real problems of communications between India and the United States. In the former category, he conceded that America had far to go in the field of race relations—although he never would appreciate how deeply the race issue affected Indian opinions of the United States. In the area of poor communications, Henderson remarked that Washington's failure to match its largesse toward Europe with commensurate assistance for Asia seemed conclusive evidence to Indians that America did not care about the welfare of the Indian people. Such aid as the United States did appropriate for underdeveloped nations—the Point Four program, for example—commonly drew criticism as "a grandiose, empty gesture made for the double purpose of convincing undeveloped peoples of United States interest in them and of gaining extensive influence over such peoples at relatively little cost." In this context many Indians were convinced that America was "engrossed in a world-wide struggle for the maintenance of the system of free enterprise" and in the process was attempting to subvert India's attempts at national economic planning.

Henderson described Indian resentment at the United States on these issues and others as so widespread as to be undeniably unrehearsed. Disappointingly, there was little Washington could do to alleviate the trouble. "We have not arrived at the present stage of our relations overnight," he observed. Nor did the situation appear likely to improve soon. Some of Nehru's opponents contended that if India wished to develop economically, it must not alienate countries, especially the United States, that could provide assistance. But this group was small and lacked influence. Even as he pointed out the existence of "islands of sincere friendliness" toward the United States, Henderson felt compelled to offer a warning. "It would be a mistake to dismiss the rising tide of Indian unfriendliness as a temporary phenomenon. This tide is likely to rise higher as Indian economic and political difficulties increase. Even if world events later should check its flow and cause its recession, the after-effects will probably be felt for years to come."[22]

14

Their Finest Hour

The tense relations Henderson described in April 1950 grew tenser still with the outbreak of the Korean war two months later. Following by hardly half a year the communist victory in China, the North Korean invasion seemed to Americans further evidence of the inherent aggressiveness of Marxism-Leninism. At the same time it provided the occasion for another major step in the construction of the American empire. In 1949 the United States had sponsored the formation of the Atlantic alliance, which pledged American lives and resources to the defense of Europe. NATO established something less than an American protectorate over the western part of the continent, but in conjunction with America's continued dominance of the world economy, the alliance gave Washington enormous influence over the actions of its allies—as the Suez crisis of 1956 would demonstrate. Yet while NATO, as befitted the first American venture in peacetime alliances in a century and a half, was narrowly and circumspectly drawn, the principle of collective security on which it rested had far broader applicability. Taking their cue from the failure of appeasement during the 1930s, American officials—with Henderson in the vanguard—argued that countenancing communist aggression anywhere endangered peace everywhere. Consequently the North Korean invasion was a test of America's resolve. The Korean peninsula, while not strategically insignificant, mattered far less from a military perspective than from the standpoint of America's credibility. Would the United States really support collective security, when support required American blood? By answering the question affirmatively, the Truman administration brought together two strands of American foreign policy—the globalism inherent in the Truman doctrine, and the willingness to use force the commitment to NATO implied. In doing so, the administration set the United States firmly on the path to acting as the policeman of the noncommunist world.

Henderson could conceive no better role for America. Since his assignment to Moscow in the 1930s he had been arguing that communism posed an abiding danger to the principles the United States stood for. Since he took charge of NEA in 1945, he had been calling for an expanded American policy of opposition to the Soviet Union and its agents. To Henderson, the Korean war stripped the issues of international politics to their fundamental form, or at least as near to fundamentals as they would ever get in a messy world. Admittedly, Syngman Rhee did not embody democracy, and perhaps his troops had committed some provocation. But the essential facts were straightforward. The communists had undertaken aggression in an effort to force their system upon a people who did not want it. In going to the aid of the South Koreans, the United States acted out of no narrow self-interest. American military leaders had repeatedly warned against a land war in Asia; Dean Acheson had declared Korea beyond the American defensive perimeter. In sending its sons to fight and die in Korea, the United States was vindicating the idea that small countries as well as large ought to be free from the threat of domination by those bent on conquest. To Henderson's thinking, Americans had never behaved more nobly.

II

Indian leaders had their doubts. Henderson encountered India's reservations during the first days of the fighting in Korea, as he sought to secure Indian support for UN action to halt the North Korean advance. As soon as the UN security council took up the Korean question Henderson called on Bajpai to explain that Korea was not simply a local problem but "a UN and world problem." The response of the international community to the situation in Korea, he said, might well determine the course of global events far into the future. The disruption of world peace by the North Koreans was "very serious." To allow the aggression to go unopposed would be "still more serious."[1]

Bajpai told Henderson that the Indian government found the Korean issue "extremely difficult." On one hand, India desired "to throw its moral weight against aggression." On the other, Indian approval of the security council resolution to send troops to Korea "might initiate a chain of events which would have unfortunate consequences in Asia." Citing one possibility, the secretary-general remarked the "most precarious" position of Burma with respect to China. It was common knowledge that Chinese Nationalist forces had found refuge in Burma and that Beijing was looking for an excuse to root them out. For India to ask the Burmese government to adopt a belligerently anticommunist posture would put Rangoon on the spot, leaving Burma to choose between following India's advice and protecting its own shaky sovereignty. Further, Washington had muddied matters considerably by linking the Korean situation to troubles in Taiwan and Indochina. Truman's announcement of naval patrols of the Taiwan strait and increased aid to the French in Indochina made the defense of Korea appear an exercise in imperialism.[2]

Henderson, having received unnecessary encouragement from Washington to "exert maximum influence" on New Delhi to gain India's support of the UN resolution, replied with an impassioned defense of the American position. He told Bajpai he had never been "more proud of being a servant of the Government of the United States" than he was at the present moment. "After having just gone through one terrible war, the United States, in spite of its ardent desire to remain at peace, had taken a courageous step for the purpose not of defending its own territory but of showing aggressors and the world at large that it took its UN obligations seriously." Did India take its obligations as seriously? Would Indians in years to come look proudly on their government's actions if it failed to take a similar stand? Henderson suggested that "hesitation and wavering at this historic moment" would encourage aggressors to continue with "an aggressive program which would inevitably result in world war." By contrast, "positive and speedy action" by India would have a "tremendous influence among all the peoples of Asia." The choice, he perorated, was not between two ideologies or two alliance systems, but between aggression and the ideals of the UN.[3]

The following day, Henderson raised the matter with Nehru directly. To Henderson's surprise he found the prime minister "exceptionally friendly and understanding." After listening to Henderson's explanation of American policy, Nehru said he did not fully agree—like Bajpai, he mentioned Taiwan and Indochina—but considering the pressure of events, he could not be "over-critical." Even so, the ambassador must realize the constraints under which his government operated. He and his colleagues had already come under attack as "tools of the Anglo-American imperialists." Although this factor did not determine his government's actions, neither could it be ignored. He described India's attitude of nonalignment between the two superpowers, an approach it shared with Burma and Indonesia, and he said his government must be careful not to act without consulting India's neighbors. Finally, even if India sided with the UN in Korea, it could provide only moral support; his government had no military forces to spare.

Henderson assured the prime minister that moral support would be just fine. At a time when Asian opinion had yet to solidify on the Korean issue, Indian backing for the UN would "immeasurably strengthen" the cause of peace. On the other hand, should India, "the most powerful and influential free country of Asia," fail to demonstrate that it took collective security seriously, the principles of the UN would become "meaningless" as far as Asia was concerned.[4]

In the event, Nehru did choose to throw India's support behind the UN. The Indian government approved the security council resolution, although New Delhi simultaneously declared that this action did not connote any change in India's basic foreign policy.

Henderson learned several hours later that an incipient split in Nehru's cabinet had been largely responsible for the decision. Bajpai, whose views were sometimes not far from Henderson's, had threatened to resign if Nehru delayed further in announcing for the UN. The secretary-general told the ambassador that the rest of the cabinet had accepted the decision with relatively little debate, despite a feeling that this move might lead to a shift in India's international position. His government hoped it could continue its policy of pursuing friendly relations with all countries regardless of ideology, but only time and the reactions of the great powers would determine whether this was possible. For his part, Bajpai said he aimed to refocus India's policy, away from one embracing "nonalignment" and toward one more accurately characterized as "independent"—a policy motivated "solely by India's ideas and objectives." Regarding his influence in getting Nehru's approval of the UN measure, Bajpai said in a not unsatisfied tone that he had had a good day, that he had made a "notable contribution to world peace."

Henderson reported this exchange with pleasure, although he cautioned Washington against premature optimism. He told Acheson that while Nehru's decision marked "a distinct step forward," it would be a mistake to assume that he would automatically take further steps in the same direction. The prime minister remained deeply opposed to the United States on Taiwan and Indochina. "It is not impossible that he will give vent at some appropriate or inappropriate time to his feelings by a critical outburst."[5]

Through the summer of 1950, Nehru took the position that the UN should concentrate less on throwing back the North Koreans than on preventing a wider war. To this end, New Delhi advocated bringing the Russians and the Chinese into the Korean debate. In the first part of July, Bajpai told Henderson he had directed the Indian ambassador in Moscow to urge the Soviets to return to the security council, which they had left to protest the council's refusal to seat Beijing—causing their failure to veto the UN resolution for help to South Korea. At the same time, the ambassador was to tell Soviet officials that India supported the Soviet stance on China. In addition the ambassador should attempt to persuade the Russians to use their influence with the North Koreans to get the latter to call off their offensive and withdraw across the thirty-eighth parallel. Finally, he should pass word that India would be willing to mediate in efforts to bring the United States and the Soviet Union together to discuss Korea and other matters of common concern.

The Russians had rejected the plan. Bajpai said Moscow's deputy foreign minister, Valerian Zorin, had replied that the Soviet Union would not return to the security council except accompanied by Communist China, that his government would not negotiate with the United States while Americans were killing Asians, and that it would not accept Indian mediation because India, having approved the security council resolution, was not neutral in the matter. At this point in his conversation with Henderson, Bajpai expressed disappointment at the failure of the Indian ambassador to rebut Zorin's remarks, not only about India's unneutrality but about Americans killing Asians. After all, the North Koreans had started the killing and continued it still. Besides, it surely afforded little comfort to dead South Koreans and their relatives that the killers were fellow Asians.[6]

This meeting confirmed Henderson's belief that in Bajpai he had an ally on the Korean issue. Consequently he read with concern an article in the *New York Times* that threatened to damage the alliance—not to mention his more tenuous position with Nehru. A few days after the Indian government announced its approval of the security council resolution, *Times* reporter Arthur Krock gave Henderson credit for pressuring the Indian government into voting as Washington wished.[7]

"Bajpai was deeply irritated," Henderson commented after speaking with the secretary-general. "I am sure Nehru is furious." Bajpai had denounced the Krock article as undermining the common cause of India and the United States by playing into the hands of leftists and others in his country who blasted the Indian government as a tool of the Americans. Henderson agreed that reports like Krock's did neither side any good. He said that of course he understood that American pressure had nothing to do with India's decision—that approval of the council's resolution reflected only the "merits and logic of events." He assured Bajpai that he too was "deeply distressed" at the implications of the article. But he could see no easy answer to the problem. American papers printed more or less what they pleased, and for the administration to issue a formal denial might simply draw attention to the matter and compound the difficulty. Bajpai granted that the ambassador was probably right, but he could not help adding that it was a source of continual exasperation that the American press gave credit—or blame, for that matter—for everything that happened anywhere, to the United States.

After further reflection, Henderson decided that something had to be done to curtail the damage to U.S.-Indian relations. He suggested to Acheson a news release to the effect that any coincidence between the ambassador's talks with Indian officials and India's announcement in favor of the resolution was entirely accidental. India made its decisions on its own, guided by events rather than American arm-twisting.[8]

A few days later a State Department representative took advantage of a news conference to make essentially the statement Henderson suggested. The affair blew over. "I do not believe that any great harm has been done," Henderson remarked shortly afterward, "although at the time I was extremely concerned."[9]

III

As Henderson returned to Bajpai's good graces, the secretary-general acted as a conduit for messages between China and the United States. Bajpai showed

Henderson cables from Ambassador Panikkar in Beijing, and on occasion he provided Henderson copies. Henderson recognized the value of these communications, but he also understood that they placed him in an uncomfortable position. India, in the eyes of many Americans, was less than reliable in its opposition to communism, and its continuing efforts to mediate between the west and the communists, especially at a moment when communists were shooting Americans, seemed downright appeasing. The McCarthyist attack on the State Department had recently begun, and while Henderson did not hesitate to risk criticism over issues he believed in, such as the need for vigilance against the Soviets, and the dangers of backing Zionism, he had no intention of being smeared for something he did not believe in—in this case, Nehru's efforts to play peacemaker.

"I am somewhat concerned," he wrote Acheson, "lest I find myself becoming gradually involved in a GOI [Government of India] effort of mediation." Henderson told the secretary of state he had made clear to Bajpai that such remarks as he offered in response to Panikkar's cables were strictly personal and did not represent official American replies. But he was, after all, the American ambassador, and inevitably his comments carried weight. "I hope the Department will indicate at once if it does not approve of what I have done so far. If it believes that it would be wiser for me to pass on further approaches by GOI without so much comment, I would be grateful it it would so inform me."[10]

Acheson preferred that Henderson keep the lines of communication open, and during the rest of the summer the ambassador continued to eavesdrop on Beijing and New Delhi, kibbitzing when Bajpai requested an opinion. Although this afforded him considerable insight into India's intentions, he suspected he was not hearing the whole story. He was reasonably certain that Bajpai knew more than he was revealing, and he guessed that Nehru was keeping the secretary-general himself in the dark on certain matters. "My present impression," he wrote in the middle of July, "is that Indian policy at this juncture is somewhat confused and working at cross purposes." Bureaucratic politics aside, it seemed that the Indian government's overall approach to the Korea problem evidenced a "confused state of mind" on the part of India's leaders. Their support of the security council resolution indicated a desire to be counted on the side of collective opposition to aggression, but their simultaneous attempts to bring the Soviets and the Chinese to the bargaining table—"even though such mediation might involve appeasement"—reflected an inclination to curry the favor of the communists.

Henderson suggested to Acheson that a letter from the secretary or President Truman reiterating the philosophical basis for American actions in Korea and elsewhere in Asia would have a beneficial effect. "A message of this kind might disturb Nehru," Henderson granted, "because he would learn from it that some of his ideas about mediation are not likely to bear fruit. On the other hand, it seems to us that he is entitled to learn from the highest quarter more about our position and what our intentions are than he apparently knows at present." For such a message to produce the maximum effect, it should be sent soon. And it must be very closely held. Any leak could be "disastrous."[11]

Acheson approved the suggestion and directed Henderson to produce a draft. Henderson's note, the secretary indicated, should also serve as a reply to a letter from Nehru calling on the United States to support the admission of China to the UN, the return of the Soviet Union to the security council, and the initiation of talks among the Americans, Chinese, and Soviets.

Henderson prepared a detailed brief for the American position on these matters. The United States, he explained, opposed China's admission to the UN as "improper if not immoral," on grounds that Beijing continued to flout the principles of the UN charter by engaging in activities calculated to foment armed uprisings in other countries and hatred between nations. Regarding the Soviets, they could reassume their seat whenever they wished, but because it had been their decision to leave, they—or the Indians, for that matter—could hardly expect the United States to beg them to return. As for negotiations with the Russians and Chinese, whatever merits such talks might have with respect to international affairs generally, they would cloud the Korea issue. "There is for the present only one solution to the problem of Korea, and that is for the North Koreans to withdraw from the Republic of Korea, and if they do not withdraw for them to be driven out by the combined forces of loyal members of the United Nations."[12]

Acheson edited Henderson's letter for length, to match Nehru's relatively short note, and he softened the tone somewhat to prevent a new round of stale argumentation, but he let stand the essential thrust of the message, and he authorized Henderson to deliver it.[13]

Henderson did not expect to convince Nehru, and he did not. The American and Indian governments continued to differ. Through the middle of September the two sides had little constructive to say to each other. Henderson's summary of a conversation with Bajpai in August captured the mood. "He listened to various points as outlined by me rather listlessly. When I asked if he would like to have notes made of them he replied in the negative. If he had notes he would feel compelled to discuss them with the Prime Minister and he did not believe the matter was worth pursuing to that extent."[14]

IV

The September 15 American landing at Inchon, which immediately reversed the momentum of the war, injected new life—and fresh problems—into U.S.–Indian relations. The marines had hardly hit the beaches before James Webb, the acting secretary of state, cabled Henderson to say that because the war had reached a critical stage, it was of the "utmost importance" for the Chinese to know they must stay out of the fighting. Accordingly, Webb instructed Henderson to approach Bajpai with a request that the secretary-general forward this warning to Panikkar and thence to the Chinese leadership. Webb added, needlessly, since Henderson could figure this out on his own, that the message would carry greater weight if relayed as representing India's own views and not simply those of the United States.[15]

The New Delhi–Beijing pipeline, of course, transmitted messages in both directions. No sooner had Bajpai agreed to pass along the American warning than he handed Henderson a cable from Panikkar. At this time, the Indian ambassador expressed no particular alarm. In Panikkar's conversations with Zhou Enlai, the Chinese premier had reiterated his country's peaceful intentions, leading Panikkar to predict that China would not interfere in the conflict to rescue the North Koreans. The Chinese, Panikkar remarked, had seen what American bombing had done to Korea's cities, yet they were making no attempts to fortify their own cities or educate their people in civil defense. From this negative evidence Panikkar concluded that direct Chinese participation in Korea seemed "beyond the range of possibility," unless the Soviet Union intervened and a third world war erupted.[16]

As the UN forces approached the thirty-eighth parallel, however, the Indians began to get nervous. Bajpai, denouncing "loose, harmful talk" of an invasion of North Korea, told Henderson he hoped the counteroffensive would capture the North Koreans below the border. Although Bajpai reaffirmed that India attached great importance to a free and unified Korea, he said his government believed that extensive military action north of the frontier would hinder the cause of peace.[17]

Four days later, Bajpai explained to Henderson that the prospect of a UN invasion of North Korea posed a most difficult problem for India. He thought that there existed real danger that the Chinese would enter the fighting if UN forces crossed the parallel. "A world war might result," he said. Reminding Henderson that the original security council resolution had called simply for the expulsion of the North Koreans, he suggested that once the invaders had fled, the military phase of the conflict should be considered over. Negotiations might then begin.

Speaking without specific instructions, Henderson predicted that the UN would not settle for such an anticlimactic result. The ambassador believed the "general opinion of the free countries" was that the artificial division of Korea had become intolerable. Only a plebiscite could determine the future of the country, and since the communists had steadfastly refused to participate in any such referendum, or even to allow UN officials into their territory, there seemed no alternative to force. Did Bajpai have a better idea? In calculated exasperation, Henderson added that countries—he did not have to identify India by name—whose citizens were not dying in Korea appeared to find it easier to ponder the situation "leisurely and philosophically." In the United States, casualty lists were lengthening and feelings were running high.

Bajpai responded that he could understand that Americans might view the Korean situation more emotionally than Indians did. Indeed, he admitted, countries like India owed a debt to the United States for bearing so much of the burden of fighting. Even so, with the future of Asia and the world in the balance, this was no time for emotion. The UN would be making a serious mistake to send its forces across the border without giving the matter a thorough airing.

Henderson was surprised, and therefore encouraged, that Bajpai did not reject outright the idea of an invasion of North Korea. Characterizing the secretary-general's attitude as "extremely cautious," Henderson commented to Acheson that it indicated that Nehru had not made up his mind on the subject. The ambassador found it hard to believe that the prime minister would actually support an invasion. At this stage Henderson could only guess whether the Indian delegate at the UN would vote no or simply abstain. Henderson suspected that Nehru would want to retain his freedom "to criticize the U.S. for developments which might occur if UN forces should enter North Korea." Nor did Henderson put it past the prime minister to exploit this moment of confusion to complicate matters further by reintroducing the issue of China's representation in the UN.[18]

Nehru's intentions soon grew more evident. Bajpai told Henderson that his government agreed with the position publicly taken by the Americans that the security council resolution of June granted UN forces the right to enter North Korea. But a narrow legalism in this case might obstruct the pursuit of peace, and therefore India recommended that the UN, before sending troops across the parallel, call upon the North Koreans to stop fighting and accept a plebiscite. Anticipating the obvious objection, Bajpai suggested placing a strict time limit on the offer so the communists would not simply procrastinate and regroup. As to a specific resolution authorizing the crossing of the parallel, the secretary-general said his government would not oppose it. But whether it would vote affirmatively or abstain depended on future developments.[19]

Henderson thought Bajpai's statement marked a "distinct shift" for the better in India's position, but before the ambassador could learn what this shift might entail, new reports from Beijing changed the situation once more. On October 3 Panikkar informed Bajpai that Zhou had said China would enter the war if American forces crossed the parallel. The Indian ambassador interpreted the Chinese announcement as entirely honest, declaring that the Chinese decision was final and that if the Americans ignored Zhou's warning, the war would certainly widen.[20]

A grim Bajpai asked Henderson for his thoughts on this latest development. Henderson acted unimpressed. He replied that the Chinese démarche lacked "legal or moral justification." He added that it cleverly played on the divergent

views among the democratic nations and put the UN in the unfortunate position of apparently having to choose between losing the military initiative,
gained at great cost, and risking a world war in "extremely unfavorable circumstances." Henderson rejected Bajpai's earlier suggestion of an offer of a ceasefire
leading to a plebiscite, contending that however little time they were given, the
communists would use the respite to reinforce themselves. Besides, once the
UN troops stopped at the border, a subsequent decision to enter North Korea
would be nearly impossible politically. The communists would portray such
entry as an invasion, and much of the world would accept the portayal.

Other costs, Henderson continued, made halting at the parallel even more
undesirable. The international community might have to garrison South Korea
indefinitely, and the military initiative in Asia would rest with the enemies of
peace. If China's threats prevented collective action in the defense of Korean
freedom, the communists would adopt similar tactics toward other small
nations. Should the world allow a situation to develop where joint efforts to
defend weak countries foundered in the face of threats by the great powers,
"then it might as well be frankly admitted that all efforts to preserve peace by
means of collective security had been in vain and that international relations
were to be governed by force—not by any code of international morality."[21]

Bajpai took this lecture well enough, but the strain of events told on both
men. The next day the secretary-general read a cable from Panikkar describing
Zhou's approval of a recent statement by Nehru, which placed principal blame
for the destruction of Korea's cities not on the communists but on the "saviors"
of Korea—implicitly, the Americans. Henderson responded with an atypically
sarcastic comment, provoking an instantaneous reaction. As he described it:
"Bajpai, who is hot tempered, flushed and told me my remark was unnecessary—that it appeared to reflect on Nehru and it was his duty to defend his
Prime Minister." Although Henderson apologized, he told Bajpai he found it
hard to remain placid in the face of Nehru's "sometimes subtle and sometimes
openly vindictive campaign against the United States," which was the more
damaging because Nehru was not only prime minister of India but the outstanding leader of Asia. Henderson went on to comment that he personally found
Nehru's statement disheartening because of his own "deep admiration and
respect" for the prime minister. Bajpai expressed regret for losing his temper.
The secretary-general said he found himself in a difficult spot, in that he had to
defend his chief even when he disagreed with him.[22]

The ultimate source of the tension, of course, was not Nehru but China—
precisely, the imponderable question of Chinese intervention. More precisely,
as Webb put it to Henderson: "It is not a question whether the Chinese Communists intend to intervene in the conflict"—since, Webb pointed out, Beijing
had been providing political, logistic, and military support to North Korea from
the beginning—"but only of the degree of their intervention." Both the Indians

and the Americans feared China's entry into the fighting, but they differed on the likelihood of such entry. The Indians took Zhou's threats at face value and considered the Americans reckless for tempting fate and risking escalation. The Americans thought Zhou was bluffing and considered the Indians irresolute or worse.[23]

The Americans also wondered whether they were getting the straight story from Panikkar. Livingston Merchant of the State Department's Far East division, speaking from personal acquaintance, described Panikkar as exhibiting an "innate sympathy for the Chinese Communists" as well as being "unconcealedly contemptuous of the white races, particularly the Americans." Henderson commented afterward that regardless of China's intentions, Panikkar "would have been glad to win Brownie points with the Peking Government by warning the United States not to cross the Thirty-eighth Parallel." Henderson added that in light of American suspicions of Panikkar, if the Chinese had sincerely desired to warn off the Americans they might have chosen a more credible messenger than the Indian ambassador. At the time, Henderson's doubts about Panikkar were reinforced by some comments of Bajpai's indicating that the secretary-general himself was wondering how far to trust the ambassador in Beijing, who owed his appointment to political connections to Nehru. Having suffered through the tenure of Joseph Davies in Moscow, Henderson found it easy to appreciate Bajpai's uneasiness.[24]

In view of the importance of clear communications between Washington and Beijing, Webb suggested that Henderson make contact with the Chinese ambassador in New Delhi. Because of the absence of diplomatic relations between the United States and China, the undersecretary recommended that Henderson ask Bajpai to arrange an introduction.[25]

Henderson agreed to raise the matter with the secretary-general, although he did not expect Bajpai or Nehru to be very helpful. For one thing, he said, the Indian government would not want to risk the embarrassment of a Chinese rebuff. For another, "it rather enjoys its present monopoly on communications."[26]

Henderson's request caught Bajpai by surprise. The secretary-general said he would have to think the matter over. After discussing it with Nehru, he consented to approach the Chinese ambassador. But when Beijing replied coolly he dropped the idea at once. With the collapse of this half-hearted effort at direct communication, the last opportunity to avoid a major collision between the United States and China vanished.[27]

V

The collision occurred at the end of November, when 300,000 Chinese troops swept into the camps of the UN allies, rolling up those parts of the front they did not demolish altogether and capturing the units they did not annihilate or scatter. The communist counteroffensive stunned Washington, leading Truman to assert that the United States was considering using nuclear weapons to stem the Chinese advance. The president went on to say that the decision on use rested with the military commander in the field.

The White House quickly corrected the latter statement. The decision regarding atomic weapons remained with the president, Truman's spokesman declared. But even the thought that Douglas MacArthur had his finger on the nuclear trigger sufficiently frightened the Indians that they could barely spit out their told-you-sos to Henderson and other American officials. "No! No! No!" screamed the *Times of India*. Nehru, learning that the crisis had prompted Britain's Attlee to fly to Washington, hinted that an Asian representative, namely the Indian prime minister, ought to do the same. Not surprisingly the State Department suggested that although it appreciated Prime Minister Nehru's concern, another time would be more convenient.[28]

On December 5 Henderson talked with Nehru for half an hour. The prime minister's Congress party had scheduled a debate in the Indian parliament regarding the world situation for the following day, and Henderson wanted to explain the American position before Nehru spoke. The Chinese offensive, the ambassador said, had complicated matters, but it did not alter the essential issue. Korea remained a testing ground for collective security. The American purpose there was "to discourage aggression, not to advance any selfish U.S. interest." In the pursuit of world peace the United States had expended much "blood and treasure." Now more than ever, peace was under attack. Americans did not want another general war. Although the United States was acting strictly in accord with the directives of the UN, in some member countries—Henderson did not need to specify—the press and public officials were criticizing American actions, while saying next to nothing against the Chinese. The Indian parliament was about to consider the matter. "Would the substance of the debate be of comfort to, and encourage, the forces of aggression? Would debaters under the leadership of the Congress Party concentrate on criticizing the United States for not following a Far East policy to India's liking and overlook the fact that Communist China with Soviet backing was openly attacking the forces of the UN?" The ambassador deeply hoped the prime minister with all his influence would do what he could to prevent the debate from following such lines.

Nehru replied that India was a free country with a tradition of active political debate. He could not take responsibility for everything spoken in parliament. Nevertheless he anticipated that the speakers would display restraint. The gravity of the situation called for constructive discussion, not efforts to place blame for past mistakes. He went on to say he was "sorely troubled as to what could best be done to prevent the onrush of war." True, collective opposition to aggression afforded, in the long run, the surest deterrent to war. But a house already aflame called for immediate efforts to extinguish the blaze. Fire-prevention measures could come later. The UN, he regretted to say, appeared ill-equipped to douse the conflagration in Korea. The only solution consisted in direct negotiation among the great powers, including China. The powers should first arrange a ceasefire. Once the killing stopped, they could take up the related problems of Taiwan and representation in the UN. Perhaps this plan would fail. Perhaps war was inevitable, and all that remained was for each country "to get in or keep out of war as gracefully as possible." He admitted his fear that the Chinese had already reached this conclusion—which then might become self-fulfilling.[29]

Despite Nehru's fears, the world survived the crisis of December 1950. The Truman administration chose not to counter China's manpower with America's nuclear arms, and notwithstanding the protests and eventual insubordination of MacArthur, the president insisted on limiting the conflict to Korea. Although Washington rejected Nehru's proposal for a conference on the Far Eastern situation, as the Chinese pushed toward the thirty-eighth parallel American leaders began to detect merit in a ceasefire. At the year's close, Henderson outlined American terms to Bajpai, who passed them to Panikkar and Beijing. But the Chinese preferred to keep fighting, with good reason. They captured Seoul at the beginning of January, and their momentum threatened to carry them considerably farther south.[30]

The UN forces managed to dig in, however, and by the end of the month they had even launched a new counteroffensive. Seoul changed hands for the fourth time in March; and as spring softened the iron mountains, the two sides found themselves almost back where they had started nine months before.

With the stabilization of the fighting front, relations between the United States and India regarding Korea lapsed into a fitful stalemate. The two countries did not quite agree to disagree, but neither did they act especially disagreeably toward each other.

VI

New Delhi had reason not to aggravate the relationship. In December 1950 Ambassador Pandit in Washington informed the State Department of a projected shortfall in India's harvest. For four years the northeast monsoon, which normally brings rain to the rice fields around Madras, had failed. Earthquakes had devastated Assam, floods had ravaged Punjab, and locusts had decimated grain standing and stored in other provinces. Facing widespread famine, the Indian government requested American aid.

Henderson immediately began lobbying for approval of the request. Having witnessed the famine in Eastern Europe at the end of World War I, he had no desire to see India repeat the calamity—knowing that in India famine, as everything else in that country, would assume a monumental scale. In addition, he

argued, there existed substantial diplomatic and geopolitical reasons for providing the requested aid. The American Congress would object to spending millions of dollars on a regime as refractory as Nehru's, but Henderson contended
that this very refractoriness might redound to America's benefit.

> Furnishing this food in the face of the present uncooperative attitude of the Gov-
> ernment of India would, particularly if done generously and ungrudgingly, provide
> a definite answer to charges made constantly by forces in Asia hostile to the United
> States that we use our economic power merely to forward our international policies
> and have no genuine interest in the welfare of Asian peoples. Our numerous friends
> in India who are dejected just now would be encouraged; and in the uncertain years
> ahead they and we could produce concrete evidence of our friendship for the
> Indian people.

Citing an earlier program by which the American Red Cross—his old outfit—
had supplied blankets to destitute Indian families, he said, "Every family given
a blanket knows it was a gift of the the U.S. people and feels the glow of friendship." On the other hand, a refusal by the United States to provide the
requested food would be exploited in full by America's enemies.

> Their slogans would be along the following lines: "The United States withholds
> food from starving Indians in order to force India to toe its line"; "The United
> States vents its resentment against Nehru by permitting millions of Indians to
> starve"; "The hypocrisy of the U.S. pronouncement of interest in the welfare of
> Asian peoples is revealed."

Under such circumstances India would drift in the direction of communism,
with potentially grave effects on the peace and stability of the rest of Asia.[31]

Helping India inevitably entailed supporting Nehru, at least to some degree,
and Henderson viewed this prospect with ambivalence. During the heat of the
fighting in Korea, Nehru's stock as world peacemaker had risen with various
groups in foreign countries who looked askance at America's foreign policy.
Henderson found the phenomenon most worrisome in Britain. Already disposed
to think ill of Nehru, he now criticized the prime minister for the company he
kept: the "self-named Liberals," "the usual run of fellow travelers for whom
Nehru has always had a soft heart," and "anti-American Conservatives who
still hope for a deal with the Soviet Union which while sacrificing other peoples
to the Soviet appetite would free the United Kingdom from 'American bond-
age.'"As for Nehru himself, Henderson described the prime minister as "con-
stitutionally unhappy when he is not leading some cause of downtrodden peoples—particularly of Asian or colored peoples—against real or imagined
oppression."[32]

For all this, Henderson could not deny the magnetism of the man, and with
India's grain request in the balance the prime minister was at the top of his

form. In February, Henderson visited Nehru for a discussion of the international situation.

> I found the Prime Minister in an excellent humor. In none of my previous conversations had he ever been so friendly or talked with such apparent frankness. He made use of his great personal charm and was evidently anxious to persuade. It is easy to understand how, when the Prime Minister is in such a mood, he is so frequently able to win over so many persons, particularly those without profound convictions based on their own experiences. In fact, as I listened to him I found myself rather regretful that I could not agree with him and say with all honesty that he was quite right and was, in my opinion, pursuing the policy most likely to preserve the peace of the world.

Predictably, Henderson resisted the temptation.

Nehru outlined his understanding of American policy. The United States, as he interpreted matters, held the belief that the Soviet Union possessed aggressive intentions toward both Europe and Asia; that Soviet aggressiveness and communist ideology were intimately connected; that communism operated not only by force but by infiltration and subversion; that the United States, in its own interests and those of world peace, must take every opportunity and every appropriate measure to stop the spread of communism; and that the United Nations and the system of collective security it represented should play an essential part in securing this end. Was this, he asked, an accurate delineation of the American view?

Henderson replied that on the whole it was, although he insisted that the United States did not oppose communism per se. The United States had not undertaken to prevent any nation that so desired from adopting a communistic form of government. Every country deserved the government it chose. What the United States objected to was "the practice of International Communism of forcing, by terror, threat and violence, free nations to submit to its yoke." While on the subject, Henderson emphasized that his government believed that unless the free nations banded together and made clear that they would collectively and resolutely oppose aggression, "the Soviet Union as the directing center of International Communism would continue to carry on its aggressive policies with a world war as the inevitable result."

This led Nehru to reflect on the possible outcome of such a war. The prime minister did not expect that the United States and the other western powers would lose. Neither, however, could he foresee a complete western victory. The west might defeat the armies of the Soviet Union and China and raze the industrial centers of those countries. But what would it do then? It could not effectively occupy the communist states. Soviet and Chinese partisans, scattered through the hinterlands, would continue to fight. The struggle would last for years, with the only victors hunger, pain, and human suffering. Should the "international communism" the ambassador referred to collapse during this phase, the anarchy that followed would spawn indigenous varieties of communism.

Therefore, Nehru continued, the objective of statesmen must be the prevention of the war no one would win. In particular, each side must be convinced that the other did not desire war, for suspicion itself could trigger the cataclysm. If, for instance, the western countries believed that the communists intended aggression, they might arm themselves heavily. The communists, observing this build-up, might be frightened into a preemptive attack. Moreover, in a hostile and distrustful world, limited friction and minor disputes could easily develop into a full-blown confrontation.

The government of India, Nehru said, sought to convince each side that the other was not preparing an attack. India's China policy illustrated this approach. His government did not believe that the Chinese had aggressive intentions toward other nations of Asia, although it recognized the Chinese desire to assume control of territory formerly under China's control, especially Taiwan and Tibet. He for one did not think China had invaded Korea out of a desire to subdue that country. Instead the invasion had resulted from Beijing's fear that the United States would use Korea as a base to invade China. For this reason he could see no settlement of the Korean question without reference to China. At the same time he predicted that the Chinese would consent to a "relatively fair" solution in Korea if such a solution were placed in the context of an overall arrangement for the Far East.

To Henderson's query regarding the nature of this comprehensive accord, Nehru replied that China should have Taiwan, that it must gain admission to the United Nations, that it ought to receive treatment as a great power, and that its views should be consulted on matters relating to the future of Pacific Asia. Eventually Beijing would demand Hong Kong, but for the time being it would probably not press the issue. Nehru placed particular emphasis on Taiwan, asserting that the island held no military significance for the United States unless the Americans intended an invasion of the Chinese mainland.

Henderson interrupted to point out that if Taiwan fell to the communists, they might use it as a base for assaults on Japan or the Philippines. This remark prompted Nehru to respond that the United States and the Chinese might agree to neutralize Taiwan, with each renouncing military bases there. Of course the problem remained of policing such a deal after the communists acquired sovereignty, but every international agreement involved risks.

Suppose, Henderson rejoined, that the United States recognized Communist China and supported its entry into the UN. Suppose additionally that Washington acquiesced in a communist takeover of Taiwan. Would this satisfy the Chinese? Would they not then demand that the United States withdraw from Japan, leaving an unarmed Japan to face an armed China and Soviet Union? If the United States refused to consent to such an arrangement, would Beijing veto any overall settlement? If the United States accepted it, would India wish to see Japan in such an exposed position?

Nehru said he had given considerable thought to the question of Japan and had come to the conclusion that any comprehensive settlement for the Far East must include that country. It would be a mistake to rearm the Japanese. Such action would further convince the Chinese and the Soviets that the United States was planning aggression. As he had suggested, the preparation for war might produce war. Perhaps the UN should guarantee Japan's security. In reply to another question from Henderson, the prime minister conceded that until the UN could establish a protectorate, the Japanese might be allowed sufficient weapons to defend themselves. But the quantity need not be great since neither China nor the Soviet Union would be likely to attack a Japan guaranteed by the UN. When Henderson reminded him that UN protection had not prevented the Chinese from entering Korea, Nehru reiterated that the Chinese and the Russians did not want war. They would not tempt fate by attacking an unarmed and neutralized Japan.

After discussion of other troublesome areas around the world, Henderson summarized what seemed to him the essential difference in view between India and the United States. India believed that communism had no aggressive intentions and that its motives were principally defensive. The United States considered communism inherently aggressive; where the communists were not actively engaged in aggression, their diffidence followed not from a philosophical aversion to violence but from fear that an attack would lead to a war against the free nations, which the communists would lose. The only guarantee of good behavior on the communists' part was continued vigilance and collective security.

Nehru did not object to this characterization, and the interview, the most civil and enlightening Henderson ever had with the prime minister, ended.[33]

VII

Henderson had requested this meeting to learn the latest Indian thinking on foreign affairs before he left for a regional conference of American officials responsible for policy toward South Asia. At the gathering in Colombo, Assistant Secretary McGhee explained a shift in American policy toward the area. Until now the United States had relied on Britain to keep the communists out of the subcontinent and adjacent territory. But British resources were diminishing and British leaders were increasingly reluctant to take any action that might antagonize India and provoke it to leave the Commonwealth. Chinese intervention in Korea had demonstrated a greater willingness on the part of the communists to take chances than American planners had recognized; consequently the rest of noncommunist Asia required stronger defenses. Because India continued to choose nonalignment over collective security, the United States might be forced to write off Indian cooperation and concentrate on Pakistan, a country that appeared more than willing to collaborate.[34]

Henderson objected. He granted that Nehru required special handling, but he contended that the United States would commit a serious blunder if it turned its back on India. "I am convinced," he said, "that in India, underneath the crust of criticism of the United States and of constantly expressed distrust of American policies, there is a hard core of basic friendliness for the United States and confidence in the motives which animate the American people."

Henderson overstated this part of his argument to counter the anti-India tenor of the meeting. At some level the Indian people may have felt a "basic friendliness" toward Americans, but Henderson was reaching when he referred to Indian confidence in American motives. From independence, Indians had distrusted American motives, often considering the United States a greater influence for war—as in Korea—than for peace. Having experienced the wrath of the Indian press and numerous Indian politicians, Henderson knew this as well as anyone.

But he would not let Washington write off India without a fight, and if he had to stretch the truth he would. He argued that instead of dismissing Nehru as hopeless, the United States should strive to move the prime minister in the direction of the west. To be sure, accomplishing this feat would require care and steadiness. "In endeavoring to bring about a change in the attitude of the Prime Minister, we should not at any time assume a cringing or flattering attitude which would give him the impression that he has the whip hand over us. We should not, on the other hand, take an attitude of truculence or hostility towards him or India which would strengthen the hands of those in India who are opposed to us." The United States must remember that Nehru did not make Indian policy in a vacuum. "We should always bear in mind that Nehru is not India; even now he cannot entirely ignore Indian public opinion. Events beyond his control may force him to decide to change his present tactics and methods."

American officials dealing with India should stand up forthrightly for American policies when these policies differed from India's. Henderson said he had done so recently regarding Korea, and he was convinced Nehru would have respected him less had he acted otherwise. The Indian government had tried, and would continue to try, American patience. "India's recent energy in opposing our foreign policies and in endeavoring to damage our reputation for good judgment in the conduct of international relations has been harmful to the United States." Where India persisted in such activities the American government must take energetic measures to defend its policies. But American officials should refrain from unnecessary criticism of India, and they should at all costs avoid casting aspersions on Nehru's motives. Summarizing, Henderson said,

"We should act as though we take it for granted that India's sense of international morality is similar to that of the United States and that India is basically on the side of those forces opposing aggression."[35]

The acting that Henderson recommended sometimes required considerable effort. After thorough hearings and debate Congress approved the Indian aid request, and in the summer of 1951 the grain began arriving in the stricken areas. Henderson did not expect gratitude from Nehru, but he was surprised at the prime minister's timing when, just as the first ships cleared the harbor at Philadelphia, the Indian government announced that it would not sign the recently concluded peace treaty with Japan. Nehru's spokesman declared that the continued American presence in Japan, the American trusteeship over the Ryukyu and Bonin islands, and the failure to return Taiwan to China and the Kuriles and South Sakhalin to the Soviet Union made the treaty unacceptable.[36]

The Indian announcement prompted Henderson to write a detailed analysis of Nehru's diplomacy. Although the subject obviously differed from that of his earlier assessments of Soviet foreign policy, in tone Henderson's memo echoed his 1936 report on Stalin's long-term strategy, with its emphasis on steady pursuit, though often by a circuitous route, of ends inimical to American interests. Nehru was not Stalin, but Henderson considered him equally devious. The ambassador asserted that the rejection of the Japan treaty was a "logical step on the part of Nehru in his efforts to achieve his one primary foreign policy objective, that is, the eventual exclusion from the mainland and waters of Asia of all Western military power and what he would consider as Western political and economic pressures." By opposing the American presence in Japan the prime minister aimed to increase India's leverage with "nationalistic and anti-white elements in Japan." Sooner or later these groups, if "discreetly encouraged and skillfully guided by such experienced Asian nationalist leaders as himself," would gain control of the Japanese government. When that happened they would denounce the treaty, kick out the United States, and presumably look to New Delhi for guidance.

Nehru's efforts to draw Japan away from the United States paralleled similar actions designed to pull China out of the Soviet orbit, with the ultimate goal, Henderson said, being an "Asia for Asians." To date, Nehru's China policy had borne little fruit, and the Chinese seemed as close to the Russians as ever. But Nehru was convinced that the alliance would dissolve sooner or later. When it did he would find himself as perhaps the leading figure in the part of the world that held the balance between the American and Soviet camps.

Subtlety would mark the prime minister's style.

> Nehru is not likely to move too openly or rapidly. He will not wish to arouse too much hostility or indignation in the United States. He realizes that for some time to come India will sorely need certain capital and consumption goods which only the United States can furnish.

He is not likely to disclose his real objectives. He will probably continue to try to appear as a democratic idealist primarily interested in the welfare of the down-trodden masses of India; as David who regretfully faces the materialistic and clumsy Goliath of militarism and imperialism. He will continue to endeavor to gain support in the Western world, particularly in the United States and the United Kingdom, of various non-Communist leftwing elements and well-intentioned idealists with progressive views as well as professional "liberals."

He will continue to make a special effort to charm and flatter naive Americans and Britons, who he thinks might be useful in helping to mold public opinion in his favor or in supporting policies in their countries which would facilitate the gain-ing of his foreign policy objectives. He will continue to make minor concessions and friendly gestures from time to time to the United States in order to keep down the tide of resentment and make United States officials think, "He will come to our side eventually if we are patient and handle him properly."

But Nehru would not change his overarching design until unyielding reality—some unmistakable display by China of hostility toward India, for example—forced a reconsideration. Since Chinese leaders could surely recognize "the potential value of Nehru in stimulating hatred in Asia against peoples of Euro-pean stock, particularly of the United States," such a display would be a long time coming. Henderson admitted that Nehru might not have worked out all these implications of his policies, but the ambassador contended that they fol-lowed inexorably from the actions the prime minister had already taken.[37]

Henderson's analysis—including his comment about Nehru's not having plot-ted the scenario to its end—doubtless captured an important aspect of Indian diplomacy. But what in Henderson's language sounded like a grand conspiracy was in fact nothing more than the obvious course any leader of a middle power would pursue in a world dominated by superpowers. To reduce America's influ-ence in Japan and Russia's in China—to create an "Asia for Asians"—would probably render India's neighborhood safer, and it would afford India the breathing space the country desperately needed to get to its primary task of economic development.

Perhaps the most interesting feature of Henderson's memo is his unwilling-ness to grant a role for idealism in Nehru's diplomacy. To be sure, Nehru knew the rules of hardball, as he demonstrated with regard to Pakistan and later China. But the prime minister also possessed a vision of a world beyond the cold war, a world in which smaller nations as well as great would have an oppor-tunity to progress in peace. His vision ultimately failed, as Henderson would have been the first to predict. Yet inherently flawed or not, the vision informed much of Indian policy during the 1950s. While Henderson might have been justified in rejecting the vision, he erred significantly in not granting it due weight in Indian thinking. A principal job of ambassadors is to report what for-eign governments are about. Primed by experience to see the cynical and self-serving in international affairs, Henderson missed half the story in India.

V

BUBBLE AND BOIL,
TROUBLE AND OIL

15

Nationalization and
Its Discontents

Henderson must have thought he spent half his life picking up the pieces of Britain's empire. It had been his hands, literally, into which the British had dropped the notes divesting themselves of responsibility for Greece and Turkey in 1947. He had carried much of the burden, on the American side, of dealing with the collapse of the British mandate in Palestine. In India, hangover hostility from British rule had contributed to suspicions that Washington intended an Americanized neo-raj. Now, in the autumn of 1951, he received an assignment to go to Tehran, where the British once more faced trouble.

A writer in the Bombay *Current*, bidding Henderson farewell, or good riddance, aptly summarized the ambassador's new mission as one designed "to pour water on troubled oil." The oil in question was the focus of a spectacular dispute between Britain and Iran; and the Truman White House, recognizing Henderson's skills in post-imperial salvage, as well as his familiarity with Iranian affairs, decided to send him to Tehran to try to patch things over. The rift, in fact, proved irremediable. But in ascertaining this Henderson became convinced that Iran was approaching the fate from which the United States had saved Greece half a decade before. As in the earlier case, he argued for a more vigorous American policy, ultimately calling for the overthrow of the Iranian prime minister, Mohammed Mosadeq. By accepting his recommendation and transforming Iran into an American proxy in the vital area of the Persian Gulf, the Eisenhower administration significantly advanced the frontier of the American empire.[1]

II

By the time Henderson got to Tehran the oil dispute was almost, but not quite, out of control. In April 1951 Mosadeq had assumed office on a pledge to lift the British incubus from his country. To most of Mosadeq's compatriots that incu-

233

bus took the form of the Anglo-Iranian Oil Company, holder of Iran's petroleum concession and operator of the world's largest refinery at Abadan. Mosadeq proposed the nationalization of the assets of Anglo-Iranian, and the Iranian parliament, or majlis, approved the measure.

The British government, which happened to be the majority shareholder in the company, pondered military intervention. Financial factors aside, London shuddered at the political and strategic ramifications of allowing Mosadeq to get away with what the British judged to be piracy. But Washington made clear its disapproval of military action, fearing that a British invasion would prompt a Soviet move in Azerbaijan, and the Attlee government decided, as the minutes of a crucial cabinet meeting in September put it, that it "could not afford to break with the United States on an issue of this kind."[2]

Instead the British opted for economic pressure. Anglo-Iranian shut down operations, depriving the Iranian government of sorely needed revenue. At the same time the company threatened legal action against any other firms dealing in Iranian oil, establishing what amounted to a blockade of petroleum exports. The British government appealed to the International Court of Justice to enforce what the British claimed were no-nationalization provisions of their contract with the Iranian government. When Mosadeq denied the court's jurisdiction, and the tribunal agreed, the dispute went to the United Nations.

Meanwhile the Truman administration threw a variety of troubleshooters into the breach: first Assistant Secretary McGhee, then all-purpose fixer Averell Harriman, then McGhee again, and finally Henderson. None, until the time of Henderson's arrival, had achieved any measurable success.

The heart of the problem was that the Americans were the only ones interested in a compromise solution. For the British, a tough stance against Mosadeq appeared imperative if Britain were to have a future in the Middle East. For Mosadeq, opposition to the British was both a philosophical raison d'être and the glue that held his ruling coalition together.

Henderson found himself caught between the two camps from the start. The British he already knew; Mosadeq he quickly got to know. Of the prime minister Henderson formed some distinct impressions at the outset. After just a few weeks in Tehran, Henderson had concluded that the prime minister was a "shrewd leader" of "demonstrated political ability," and at the same time a "demagogue who well understands Iranian emotions and character." Mosadeq combined "personal prejudices against the British," an "undoubted understanding of Russian intentions in Iran," and an "almost megalomanic desire to act as the champion of the people in the struggle for 'independence'." Henderson respected Mosadeq and, in contrast to his experience with Nehru, grew to like him. Describing their relationship many years after the fact, Henderson recalled,

Mosadeq was an attractive man although he was neither handsome nor elegant. He was tall and lanky; his long horselike face topped with rather disheveled gray hair was expressive like that of an actor. He had a large mouth and when he smiled, his whole face lit up and one felt drawn toward him. He liked jokes and liked to laugh at them—a trait which is always helpful, particularly when one is engaged in serious conversation. He was troubled with dizzy spells so he would remain in bed much of the time. In general I found our conversations interesting and even agreeable. During most of them he was in bed and I was sitting beside him. He was quite frank, at times, without being offensive in criticizing our policies, and I was equally frank with him. So we got along quite well, each pointing out where he felt the other was wrong.[3]

Henderson soon recognized that Mosadeq had as little interest in a compromise solution to the oil dispute as the British did. To back down from nationalization, Mosadeq repeatedly said, would be to sell Iran's soul. Besides, the only way he could maintain control of the nationalist movement in the country, against radical elements like the communist Tudeh party, was to adopt a hard line against London. The people demanded no less.

What Mosadeq's nominal boss demanded was another question. Under the Iranian constitution the prime minister served at the pleasure of the monarch, in this case Shah Mohammed Reza Pahlevi. Such, at least, was the interpretation the shah placed upon the constitution. Mosadeq had a different idea, believing that his power flowed upward from the people rather than down from the throne. In the prime minister's view the shah could no more dismiss him than the king of England could fire Clement Attlee. Had the shah been a more commanding figure—like his father, for example, a former commander of Cossacks who had threatened and connived his way to the throne in 1925 and had single-handedly established a dynasty (albeit one of only two generations), and whose independence of mind had provoked the British occupying force to send him into exile in 1941—he might have forced Mosadeq to bend. But the second Pahlevi lacked the appeal to inspire obedience and the strength to compel it. Consequently, although the shah was more favorably disposed toward compromise with Britain, Mosadeq's will rather than his prevailed.[4]

Shortly after Henderson's arrival in Tehran, the ambassador outlined the Iranian scene for Washington's edification. Treating first the shah, Henderson described him as "indecisive and weak, though well-intentioned." The young monarch—in 1951 he turned thirty, slightly more than half Henderson's age—possessed "no confidence in his own influence," or if he did, he hid it well. In the shah's favor, Henderson noted that to many Iranians he represented stability and continuity of leadership. No one could question his opposition to communism. He might have been a reformer, but he had not the faintest idea where to start. Instead he allowed himself to be stymied by what Henderson called the "landowning–merchant oligarchy," a group motivated by "complete self-interest" and constituting "one of the main obstacles to the progress of the Iranian people and to the development of the country's resources." Nor was the shah willing to challenge Mosadeq, with reason. "He is probably correct in his belief that if he should try just now to remove Mosadeq from the premiership, or if he should take any other measure which might seem to run counter to nationalist aspirations, the prestige and influence of the Crown would suffer severely and he might even be overthrown." This consequence could be disastrous, and not only to the dynasty. "The disappearance of the shah would mean the loss to the Western world of a potentially powerful anti-communist element, and the ensuing struggle for power might lead to chaos which the organized Tudeh Party would exploit."

The shah commanded the loyalty of the military, but this meant little since the army struck fear in no one. "The lower ranks are discontented and ill paid," Henderson wrote, "the junior officers reportedly are receptive to communist propaganda, and the senior officers are often incompetent and corrupt." American military advisers, dispatched to Iran under terms of a 1950 defense-assistance pact, were helping to improve the situation somewhat. Yet there were limits to their effectiveness, not the least being the fact that a strengthened army would attract Mosadeq's attention. Henderson suspected that if the prime minister came to consider the military a threat he would channel some of the country's anti-British resentment against the American advisers. Should this happen and they be forced out of the country, the result would be a "most serious blow to U.S. policy."

While the army was impotent, the Tudeh was not. Although its hard core numbered only several thousand, the party magnified its influence by working through a much larger network of fellow-travelers. These sympathizers had established what Henderson labeled "stooge organizations" and "Peace Fronts" and had been instrumental in convincing most of the Iranian people that the Tudeh was merely "an indigenous political movement advocating reforms close to the heart of the populace." At the moment the Tudeh was collaborating with Mosadeq's National Front coalition, but Henderson had no doubt that when the party accomplished its immediate task of ejecting the British it would resume the struggle for its ultimate objective: "the destruction of all remaining rivals for power in Iran." So far the Tudeh had succeeded in spreading the belief that only the west posed a danger to the country. "The average Iranian fails to see any present tangible evidence of Soviet imperialism, whereas he imagines he sees numerous signs of endeavors by British and Americans to maintain old controls and even obtain new holds on the country." For their part, the Soviets lent credibility to the radicals' allegations by a constant barrage of propaganda. "The U.S.S.R. is Queen of the airwaves in this area," Henderson wrote.[5]

Another factor, the importance of which Henderson could only guess at this point, was Islamic fundamentalism. It is unclear how much of Henderson's disdain for those he called the "demagogic mullahs" resulted from his upbringing in the house of a Christian minister. For what it is worth, the British chargé d'affaires in Tehran held a nearly identical view, calling Ayatollah Abol-Qasem Kashani a "sly, corrupt, and anti-Western demagogue." In any event, Henderson accounted Kashani and his associates "retrogressive and anti-foreign." But the American ambassador could not tell what direction their xenophobia would take. They defied classification on an ordinary spectrum of reaction to radicalism, since they occupied both ends simultaneously. For the time being the impressive energies they had tapped were directed against Britain. "Anti-British slogans, particularly those connected with the oil dispute, fit the intolerant aspects of Islam. The movement to drive out the British has gained almost the significance of a religious crusade in some quarters." Although Mosadeq had formed an alliance with the mullahs, Henderson thought the prime minister was deluding himself if he thought he could control them. Should the government decide, for whatever reason, to compromise with the British, "there would be religious fanatics ready to stir up popular emotions and to assassinate responsible officials." "Religious fanacticism," Henderson concluded, "can be used to combat communism, but it cannot be employed as a constructive force for the country's progress."[6]

Turning to issues, Henderson asserted that the oil dispute was the principal threat to Iran's stability. Politically, the contest provided ammunition to the Tudehists, who could generate enthusiasm for their cause simply by denouncing Britain. Economically, the shutdown of the petroleum industry was creating great distress. In the absence of royalty payments, bureaucrats and soldiers found themselves losing their race with inflation. The Iranian population as a whole was suffering still more, with rising unemployment combining with rising prices to squeeze household budgets.

At best the future looked cloudy. Henderson predicted that British influence would continue to decline, which would open the country further to communist influence. "Iranians, long accustomed to playing foreign powers against each other, may dangerously allow themselves to be vulnerable to Soviet penetration to such an extent that if or when they turn later to the Western world to save them from Soviet domination their position will already have become irretrievable." Yet Henderson was not prepared to count the British out entirely.

> Despite a general condemnation throughout the country, the British still have much powerful unseen support which might be effectively mobilized in certain circumstances. For instance, if as a result of an understanding attitude on the part of the British the oil dispute could be settled in a manner inoffensive to reasonable Iranian nationalist elements, or if the Russians or the communists should make a misstep in their program, the British might still make a comeback.

But they would never regain their former hegemony. And this, Henderson argued, meant that the "relative responsibility of the United States on behalf of the free world in preventing Iran from passing into the Soviet sphere has increased."[7]

In 1951, at the time Henderson took up his Tehran post, the British embassy dominated the diplomatic skyline of the capital. Sprawling across sixteen city blocks, the concrete presence of Britain in Persia—as unreconstructed imperialists like Churchill insisted on calling the country—served as a constant reminder of British influence. The American embassy, by contrast, appeared insignificant. A new building was under construction, but even after completion in 1952 it gave the impression of a high school planted somewhere in middle America. Indeed, it soon acquired the nickname Henderson High.[8]

The modest façade, however, did not fool the various contestants in Iranian politics who sought to turn American influence to their different ends, and Henderson had hardly unpacked his bags before a procession of visitors began to march through his office. All wanted to know what position the United States would adopt in Iran's conflict with Britain and in the various quarrels within the country.

Among the first to request an interview was Kashani. Henderson suspected that the ayatollah intended to use an audience with the American ambassador to bolster his standing in Tehran, probably by misrepresenting the American position vis-à-vis Britain and Iran. For this reason Henderson declined to see Kashani, instead passing him along to Arthur Richards, the counselor of the embassy. Richards afterwards described the conversation to Henderson, and the ambassador relayed the description to Washington.

"With a complete lack of modesty," Henderson wrote, "Kashani dwelt at length on his influence. He said that he put Mosadeq in power and that it was he who was keeping him in." The ayatollah explained that he had opposed Ali Razmara, the former premier who had been assassinated in March 1951 by a member of an extremist Islamic sect, "because Razmara was a British stooge." He asserted that Muslims "from India to North Africa were now under his control and would obey his every command." Yet he conceded that he could not handle the communists, at least not by himself. For this he needed the help of the United States. "Kashani said that immediate and dramatic efforts must be made to save Iran from the Communist menace." He knew that the Americans appreciated the danger communism posed to his country, but he claimed that most of their efforts to halt it were "useless" because they focused on strengthening the Iranian army instead of on improving the living conditions of the Iranian people. "What good would a well equipped and well trained army of even 250,000 do against 18 million hungry people?" he asked.

Kashani proposed two methods by which the United States might rectify its errors. The first was to lend Iran $250 million at once. Most of this would have to be in cash, to allow the Iranian government to meet its financial obligations. The rest could take the form of machinery and equipment for building dams, roads, and the like. With assistance of this nature "the population of Iran would rapidly increase, its productivity would increase at a more rapid rate, the country would be prosperous, and Communism would be thwarted." The second option available to the Americans—one that might render the first unnecessary—was for Washington to pressure Britain "to recognize Iranian legal rights and allow Iran freely to sell its oil." Kashani charged the British, accurately, with intimidating potential purchasers. The mere thought of Britain's perfidy sent the ayatollah into a rage. "He spoke with violence of the harm which the British had done to Iran," and he declared that "any impartial international group which studied the matter of compensation would decide that Iran owed the British nothing; rather the British owed Iran many millions of pounds because of its plundering over the last half century."

In passing along this summary of the conversation, Henderson added his own thoughts. "Kashani's whole approach was unrealistic," he remarked. Evidently the ayatollah thought Washington simply had to snap its fingers to get the British to settle. Trying to convince him otherwise was a lost cause. All the same, Henderson had to admit that Kashani was "a shrewd religious leader who has gained considerable influence."[9]

Kashani's belief in America's power was matched by that of the embassy's next visitor, Jamal Emami, the leader of the opposition in the majlis. Henderson later recounted his surprise when Emami opened the conversation by asking "whether I was satisfied with my present Nationalist Government of Iran." Henderson continued his summary:

> I laughed. I said I thought it was his government and not mine. He then tried to prove that United States policy was responsible for Mosadeq coming into power and remaining in power, and proceeded to stress the dangers of communism and the unwillingness of Mosadeq to acknowledge the existence of such dangers. He insisted that if Mosadeq were not removed almost immediately and replaced by someone who temporarily at least would play the role of dictator in bringing order out of chaos in Iran, Iran was doomed.

Emami went on to declare that several of Mosadeq's close advisers were collaborating with the Tudeh. From ignorance or design, the prime minister was leading Iran to disaster.

> The time had come for all groups interested in the maintenance of Iran's independence to concentrate on persuading the Shah peremptorily to remove Mosadeq. . . . The Shah should not bother about Parliamentary procedures since Parliament was helpless in the present situation. The Shah should merely announce that in order to save Iran from the political and economic chaos in which it was

drifting, he was taking the extraordinary measure of removing Mosadeq and replacing him with a Prime Minister who would restore law and order and bring prosperity back to Iran.

Emami concluded by asking Henderson whether he as American ambassador would be willing to use his influence to persuade the shah to take this necessary measure, and whether the United States government would be prepared, in the event that a new pro-western regime took power, to assist Iran over its current financial crisis.

After only four months in the country, Henderson was not prepared to lend his prestige to a coup, technically constitutional or not, and he had no reason to believe that Washington would either. He replied that he considered it improper for an American official to make commitments that might encourage any Iranian citizen to take action against the Iranian government. "I must do nothing which might justify charges being made that a representative of the United States participated in a plan or conspiracy to overthrow the government of Iran." But Henderson did not turn Emami away empty-handed. The ambassador told his visitor he considered it his duty to the United States and to world peace to do all he properly could to help Iran preserve its independence, and, whenever possible and proper, to give Iran such guidance in that direction as would be helpful. In other words—or so it must have seemed to Emami—try again later.[10]

Emami had hardly stepped out the door when General Mansur Mozayeni, who had just resigned as chief of police, arrived. Indicating the sad state into which his country had fallen, Mozayeni explained the circumstances leading to his resignation. He said that several months earlier the shah had forwarded his name for police chief, and Mosadeq had seconded the nomination. He had accepted the job on the understanding that he would answer only to the shah, to the prime minister, and to the interior minister. He soon discovered, however, that Kashani also claimed the right to issue orders. In several instances the ayatollah demanded the reassignment of certain police officials to remote parts of the country, and when he—Mozayeni—refused, Kashani grew hostile and threatening. Not long afterward, a group of "knife men" belonging to a radical fundamentalist sect were arrested for raiding a printing establishment. Over Kashani's protests, Mozayeni succeeded in seeing that they were exiled, only to have Mosadeq buckle under Kashani's pressure and bring the culprits back to Iran and release them.

Mozayeni went on to describe an incident in which his men had seized "a great mass of documents" delineating the anti-government activities of "key Tudeh agitators." With the shah's approval he delivered the documents to Mosadeq along with a plan for arresting the party leaders. The prime minister responded by ridiculing the notion of a communist threat, instead blaming the British for Iran's troubles. At this point Mozayeni had resigned. He told Henderson gloomily that his successor was "a tool of Kashani." He added that he was leaving the country because his life was in danger, but before he departed he intended to tell the shah that for Iran's safety he must insist that Mosadeq take firmer measures against the communists. If the prime minister refused, the shah must dismiss him. Finally coming to the point of his visit, he asked Henderson if he would be willing to counsel the shah to accept this advice.

Henderson responded by inquiring whether the general planned to suggest that the shah give Mosadeq an ultimatum privately or in public. Henderson thought the private route would be unwise, because "Mosadeq with his eloquence might be able to drown the Shah's request in a torrent of words and divert the conversation to other channels." Better, Henderson asserted, for the shah to prepare a public statement. Mozayeni agreed that this would be a good idea. As to the feeler regarding intercession, Henderson answered, according to the account of the conversation he sent to Washington, "I did not reply to the General's question either directly or by inference as to whether I would be willing to endeavor to persuade the Shah to accept his suggestion."

Henderson summarized his discussions with Emami and Mozayeni with the comment, "I do not wish to exaggerate the importance of the talks with Emami and Mozayeni since neither has a reputation for sound political sense, neither is likely to prove a real leader, and both have a personal bias. Nevertheless I consider their attitude and suggestions interesting as illustrative of similar thinking that seems to be going on in somewhat wide but more timorous circles."[11]

During this same period, Henderson renewed his acquaintance with Hussein Ala, formerly Iranian ambassador to Washington, later prime minister and now minister of court. At a dinner at Ala's house, Henderson spoke with the minister and his son and daughter. The children, students in the United States home on vacation, had attended that day's session of the majlis and had come away thrilled by Mosadeq's oratorical power. It was "wonderful," they said, to hear a "great Iranian patriot" put to rout the "shrill, excited pro-British deputies." Ala himself admitted that Mosadeq had won a great forensic victory and overawed the opposition. Henderson, who had reviewed the debate before coming to dinner, commented that it was unfortunate that Mosadeq in making his case had felt compelled to attack the United States. Ala agreed but pointed out that the Iranian mind had identified America with Britain in the latter's struggle to retain control of Iranian oil. Ala remarked that the British embassy had been spreading reports that the United States would back Britain to the hilt and that there existed no possibility of breaking the Anglo-American united front. The minister could not comment on the veracity of the reports, but he said that in the absence of an American denial they were assuming an air of authority. The additional fact that the United States was withholding economic aid lent credibility to charges that Washington intended to assist the British in squeezing Iran.[12]

Ala had touched on the weak spot of American policy toward Iran. As would happen repeatedly throughout the third world, the United States found itself in the uncomfortable position of having to support European imperialists against local nationalists, out of fear that lack of support would undermine American initiatives in Europe. American policy-makers recognized that opposing the nationalists tended to polarize situations, driving centrists to the left and leftists toward Moscow. In the case of Iran, standing with the British—by suspending aid that would ease Iran's economic plight and thereby diminish Iran's incentive to settle the oil dispute on British terms—not only tarred America with the anti-British brush but also threatened to force Mosadeq to extreme actions in order to outflank the radicals. Yet as long as Europe mattered more to America than countries like Iran, what could be done?

Henderson appreciated the problem, and although he recognized that it had no easy answer—perhaps no answer at all—he felt obliged to warn Washington of the trend of perceptions in Iran. As Ala's comments indicated, the United States was losing the battle to maintain an identity separate from Britain's. Before long, Henderson predicted, America would be seen as an enemy of the Iranian people. "There can be little doubt that as the financial noose tightens the outcry against the United States will become more shrill."[13]

III

The only way out of the dilemma, or so it seemed to the Truman administration, was to arrange a settlement of the oil dispute. Truman invited Mosadeq to Washington in late 1951 to talk matters over. The president tried to get the prime minister to bend, but to no avail. To a suggestion that Iran reconsider its opposition to World Court jurisdiction, Mosadeq declared that this would be "quite impossible." The Iranian people had had "quite enough of the World Court." In Washington and London, American officials attempted to convince Churchill and Anthony Eden, recently returned to office, to accept a compromise. When Eden visited America at the beginning of 1952 Acheson and Robert Lovett, now defense secretary, sharply criticized British policy. If Britain continued to hold out, the two secretaries predicted, there would be the devil to pay. Eden put them off, asserting that the key to success with Mosadeq consisted in "playing our hand slowly and carefully."[14]

The divergence between the British and American policies toward the oil dispute followed from what one of Eden's associates at the British foreign office called an "acute difference of opinion" regarding the nature of events taking place in Iran. In particular, British and Americans held opposing views of Iranian nationalism. George Middleton, second in command at the British embassy in Tehran, offered the most succinct appraisal of the two positions.

> The American view is that Persian nationalism is a potent and spontaneous force which will be an overriding force on its own account regardless of the wishes and actions of any future government. Our view is that Iranian nationalism certainly exists but that its effectiveness as a political force is largely a matter of manipulation.[15]

Middleton's superior, Francis Shepherd, who distrusted the Americans only slightly less than he distrusted the Iranians, detected a deeper and more sinister significance in the Americans' attitude on Iran.

> They retain strongly their suspicion of the colonialist and imperialist attributes of British policy in all parts of the East, and they are consequently all the more impressionable to Persian complaints, however unfounded they may be, of alleged British colonialist policy in Persia and inclined to lean over backwards to be kind to the Persians. . . .
>
> Furthermore, although they are aware that we have longer experience in this country than they have, they wish to build up their own opinions for themselves, and are cagey about accepting views or advice from us which they think may be tainted by self-interest. . . .
>
> Finally, they are certainly ambitious of increasing American influence in this country and regard us as their chief rival.[16]

Fortunately for Anglo-American relations—and for Henderson's peace of mind—Shepherd was soon transferred to Poland, where he found less to accuse the Americans of. Henderson then dealt with Middleton. The American ambassador and the British chargé d'affaires thought along similar lines and developed a close working relationship, to the extent of collaborating on reports to Washington and London. In November 1951, for example, Middleton relayed to the British foreign office the joint judgment of the two that while the Iranians had a "historic suspicion" of Russia, they had adopted an "ostrich-like attitude" regarding current Soviet intentions. Although "naturally xenophobic," the Iranians seemed capable of aiming their hostility in only one direction at a time. At the moment they were targeting the west. Henderson and Middleton also thought similarly regarding the need for a more forthcoming British policy. Speaking for both—and speaking in opposition to his superiors in London—Middleton declared that it would be "extremely dangerous to let matters stagnate."[17]

While Henderson and Middleton cooperated, Mosadeq did his best to drive the Americans and British apart. The prime minister especially sought to discredit the British in American eyes. At the end of November, as elections for the majlis approached, Mosadeq told Henderson that the Iranian people now had an opportunity to choose persons who genuinely represented them. In the past, the prime minister asserted, the British had succeeded by bribery and fraud to pack the assembly with their "puppets." Mosadeq read Henderson a note he had sent to the British government and which he intended to publish, protesting British interference in Iran's internal affairs. When Henderson commented that the note was rather strong, Mosadeq replied that he intended it to be strong. The British, he asserted, were doing their utmost to push his country to war. They were attempting to ruin Iran economically and politically. They were intriguing in the majlis and the court, among journalists and students, and they were inciting public servants to demand higher salaries. To Henderson's inquiry regarding evidence of these crimes, Mosadeq replied that naturally the British were not leaving a paper trail. But he possessed "plenty of circumstantial evidence." Mosadeq told Henderson that it saddened him to have to make such statements about a country with which the United States had friendly relations. The British, however, had left him no alternative. Henderson, perhaps thinking that the prime minister did not look particularly sad, suggested that he might have raised the matter privately with the British embassy. Oh no!, Mosadeq replied. Only the impact of public disclosure promised any chance of forcing the British to mend their ways.[18]

Mosadeq's comments, together with his own estimate of the situation in Iran, caused Henderson to reflect on the advisability of continued coordination of policy with London. During the previous eight years Henderson had come to recognize that although the American and British governments shared a general philosophy of international affairs—including, most notably, opposition to Soviet expansion—Washington and London might differ enough tactically to wind up pursuing practically contradictory policies. Henderson had felt the friction in Iraq, during the Palestine fight and in India. He saw it more clearly than ever in Tehran.

At the beginning of 1952 he wrote a long letter to the State Department explicating his views on the subject. "I do not believe that I need to try to assure you and the other members of the Department who are acquainted with my past that I am in any sense anti-British," he began. "I have believed for years and still believe that close cooperation between the British and ourselves is necessary in the present world." He continued,

> I do not believe, however, that it would pay either us or the British for this cooperation to be based on a misunderstanding of a given situation or on a glossing-over of the unpleasant facts of a given situation. For the British to take the position that with their century or so of experience they are better able to analyze developments which are taking place in this area, and to decide what policy to apply towards developments, in some respects may be right. We have at times found that their

policies are wiser than ours. On the other hand the British have on occasions been over-optimistic regarding their ability to deal with certain situations in accordance with nineteenth century formulas and the result has been that those situations have deteriorated rather than improved.

Henderson thought Iran demonstrated the latter tendency. "I am afraid that at the present time the British policies with respect to Iran are of a destructive rather than of a constructive character." He conceded that the British were under great provocation. "The Iranians have been most disgustingly unreasonable and disagreeable." But this did not excuse London's mishandling of the situation.

Typically, Henderson advocated a more forceful American policy. To some extent he agreed with British officials who blamed Iran for Iran's troubles, but he found alarmingly shortsighted Britain's desire to let Iran slide toward disintegration. The United States, he said, could not afford "to sit on the sidelines and chuckle while the Iranians, with shouts of religious and national fervor, proceed to ruin themselves." He added, "It seems to me that efforts should be made to help the Iranians extricate themselves from the mess in which they have put themselves." Henderson considered the Truman doctrine an appropriate model for American action in Iran.

I can well remember how in February 1947 the British gave us notice that they would no longer be able to carry the Greek and Turkish burden. It was clear to us that Greece and Turkey would be lost unless we should take over. We therefore did and although we have made a considerable number of mistakes and although we have expended much money, effort and material in Greece and Turkey, our efforts have been amply rewarded and these two countries are probably among the brighter spots in this part of the world.

Unfortunately, at the time the British threw Turkey and Greece into our laps they did not include Iran. They wanted to continue to call the tune in this country. I do not believe that it is entirely unfair to say that one reason they wished to call the tune was because they were making a considerable profit from this country. They are no longer making this profit and I do not believe unless they change their basic attitude they are going to make any profit for a good many years.

It seems to me therefore we might well consider whether or not the time might be at hand for the British to at least let us have a free hand in an endeavor to save Iran.[19]

IV

At the time Henderson offered these comments, Iran seemed more in need of saving than ever. Fighting had broken out between the religious right and the Tudeh, and although disturbances of this sort had marked Iranian politics for some while, Henderson detected an important change in the government's response. Formerly the police had worked to restore order. In this case they seemed "so disorganized that they, in fact, became part of the mob." Whether this represented a deliberate decision on Mosadeq's part, Henderson could not say, but he did predict that "overt political activity would henceforth be confined to the streets."[20]

In addition, Mosadeq's grip on the National Front seemed to be slipping. During his visit to America in the late autumn of 1951, Mosadeq's rivals in Iran had made good use of his absence. Kashani, ostensibly still an ally, managed to install his own lieutenants in key positions. The religious radicals then employed their expanded influence to launch an attack on the communists, and otherwise maneuvered against present and possible future foes.[21]

At the end of January 1952 Henderson offered an assessment of the current condition of Iranian politics. Having received word that the British were sending a special envoy to Washington to brief the State Department on Iran, Henderson wanted to ensure that Acheson receive an independent appraisal. At the moment, Henderson said, Mosadeq remained in control. The Iranian people could not get enough of blaming Britain for their troubles, and the prime minister was still riding the wave of resentment. He had scheduled elections, which almost certainly would be rigged. The results could be expected to strengthen the National Front. To some degree the new strength would not be the prime minister's, reflecting instead an increase in the number of assembly seats for Kashani's faction. The Tudeh, although underground, would run candidates through various front organizations. The communists' success would depend largely on the economic situation: as conditions deteriorated, the left's chances improved. As for the shah, not much was being heard from him. Although he retained the loyalty of the army, unless he asserted himself more forcefully and—perhaps more to the point—convinced the troops they would continue to receive their paychecks, his influence might collapse entirely.

Looking to the future, Henderson argued that everything hinged on the economy. The economy in turn depended on the outcome of the oil dispute and the possibility of foreign assistance.

If there is no resumption of oil revenues or no Western budgetary aid, xenophobia will grow against the West, which may show itself in acts designed to eliminate varied Western activities in Iran, including U.S. military missions, Point Four, and Western cultural, business and educational institutions. Internal unrest and weakening controls over state administrative and armed forces will accelerate the progress of Tudeh infiltration and influence.

Before Government finances reach bottom it is quite possible that Iran will make desperate efforts to strengthen its financial position with aid from or trade with the Soviet bloc. Such action would make the ultimate solution of the oil problem more difficult and would increase the danger of ultimate Soviet domination of Iran.

Even economic assistance would provide at best a stopgap. Henderson asserted that western aid, unless far larger than seemed likely, would not materially raise the Iranian standard of living. And once having elevated hopes, it might well produce more disillusionment than benefit. That left a single answer. "The resumption of Iran's income from oil is indispensable in the near future if the country is not to be lost to the free world."[22]

At the end of January, when he drafted this analysis, Henderson thought Iran's condition would decline slowly, but events of the next few weeks suggested that the patient was sicker than he thought. Mosadeq, the "wily old master," appeared to be losing his touch. Already several important members of the National Front had broken away, dissatisfied with their share of the spoils of office. Others, notably Kashani and Hussein Makki, who styled himself Iran's oil expert and Mosadeq's heir apparent, remained within the coalition but were demanding a greater say in charting the direction of government policy. Rumors had it that Kashani was dabbling in anti-Mosadeq intrigues. "It seems apparent," Henderson remarked, "that the National Front coalition is unstable and should not be expected to endure indefinitely." The ambassador was not yet predicting an imminent demise, for he believed that so long as Mosadeq, Kashani, and Makki—"the three outstanding personalities in the Government"—remained united, the front would hold. But if they split, the end might come rapidly.

Henderson identified three potential causes of defection. The first was economic despair resulting from failure to revive the oil industry. Mosadeq would disguise this failure as long as possible, but eventually he would face a schism between radicals who would oppose any reasonable settlement with the British and moderates who preferred half a barrel to none. A second disruptive force was the approaching bankruptcy of the government. Mosadeq had demonstrated substantial ingenuity in eluding his bill collectors, and room for further improvisation—dipping into the country's gold reserves or borrowing against future oil sales—remained. But when the paychecks started bouncing, the coalition would probably crumble. The third and most unknowable element was the shah. Should he "change from his habitual vacillation or drop his policy of awaiting the play of other forces which might painlessly dispose of the Mosadeq Government without his intervention," the monarch could put the National Front to flight. As yet Henderson detected no evidence of such an awakening.[23]

In the middle of March, however, Henderson sensed a slight stirring. According to Ala and others close to the court, the shah was finally getting fed up with Mosadeq's dithering on the oil issue. He had decided, Henderson's informants said, to give the prime minister and the majlis a few more weeks to settle the affair. If they failed, he would dissolve the assembly and appoint a new prime minister.

Henderson had doubts. "It will be recalled," he wrote Acheson, "that friends of the Shah and the Shah himself during the last twelve months have been stating that when the proper moment comes the Shah will intervene to save the country." So far the monarch had never discovered the proper moment. Henderson, being a bull-by-the-horns type himself, found the shah's temporizing exasperating, but he had to grant that there might be a certain method to the maddening slowness. Although the constitution granted the crown the right to dismiss the parliament and fire the prime minister, the majlis and Mosadeq almost certainly would not go in peace. Mosadeq, in fact, might even be hoping the shah would make him a political martyr. Better, for the time being, to leave Mosadeq to face the troubles he had brought upon the country—to let him "squirm," as one of the shah's advisers had explained the strategy to Henderson.[24]

To get a clearer fix on the situation, Henderson visited the shah himself. The young man was in good spirits. "He seemed much more cheerful and self-confident than during our previous conversations, although obviously still concerned regarding the future." The shah defended his decision not to take action against Mosadeq, saying had he done so popular opinion would have been against him, Mosadeq would have been a greater hero than ever, and the new government would have had to impose its will by force. But the situation was changing. Where three months earlier Mosadeq's and Kashani's mobs had ruled the streets, now those two demagogues were alive "only because my army is protecting them."

Although Henderson thought the shah was overstating the case, he chose not to contest the point. Instead he asked the shah to account for the change in the popular mood. The monarch replied that Mosadeq had gotten "carried away by his own popularity." He had promised too much. He was "still making promises but they were becoming progressively less convincing." Most "thinking people" now believed that Mosadeq had no constructive program. Henderson inquired what would happen if the prime minister proposed vigorous measures to strengthen the economy—higher taxes, for example, or import restrictions. The shah dismissed the possibility. Mosadeq and his henchmen were "temperamentally and technically unable" to launch reforms of this nature. Besides, the mere suggestion of such a plan would shatter the National Front, which continued to rely for its unity on the expectation that an oil settlement would deliver Iran from its trials. No, the shah said, Mosadeq could not long postpone the day of reckoning. Soon he would have no choice but to resign. He—the shah—would not even have to intervene, and barring a direct threat to national security he would not do so.[25]

This was what Henderson had been afraid the shah would say, and the ambassador was not encouraged. Nor was he heartened by an approach from Mosadeq a few days later—although he was not exactly disheartened either, since the prime minister circled his apparent topic several times without ever landing. Mosadeq recited a list of reasons why the United States should provide economic aid to Iran. First, the American government had long been promising to help Iran but had not followed through. Second, because the United States was compensating Britain for revenues lost as a result of the oil dispute, fairness dictated doing the same for Iran. Third, the United States was committed to peace, as it was demonstrating at great cost in Korea. A civil war might erupt in Iran, between rich and poor, unless America provided aid. Fourth and finally, without American dollars Iran would have to curtail imports from America, which would have a serious impact on the American economy. Mosadeq concluded—to Henderson's surprise—by *not* asking for American aid. Instead he declared that the United States had missed its opportunity. Iranians were a "proud and sensitive people," and they would not beg for charity.

Henderson rebutted Mosadeq's argument point by point. Far from ignoring or short-changing Iran, he said, the United States had assisted the country for years, politically and economically. Perhaps the level of financial aid had not been what some Iranians desired, but neither had it been insignificant. The American government was not compensating Britain for lost oil revenues. If the prime minister was referring to the Marshall plan, that program antedated the petroleum dispute and had nothing to do with it. As to the argument that the United States should compensate Iran for lost revenues, Henderson reminded Mosadeq that Iran, not the United States, had made the decision to nationalize Britain's oil assets. The ambassador denied any connection between American action in Korea, undertaken at the behest of the UN and designed to repel a cross-border invasion, and a future class war in Iran. Regarding the loss of the Iranian market for American exports, Henderson respectfully submitted that the American economy did not exactly live or die on the basis of what Iran purchased. Besides, American exporters did not make American foreign policy.

Henderson asked Mosadeq how he aimed to solve Iran's economic problems without foreign aid. "He seemed somewhat at a loss," the ambassador recorded, "but eventually he said he supposed he would recommend to the new Majlis stringent cuts in the budget." To Henderson's observation that budget cuts would entail laying off government employees and intensifying public pain, Mosadeq replied that anything was better than submitting to the British. Furthermore, if he accepted a compromise, the communists would seize the nationalist banner and cause far more trouble than would government layoffs. He added—"half jokingly," Henderson thought—that if he sliced the budget the consequent disorders might cause the United States to reconsider and extend aid after all.

The obvious conclusion to be drawn from this performance was that Mosadeq wanted American assistance without hazarding a direct request. But, with Mosadeq, Henderson could never be sure whether the obvious conclusion was correct. Of one thing, however, the ambassador felt confident: that the shah's prediction that Mosadeq would be forced to resign soon was mistaken. "Neither by word nor bearing," Henderson wrote, "did Mosadeq indicate that he had any intention to resign. . . . On the contrary, he seemed to be as full as ever of fighting spirit."[26]

Pulling together the various strands of evidence at his disposal, Henderson offered some tentative prognostications. With the disclaimer that "so many political and economic factors are involved that even the best informed and most experienced Iranians appear confused regarding the present situation and extremely hazy regarding what the future holds," the ambassador gave his reasoned guesses, in order of probability:

a) Mosadeq will continue in office at least for several months, drawing on all available sources, including the printing press, to enable the Government to meet its payroll and most pressing current bills.

b) The Majlis or the Senate or both, under pressure from the Shah, will invite Mosadeq to resign.

c) Mosadeq will insist on retiring.

d) The Shah, without reference to the Majlis or the Senate, will demand Mosadeq's resignation.

e) Some form of military coup, with or without the knowledge of the Shah, will overthrow Mosadeq.

f) The Majlis on its own accord will accept the pro forma resignation which it is customary for the Prime Minister to proffer to the new Majlis.

Regarding the consequences of the scenario at the top of his list, Henderson predicted that Iran's economic position would continue to degenerate, inducing considerable unrest among the populace at large; that the National Front would attempt to blame the country's distress on the "Anglo-Americans"; and that amid the turmoil the Tudeh would gain adherents and influence. When this occurred, Henderson believed, the likely outcomes would reduce to two:

a) The National Front Government may develop into a dictatorship or may be replaced by a strong anti-Western government which will take measures of a confiscatory and revolutionary nature to obtain needed funds, or

b) The Shah or the Senate or the Majlis, or some military group, may intervene and may bring in a Government which with the assistance of the West will endeavor to restore emaciated Iran to a semblance of vigor.[27]

16

The Wily Premier, the Angry Ayatollah, and Hamlet on the Peacock Throne

Nineteen-fifty-two was a quiet year in American foreign affairs, certainly by comparison with those that came before and those to follow. The "great debate" of the previous spring had ended with Congress agreeing to send American troops to Europe, thus putting the "O" in NATO and physically committing the United States to the defense of the continent. The Marshall plan was working; after an American jump-start the economy of Western Europe was beginning to hum. In the Far East the triple shocks of 1949–1951—the communist victory in China, the North Korean attack on South Korea, and the Chinese intervention in the Korean war—still reverberated, but stalemate, as in the truce talks at Panmunjom, had become the predominant motif. After the ratification of the peace treaty with Japan, American responsibility for governing that former enemy ended. Although the French were encountering increasing difficulties in Indochina, real crisis there remained two years distant. The Middle East experienced business as usual, which meant trouble for the British. In 1952 Britain's greatest source of concern in the area was Egypt; London was trying to negotiate an extension of its treaty granting base rights at Suez. A colonels' revolt in July did not help matters, but for the moment the new head man, Gamal Abdel Nasser, seemed not entirely immoderate. Only later would his radicalism become apparent, and not until 1956, when he followed Mosadeq's example and nationalized the Suez Canal Company, would he spark the emergency that would bring the United States fully into Middle Eastern affairs.

In 1952 Henderson turned sixty. After three decades in the foreign service he was at the pinnacle of his profession. He certainly looked the diplomat.

Much earlier his dark hair had receded, adding height to an already impressive forehead; now the strands were nearly white. His clipped mustache retained a few black flecks, but not so many that an observer didn't have to look twice to be sure it was there. His strong nose and firm chin, and the mouth that habitually settled into a straight line, indicated a man sure of himself, or at least of his mission. Calm eyes met visitors straight on. He was neither tall nor short, paunchy nor particularly trim. His carriage was erect without being military; he wore a morning coat well. He drank sparingly, smoked unobtrusively. In the State Department and foreign service he was regarded with the greatest respect. Superiors valued his judgment, recognizing that it came from a person of unflinching integrity. Peers judged him the consummate career officer, a man who did not allow political considerations to color his advice, whose steady advancement owed to solid work and devotion to duty. Subordinates looked up to him as a model of what they might become.

Henderson's professional stature lent weight to his reporting from Tehran, which through the course of 1952 described a situation proceeding from bad to worse. This decline surprised Henderson, since when the year began he had not thought matters could deteriorate much more without provoking someone to do something about it. Of the two alternatives he had described in March— a descent to dictatorship under the National Front or elements even more anti-western, or a move by the shah and his supporters to restore vigor to Iran— there was no doubt which he preferred. Accordingly he sought to nerve the shah for action. At the same time he continued to work with Mosadeq to ensure that when and if the monarch decided to save Iran, there would be a country left to save.

During an interview in April, the shah seemed as confident as ever that Mosadeq would pass from the scene, although he was less sure that the prime minister's departure would come soon. The popularity of the National Front, the shah said, was "continually ebbing." Eventually Mosadeq's coalition would give way to a government more cooperative with the free world. Meanwhile the influence of the crown continued to increase. A few months ago Mosadeq was treating him as a cipher, but just yesterday the prime minister had made a point of boasting that his cabinet enjoyed royal support. When Mosadeq was "thoroughly discredited," he—Mosadeq—would retire. Until then, it would be a mistake to move against him.

For now, the shah continued, the United States might help matters by resuming assistance to the Iranian military. American aid had lapsed after Mosadeq refused to allow as much supervision as the American Congress required. The shah said he understood that Mosadeq's obstinacy made it difficult for the American government to provide aid, but he declared that a strong Iran would prove a valuable asset to the United States in the Middle East. Likening his country to Greece and Turkey in 1947, he asserted that it would be a tragedy if Washington could not find some way to prevent Iran from sliding into chaos.

Henderson had checked into the matter of military aid and was forced to reply that the law admitted no exceptions: unless the Iranian government allowed supervision the United States could ship no arms. Henderson also had to point out that time was working against American assistance. The current appropriations bill would terminate at the end of June, and unless Tehran made a satisfactory application in the next few weeks the funds earmarked for Iran would go elsewhere. Reopening the spigot might take as long as two or three years.[1]

With the shah seeking military aid, Mosadeq, in his roundabout fashion, continued to angle for economic help. Near the end of April the prime minister called Henderson in to discuss a matter of "the gravest importance." The British, he asserted, were making every effort to prevent the meeting of the new majlis, to overthrow his government, and to install a "stooge" in his place. During the last several days British agents had approached recently elected deputies to try to persuade them to boycott the assembly's convocation. This would block a quorum and paralyze the legislature. It would also starve the government of funds. Mosadeq added that the British might undertake some form of covert operation to topple him.

The prime minister went on to say he had discussed this matter with the shah. He had told the monarch he would fight the British to the last. He was old and fatigued. He would like nothing better than to retire. But duty came first. Should a pro-British government assume power, it would not survive long and would inevitably be succeeded by a communist regime. The National Front embodied the hopes of the youth of Iran. If the National Front disappeared these hopes would be shattered and the young people would turn to communism. Mosadeq said that in the event that the British succeeded in preventing a quorum he would call on the council of ministers to pass emergency measures to cut spending and raise taxes. But even this strategy, he confided, would not solve Iran's problems, since fresh revenues would not enter the treasury for at least two months. Meanwhile the government had to meet its obligations. Idled oil workers in the fields near Abadan were already feeling the pinch. By a "great effort" he had succeeded in finding five million tomans to alleviate the shortfall, but he needed at least ten million more in the next few days and had "no idea" how to obtain them.

Actually he did have an idea, as Henderson learned. Mosadeq said he once had resolved never to ask the United States for financial aid. His pride made it difficult even to mention the subject to the American ambassador. But in this emergency he must put personal feelings aside and do everything possible to rescue his country. He was not asking for new assistance, he declared. He would, however, appreciate Henderson's bringing to Washington's attention the parlous state of Iran's economy, and if the American government could divert already appropriated technical-assistance funds to his treasury, "Iran might be saved."

Mosadeq said he hoped the Americans had learned from their mistakes in China, where a failure of aid had led to a communist victory. Doubtless the British would object to American assistance, for they relished the prospect of rioting and disorder as possibly leading to his personal defeat. It was common knowledge in his country that the British hoped to partition Iran, claiming the southern half for themselves and granting the northern provinces to the Soviet Union. Immorality aside, such a scheme was "extremely short sighted." As Korea demonstrated, the communists did nothing by halves, and they would not settle for half of Iran without a major war. For himself, if he had to make a choice between the Russians and the British, "it would not be the British." Lest Henderson miss the slightest significance of what he was saying, Mosadeq asserted that any pro-British or pro-Soviet government in Iran would surely eject all American officials and attempt to abolish American influence. The British and the Russians differed on many subjects, but they concurred on the necessity of shutting out the Americans. To add emphasis to his claims of imminent danger facing Iran, he waved a sheaf of documents, which he said had just arrived from the north and indicated that the Russians were massing troops and supplies along the border.

Henderson told the prime minister he was mistaken regarding Britain. The British government and most responsible persons in Britain had no desire to take over Iran or to see any other country do so. As to British attitudes toward the United States, a few resentful types might like to see American influence eradicated in Iran, but the majority of informed individuals welcomed American help in keeping Iran free. The ambassador said he would pass along Mosadeq's suggestion regarding the transfer of Point Four funds, while making clear that the prime minister was not asking for aid. He did not know whether such an arrangement would prove possible. He assured Mosadeq that the United States government remained "deeply interested in the maintenance of Iranian independence," but he pointed out that the American people had difficulty understanding why they should support Iran when Iran could solve most of its financial problems simply by settling its oil dispute with the British.[2]

Thinking about the conversation afterward, Henderson took Mosadeq's semi-request for American aid as an indication that the prime minister, more beset than ever by the country's economic woes, recognized he was in trouble. Otherwise he would not risk the charges of selling out to the Anglo-Americans that the radicals would surely raise. Mosadeq's acquiescence in the continued stationing of American military advisers in the country seemed further evidence in this regard. To a certain extent, Henderson was pleased to inform Washington, Mosadeq's modest change of heart was the result of pressure from the shah—or at least from the shah's followers. Although the Iranian senate had never been a source of Mosadeq's strength, largely because half its members were appointed by the shah, the upper house recently had turned "clearly hostile." Emboldened by this development, and by other signs of the prime minister's weakness, newspapers previously supportive or neutral had gone over to the opposition. Even within Mosadeq's cabinet, five ministers were reported to have advocated the shah's acceptance of Mosadeq's pro forma resignation. "It has now become popular to blame the Mosadeq Government for current conditions."[3]

A short while later, Henderson received a visit from Hussein Ala, who said he needed to talk in "utmost confidence." Ala told Henderson he was worried that the shah would let this wonderful opportunity to destroy Mosadeq slip through his fingers. Until recently the crown's policy of letting the prime minister hang himself had seemed to be working, and while Mosadeq had proved unable to rectify Iran's problems, the shah had gained strength and popularity. But Ala said he feared that the shah was waiting too long. The country's economic troubles might engulf not only the prime minister but the shah as well, and what the populace had applauded as masterful inactivity on the part of the monarchy was increasingly being viewed as timidity.

Ala asked Henderson for advice. Mosadeq was about to go to the Netherlands to address the World Court. Should the shah demand his resignation before he left Tehran? Partly answering his own question, Ala pointed out that if the World Court ruled against Iran on the matter at issue, the crown and the successor premier inevitably would receive much of the blame. Should the shah let Mosadeq leave for the Hague and replace him in absentia? This might appear cowardly. Suppose the shah waited for Mosadeq's return to fire him. The prime minister, who had plenty of spies, might get wind of the plan and stop off in Switzerland or some other third country for weeks or months. The country could go bankrupt waiting for him to come back. Of course, no matter how the shah sacked Mosadeq, Iran's economic problems would remain. Ala commented that it was "quite clear" that American assistance hinged on a settlement of the oil dispute. Unfortunately the British showed no signs of moderating their demands. Indeed, Ala said he detected a trend in the opposite direction, and as Iran's financial crisis deepened the British would raise their demands further. To replace Mosadeq might only convince them that Iran was weakening. The minister concluded by soliciting Henderson's opinions, assuring him they would be treated on a "confidential and personal basis."

Henderson replied by asking if the shah had in mind a particular successor to Mosadeq. Ala mentioned several names, the most prominent being Ahmad Qavam, the former prime minister and the apparent favorite among the old-line politicians; and Allahyar Saleh, the leader of a moderate faction within the National Front. Ala had little good to say about Qavam, but he liked Saleh, as did the shah. Saleh was a man of integrity, Ala asserted, and the shah particularly appreciated his courage and determination in dealing with Kashani.

While Henderson did not know Qavam well, he had reservations about Saleh, who had been agitating for the expulsion of American military advisers. In itself, this was not an issue of primary importance, and Henderson assured Ala he did not judge Iranian politicians solely on their attitude toward American advisers. But he worried that Saleh's stance on this question might be symptomatic regarding more-basic matters affecting Iran's future. On various occasions in the past Saleh had displayed "appeasement tendencies in dealing with Russia and international communism." Further, there was little reason to believe that Saleh would be any more amenable than Mosadeq to a compromise on the oil dispute. The ambassador granted that he knew less of Saleh than he should, and he said—as unconvincingly, no doubt, as he intended—that he did not wish to prejudge the man.

Ala then asked Henderson what sort of final settlement the British might be willing to accept. Henderson sidestepped the question. He replied that he had no specific knowledge. Why didn't Ala ask the British himself?

By the time the conversation ended, Henderson had lost what small hope he had allowed himself that the shah might be nearing a decisive move against Mosadeq. Perhaps this was just as well, if the shah could not come up with a better substitute than Saleh. Henderson did have to concede that there existed some basis for Ala's fears that Britain would respond to Mosadeq's ouster by squeezing harder. He commented to Acheson, "I cannot therefore brush them lightly aside and suggest that he tell the Shah that he should get rid of Mosadeq now."[4]

II

Less than a week later, Ala was back. Mosadeq, evidently aiming to preempt action by the shah, had requested a conference with the monarch. As Ala described the conference, the prime minister had begun amicably but suddenly launched into a tirade against the court and the army for interfering in the country's politics. He charged the queen mother and other members of the royal family with intrigues against him, and he insisted that the shah take measures to stop these illegitimate maneuvers at once. The shah, according to Ala, had "heatedly denied" the accusations. On the contrary, the shah said, the intriguers were Mosadeq and his supporters, particularly Kashani, who had fixed elections and used terrorist tactics to intimidate their opponents. Mosadeq himself, he continued, had personally slandered Iranian statesmen of irreproachable character. Perhaps surprised at this vigorous counterattack, Mosadeq had backed off, saying he had not meant to imply that the shah was responsible for the failings of his followers. The shah pressed forward. He declared that he himself had been studiously correct in his treatment of Mosadeq, despite deep doubts about the prime minister's policies. He added that Mosadeq had yet to explain how he intended to address the government's overwhelming financial problems.

Ala asserted that this confrontation had thrown the prime minister into a depression, and that the shah desired to capitalize on his advantage. The monarch had told Ala that steps must be taken in the near future to replace Mosadeq. To this end he had directed Ala to call on Henderson and ask the ambassador's advice. In particular, he needed to know the attitude of the American government toward Mosadeq. Did Washington want him removed or not?

The question did not take Henderson unawares, and the ambassador framed his reply with care. He said that the United States considered Iran a friendly country, and that it had always endeavored to treat Mosadeq as it would the prime minister of any friendly country. But this did not imply American support for Mosadeq as a politician. The shah certainly knew of the problems Mosadeq's attitude had created in Iranian-American relations. Mosadeq himself knew, since he—Henderson—had never hidden from the prime minister his opinion that the National Front was leading Iran in a dangerous direction. Henderson commented that some months earlier he personally had reached the conclusion that as long as Mosadeq remained in office the oil problem could not be solved, and that without a solution on the oil front Iran could not rectify its domestic and international affairs. Henderson professed great admiration for Mosadeq's better qualities and declared—without batting an eye—that Washington held him "in high esteem." All the same, the prime minister's unreasonableness on the oil issue was "causing harm to Iran and to the entire community of free nations."

Ala thanked Henderson for his forthrightness. The minister of court reiterated that the shah had decided that the time was near for a change in government, but he said the shah did not wish to act without consulting the ambassador. Could Henderson, "personally and confidentially," offer counsel regarding how to proceed?

Following a disclaimer that he was not sufficiently acquainted with the situation to make a specific recommendation, Henderson proceeded to do just that. He suggested that the shah take no step until Mosadeq presented Iran's case before the World Court. But as soon as the prime minister completed his argument, the shah should demand that he return to Iran at once. By this act alone, the shah would seize the initiative. If Mosadeq balked, the shah would have grounds to fire him. If he came home as directed, the shah would have reaffirmed his right to give Mosadeq orders.

Ala expressed appreciation for this advice, and he went on to offer another pitch for Saleh as a successor to Mosadeq. The shah, he said, had met recently with Saleh and was more favorably impressed than ever with his "reasonable attitude and high ideals." Saleh had spoken quite critically of Mosadeq, which the shah found reassuring.

Henderson refused to budge from his earlier position. Saleh, he said, had a reputation for hostility to the west and especially to the American presence in Iran. This being the case, he worried that Saleh "might lead Iran to ruin" even faster than Mosadeq was doing. It lay entirely with the shah, of course, to choose whomever he desired as prime minister. But he should be warned that the United States would find it difficult to cooperate with Iran if he selected Saleh.

Although Ala must have left this meeting with quite a different view, Henderson in his report to Washington contended that he had "leaned over backwards not to become involved in Iranian internal affairs." Maybe he believed that his avoidance of a point-blank ultimatum to fire Mosadeq constituted non-involvement, or that by not explicitly vetoing Saleh he was simply offering a disinterested opinion. But Ala doubtless had no difficulty discerning that the American ambassador, and presumably the American government, wanted Mosadeq out and someone besides Saleh in. Perhaps rereading his own summary, Henderson felt it necessary to defend his action. He wrote to Acheson,

> I responded as I did to Ala's questions during the conversation because I believed that if I had not done so both the Shah and Ala would have obtained the impression that I am unwilling to talk frankly with them. As a result my present relationship with them might have been adversely affected. Furthermore, it seemed to me that I should not be evasive when a trustworthy emissary of the Shah approaches me in this fashion.[5]

Even as he did his best to undermine Mosadeq, Henderson thought it politic to keep in touch with the prime minister. At the beginning of July he paid the old man a visit. "He was in an extremely depressed mood," Henderson noted afterward. The majlis had just elected an opponent of Mosadeq to its presidency, and the prime minister, characteristically, attributed the election to British machinations. Under the circumstances, he told Henderson, he must tender his resignation. He went on to say that the parliament faced two alternatives. On one hand it could decide that Iran simply must get the oil flowing again. If it opted in this direction it would authorize the formation of a new government willing "to exchange Iranian independence for oil revenues." Mosadeq said he assumed that the United States knew that this would mean the return of British rule and the end of American influence in Iran. Moreover, the British, although strong in Iran, were weak internationally, and therefore at some point a government of British puppets would be compelled to "appease the Russians." On the other hand, if the majlis chose independence over oil lucre, it would support his government in its efforts to balance the budget and develop resources other than oil. In this case Iran would welcome the continued interest of the United States.

Lest Henderson somehow miss his drift, Mosadeq went on to assert that the Americans were defeating their own anticommunist purposes by allying themselves so closely with Britain. British policies were designed "either to dominate Iran or to strangle it." Whatever success Britain achieved in regaining control of Iran would not last long. Sooner or later they would be driven out, but this time by communists, not nationalists.

Earlier Mosadeq had inquired about reports that Washington was considering a plan to purchase Iranian oil on behalf of Nationalist China. Henderson took this opportunity to say that the reports were false, and he added that he could foresee no circumstances under which the United States could buy Iran's petroleum. To do so would result in a sharp deterioration of relations between the United States and Britain, which in turn would have "unfortunate consequences for the whole free world, including Iran." Mosadeq responded that he had not seriously credited the reports. Even so, if Iran stood firm, eventually it would break the British blockade. The world would beat a path to Iran's door.

As Henderson got up to leave, Mosadeq—possibly forgetting his earlier offer to resign—said that if his government fell, it would "go down fighting." Many difficulties lay ahead, for the British had succeeded in their efforts to infiltrate the parliament. He would accept whatever fate Allah had in store for him.[6]

III

Allah may or may not have been involved in the chain of events, collectively designated 30 Tir (July 20), that led to a resurgence of Mosadeq's power. Hoping to neutralize the shah, the prime minister demanded control of the war ministry. The shah resisted, appointing instead a close ally of the court as war minister. Mosadeq then resigned the premiership, and the shah appointed Qavam in his stead. Qavam would have had trouble enough dealing with Mosadeq's secular supporters, but he also alienated Kashani and the religious radicals by declaring that although he respected the teachings of Islam, he would "divorce religion from politics" and "prevent the dissemination of superstitious and retrogressive ideas."[7]

At this point the National Front took its case to the streets. Ignoring a ban on public gatherings, thousands of Mosadeq's and Kashani's partisans, with a strong admixture of Tudehists, packed the majlis square. When the crowd attempted to break a cordon of police and army troops surrounding the parliament building, a pitched battle broke out, leading to the deaths of several demonstrators. As per custom, the bodies of the fallen were carried shoulder high to the nearest mosque, creating an even greater uproar. Kashani issued a *fatwa*, or religious order, calling on the armed forces to join the people in the struggle against the government. Although ayatollahs in the past had used the *fatwa* against foreign foes and infidels, Kashani's call marked the first time the political struggle against the shah had taken such an explicitly religious turn. Facing the prospect of widespread insubordination for conscience's sake, the shah backed down, ordering the army to the barracks and dismissing Qavam. The majlis met in secret session and voted unanimously to reinstate Mosadeq, and the shah bowed to the decision.

Mosadeq's return, however, did not immediately end the rioting. The Tudeh, chanting "Down with the Shah! We want a People's Republic!," continued to engage anticommunist forces. Eventually the leftists abandoned the fight, but not before some two hundred persons were killed and the Tudeh and Mosadeq's party entered into a tacit agreement to deescalate their political and propaganda war against each other.[8]

Henderson found the crown's capitulation depressing but not entirely demoralizing. "In the foreseeable future," he wrote, "the Shah is likely to be a negligible political factor in Iran." Yet he continued, "We do not believe we should throw up our hands while the Iranians rush by in a mad and suicidal career like so many million lemmings." The United States might still manage to turn Iran from its self-destructive course, although this would require "prompt and radical changes" in American policy. Until the dust settled he would not advocate specific actions, but he did think that Washington now more than ever should seriously reconsider its deference to London on the Iran issue. Even "the most sensitive and suspicious Britishers" would have to admit that until the present the United States had backed Britain's policies toward Iran "to the hilt." "They cannot say that the failure of these policies was due to lack of United States cooperation." This cooperation, however, had failed to produce a settlement and had earned America only the condemnation of the Iranian masses. The Tudeh had succeeded in playing upon "the natural suspicions of Iranians regarding foreigners, perhaps justified by centuries of experience," and had created a climate for widespread belief in "malicious rumors propagated against Americans."

Although Mosadeq remained atop the government, the extent of his control was open to doubt. Henderson questioned the prime minister's ability to ride herd on those of his associates who had tasted power. Kashani certainly had been strengthened by the week's events, and the ayatollah might well decide at some point that he no longer needed Mosadeq. In addition, the prime minister and the Tudeh evidently had come to an understanding, even if Henderson could not discern the nature of the agreement; yet the communists likely intended "to ensnare and corrupt National Frontists at various levels." As for the army, the principal countervailing force, the officer corps as a whole had not been completely undone by the response of the rank and file to Kashani's *fatwa*, despite the fact that the spirit of certain units had been "shattered."

A crucial question involved Mosadeq's current attitude toward the United States. Since the recent disruptions, Henderson had deemed it wise to steer clear of the prime minister, believing he should wait to be summoned. Eventually the call would come, most probably when Mosadeq ran out of ways to put off facing the economic crisis. Because he now had nearly complete control of the government's financial and economic policies, he could postpone the day of reckoning. "He should be able to print enough bank notes to finance the Goverment for months." Still, prudence demanded preparation for the inevitable moment when the prime minister again would raise the issue of American aid. Much would depend on the circumstances. If Mosadeq appeared friendly, Henderson said, he would tell the prime minister that now that the National Front had crushed all opposition he could afford to adopt a conciliatory line on the oil question. "I might point out to him if I find him in a tractable mood that if Iran is to make real progress internally under his Government it should take an attitude which will help to remove suspicions and misunderstandings which are adversely affecting close cooperation between Iranians and the peoples who are their natural friends." Henderson added, "I hope that such an approach will have more effect than some of my previous efforts at persuasion."[9]

Four days later Henderson got his chance. Without directly requesting an interview, Mosadeq indicated he wished to speak to the ambassador. Henderson began his side of the conversation by expressing concern at the large number and vitriolic character of anti-American rumors flooding Tehran. It would serve no purpose, he said, to try to deny them all, but some implied or declared explicitly that he or members of the embassy staff had colluded with Qavam, and these demanded a response. Henderson asserted that the United States had not meddled in Iranian affairs and did not intend to do so. It had neither favored nor opposed Qavam for the premiership. Internal political decisions of this nature were matters for Iranians alone to decide. Stories that the embassy had intrigued to effect Mosadeq's resignation and Qavam's appointment were so ridiculous that he failed to understand how "intelligent Iranians, even in this emotional period, could credit them." Such tales must have been circulated "either by people who had completely lost their ability to reason or by enemies of Iran who desired to undermine Iran's independence by impairing relations between Iran and the United States."

Henderson went on to describe the nature of Qavam's approaches to the embassy. In a discussion the day after his appointment as prime minister, Qavam had said he hoped to achieve an oil settlement that would not in any way sacrifice Iran's rights or sovereignty; but in the meantime his government needed emergency assistance. Henderson told Mosadeq he had not encouraged Qavam in the matter. He did not know whether the United States government had funds available. If not, Congress would have to pass new legislation, and Congress had adjourned until after the November elections in the United

States. Even if the money were available, the State Department would insist that any aid not give offense to public opinion in the United States or the United Kingdom. It would not serve the interests of the free world, Iran included, for American assistance to Iran to lead to antagonism between Washington and London. Henderson said he had warned Qavam that American taxpayers were not prepared to underwrite any country indefinitely. The prime minister would improve his chances of receiving American aid by putting Iran's economic house in order. Should the prime minister undertake necessary reforms, he—Henderson—could recommend to Washington that Iran receive assistance for a limited time. He could not, however, know how Washington would treat such a recommendation.

Mosadeq listened quietly to Henderson's exposition, thanked the ambassador for conveying this confidential information—and then launched into a scorching attack on American policy. He charged the United States with pressuring Qavam to bow to British demands and thereby allow the return of British rule. Henderson interrupted to say that such an allegation was "entirely unjustified." The American government had never suggested to Qavam or anyone else that Iran should sacrifice one iota of its sovereignty. The United States had consistently held that Iran and Britain could reach a reasonable settlement that would not in any way diminish Iran's rights. Furthermore, he had never pressed Qavam to settle with the British. He had simply informed the prime minister of the circumstances under which the United States might be willing to extend aid. In this he told Qavam nothing he had not told Mosadeq earlier.

Mosadeq declared that the United States favored Qavam over himself. The American government had indicated that it would approve Qavam's request for aid after it had repeatedly refused his own. Henderson denied this as well. The United States did not play favorites, he asserted. Washington's response to Qavam's request owed solely to the nature and circumstances of that request. In any event, he—Henderson—had recommended in the past and was prepared to recommend in the future that Washington grant aid to Mosadeq's government provided the prime minister explicitly request such aid and indicate an intention to pursue sensible and sound policies.

Mosadeq next alleged that Henderson, by excessive friendliness toward Qavam, had encouraged Qavam against him. Henderson responded that whatever friendliness he had shown Qavam followed from the fact that the man was, at the time, prime minister of Iran. An ambassador's job included the cultivation of amicable relations with the leaders of the host country, whoever those leaders happened to be. Henderson pointed out that he had incurred criticism in the past for being friendly with Mosadeq. So long as he headed the American embassy he would do his best to get along with anyone who held the premiership.

Mosadeq proceeded to denounce American foreign policy more broadly. Henderson summarized:

> He said the United States had no diplomacy. The United States in the Middle East was merely an agent of Britain. The manifestations of anti-Americanism witnessed during recent days had shown how great had been the failure of so-called U.S. diplomacy in Iran. The United States had given a billion dollars of aid to Turkey and yet when Iran was bankrupt and on the verge of communism, it had refused financial assistance, first, because it feared that if Iran should be able to operate its own oil industry, U.S. oil interests in Saudi Arabia and elsewhere might suffer and, secondly, because it was afraid of British displeasure.

When Henderson replied that American petroleum interests did not control American foreign policy, Mosadeq asserted that even the British were saying the United States was backing Britain against Iran for fear that a compromise settlement would destabilize American concessions in other countries of the Middle East. Henderson again denied the significance of oil in American diplomacy, adding that whatever the case on this score, the countries in which Americans held concessions hardly appeared likely to follow the path Iran was taking.

After Henderson commented that the free world would suffer from a rift between the United States and Britain, Mosadeq "began to chant that Iran would prefer to go Communist than for the United States and the United Kingdom to have differences of opinion with regard to it." When Henderson managed to break in, the ambassador said the choice the prime minister described was no choice at all: if significant misunderstandings arose between the United States and Britain, Iran would fall to the Soviets as a side effect.

Mosadeq then charged the American judge on the World Court with being an agent of American diplomacy. Henderson responded that American judges neither at home nor abroad received instructions from the administration in power, but he added that he doubted that anyone not educated in the United States could understand the depth of the American commitment to an independent judiciary. Henderson said he had heard that Mosadeq himself hoped to separate the judicial branch from the executive in Iran. If he succeeded, perhaps someday he would appreciate how Americans felt on the issue.

The prime minister turned his guns on the Iranian army. It was now "hated by all Iranians." At a moment of imminent peril from the communists, the army had abandoned its position as a stabilizing factor in Iranian politics. Under the orders of Qavam, "who was a British agent," the army had fired on and killed hundreds of Iranians. The people knew that the army was a "tool of the British." To Henderson's query whether he intended to try to restore the prestige of the military, Mosadeq replied that it was too late. "Nothing can save the army." In fact, the officer corps had become positively dangerous to Iranian freedom. Officers and men, "humiliated at their present unpopularity," might try at any moment to regain public favor by taking the leadership in a communist revolt.

Henderson did not respond to this astonishing statement, but he did ask if the prime minister could clarify his position regarding American military aid and the American advisory mission in Iran. Mosadeq replied that he did not wish to discuss the matter at present. Henderson inquired whether the head of the American mission should call on the prime minister, now that the prime minister also headed the war department. Mosadeq answered that he would be glad to speak to the American commander, "so long as the military mission is here." Henderson could not tell whether this last phrase had special significance.

As the ambassador took his leave, Mosadeq said he hoped Henderson would not interpret amiss the directness of his comments. It was his practice, he said, to talk on a personal basis rather than as the prime minister to an ambassador. He simply wanted Henderson to know that his country and his government confronted "great danger" and that he could not understand America's inaction after Henderson's repeated assertions of American interest and friendship. Henderson said the United States was trying in many ways to help Iran; to which Mosadeq replied with a laugh that if the Americans really were attempting to assist they were doing a good job hiding their helpful activities.

Henderson found the two-and-a-half-hour session "exhausting and discouraging."

> As I listened to him I could not but be discouraged at the thought that a person so lacking in stability and clearly dominated by emotions and prejudices should represent the only bulwark left between Iran and Communism. As during several previous conversations, I had the feeling at times that I was talking with someone not quite sane, and that therefore he should be humored rather than reasoned with.
>
> On occasions he resorted to such wild exaggerations and extravagances that it seemed almost useless to talk further. At one point I almost decided to abandon our conversation when he repeated again and again in a monotone that "Iran would never, never want the United Kingdom and the United States to have any differences over it. Iran would prefer to go Communist rather than cause any trouble between the United States and the United Kingdom."
>
> There were periods during our talk when he seemed lucid and sensible. The general impression he left was, however, one of deterioration. I have noticed in the past that in the evenings he is likely to be more tired and to have less control over his emotions. I can only hope that his behavior last evening was due to the strain of recent events and fatigue and does not indicate serious degeneration.[10]

As it turned out, fatigue—or intentional theatrics—accounted for Mosadeq's emotional behavior. Less than forty-eight hours later Henderson had another session with the prime minister. "I found him more composed than usual. Although showing emotion at times, he did not engage in extravagances and he made relatively few wild statements. He was apparently endeavoring to restore personal relations of a cordial character."

For his part, Henderson said he was unhappy at the pessimism the prime minister had displayed at their meeting two days before. He told Mosadeq he considered him Iran's principal defender against communism. He hoped the statements the prime minister had made to the effect that Iran was drifting inevitably toward Moscow did not represent his "sober views." In light of the courage the prime minister had exhibited in past crises, he found it difficult to believe he held such bleak opinions now.

Mosadeq denied he had exaggerated the danger. The situation with the army was "almost hopeless." He had been looking for a reliable chief of staff but could find no one who was both capable and trustworthy. Most Iranian generals were "venal and outmoded." Officers of lower rank, while perhaps qualified in other respects, lacked the stature a staff chief required. He agreed with Henderson that he constituted the "only effective barrier in Iran against Communism." The different factions within the National Front were so jealous of each other that they would listen to no one but himself, and if he relaxed his grip he did not know what would become of the alliance. He did not have to tell the ambassador of Iran's economic distress. The government could print money, but this would lead to inflation and a further decline, which the communists would quickly exploit. For now, he had the Tudeh under control. For how much longer, who knew?

Mosadeq and Henderson waltzed around the issue of economic aid for most of an hour. As before, the prime minister asserted that the communists would take over Iran if American dollars did not come to the rescue, without directly asking for assistance. Henderson reiterated that the United States would happily provide financial help if Iran compromised with Britain on the oil question.[11]

IV

But Mosadeq had no intention of backing down, and neither did the British. George Middleton, surveying the aftermath of 30 Tir, and following an audience of his own with the prime minister, came to conclusions nearly identical to Henderson's first reactions. The British chargé d'affaires asserted of Mosadeq, "His megalomania is now verging on mental instability." London agreed. Hoping to push Mosadeq, or at least Mosadeq's government, over the edge, Eden and Churchill refused to moderate their opposition. At the same time they worked more assiduously than ever to prevent the Americans from coming to Mosadeq's rescue.[12]

Henderson set himself against the British policy. The American ambassador had no interest in preserving the assets of the Anglo-Iranian company, especially if preservation meant risking the loss of Iran to the communists. He argued emphatically that should Washington continue to defer to London and London continue to stonewall Tehran, the United States would see its already tenuous influence in Iranian affairs vanish entirely. As Iran's economic problems deepened, there would develop "increased tendencies to blame the West, particularly the United States, for all of Iran's ills." The Tudeh would demand the withdrawal of the American military mission. Mosadeq might feel forced to comply. The radicals would probably also turn their attacks on American Point Four advisers. "As the agitation against the West increases, Iranians with Western background and sympathies will be compelled to hide their real feelings or be gradually eliminated from public life." Although the government might attempt to steer a neutral course between east and west, the communists probably would emerge the winners. "The nationalists may realize that they are leading Iran into the Communist camp in time to shift their policies and rally the country against Tudeh infiltration and Soviet pressures. It is more likely, however, that this realization would come too late."

On the other hand, should Washington manage to persuade the British to settle, and should American aid begin to flow, Iran's downward spiral might be halted. Even this course offered no guarantees, though. The situation might already be beyond repair. At its most accommodating, the United States might find itself spurned and reviled. Hostility toward the west, now increasingly directed against America, had sunk its roots so deep in Iranian society that "the most conciliatory and friendly gestures on the part of the West may not be able to prevent Iran from proceeding along its suicidal course." Mosadeq might moderate his language about the perfidy of the Anglo-Americans, but he would not live forever. If his health remained good, he might fall to an assassin's bullet, as his predecessor Ali Razmara had. Should this happen, the communists would be the ultimate beneficiaries.

Henderson considered assassination a real possibility. The prime minister had already been the object of murder attempts. Furthermore, Mosadeq's principal ally in the National Front, Kashani, was also his chief rival, and considering the ayatollah's conviction that he had Allah on his side, Henderson did not doubt that Kashani could rationalize killing as a means to an end. Indeed, Kashani had recently engineered a vote in the majlis pardoning Razmara's assassin, and he was known to have reached an accord with a terrorist group responsible for the deaths of several Iranian officials.[13]

Although Henderson at this time judged Kashani a greater immediate threat to Mosadeq and the stability of Iran than the Tudeh, he believed that the communists eventually would profit from a takeover by the religious right. Throughout this period, Henderson consistently underestimated the seriousness and capacity for positive action of the Islamic fundamentalists. Henderson could appreciate the force of secular Iranian nationalism, which had its counterparts, if not its models, in European and American history. Communism, too, he could understand, having studied its operation in Europe and elsewhere for more than thirty years. But the strength of radical Shiism as a transforming influence in Iran was essentially lost on him. Having been reared in a culture that drew a sharp line between church and state, he had difficulty accepting the legitimacy—or recognizing that rational Iranians could accept the legitimacy—of political religious leaders like Kashani. Thus he tended to dismiss Kashani as a demagogue, distinguishing him from those he described as the "really religious clerics"—that is, the nonpolitical ones.[14]

Of course Henderson was neither the first nor the last western observer to sell the mullahs short, but in his case, although he missed the staying power of the movement they led, he did not mistake their capacity for challenging the status quo. From personal observation he knew that Mosadeq definitely felt challenged. During one discussion of Kashani's growing influence, the prime minister commented that whenever he went to sleep he never knew if an assassin would strike in the night. Early in August, immediately following news that the majlis had named Kashani its president, Henderson visited Mosadeq to get his reaction. The prime minister appeared not to know what Henderson was talking about. Certain members of the majlis, he said, had discussed a Kashani candidacy, but no election had taken place. In Henderson's presence the prime minister made a telephone call to verify his statement, only to discover that the ambassador was right. "Mosadeq was obviously shocked," Henderson wrote. "For a moment he seemed to forget my presence and did not seek to hide his distress and agitation. He fell back on his bed and closed his eyes. I thought he might lose consciousness." When he regained his composure Mosadeq asked Henderson what he thought of the matter. "In an effort to ease the tension I laughed and said I was merely an observer and was not entitled to any views." Henderson did remark that the election would affect him personally in that he would now have to call on Kashani and establish official relations with him. "Mosadeq made a wry face and said my visit could be courteous and short. I again laughed and said I was not sure how short a conversation with Kashani could be."[15]

Henderson soon found out, and discovered that the answer was not short. In mid-August the majlis president treated him to what Henderson described as a "long harangue" on the sins of the west generally and the United States in particular. It was the same old story, and Henderson had little of substance to pass along from the conversation. The interview chiefly served to confirm Henderson's earlier opinion that the ayatollah was "inherently unstable."[16]

Indeed, to Henderson's way of thinking, Kashani made Mosadeq look like a chunk of bedrock. The ambassador declared to the State Department that Kashani was "clearly trying to undermine the influence of other rival leaders and make himself the complete master of the National Front and of Iran." For the moment the ayatollah was accepting a position subordinate to Mosadeq, but Henderson had no doubt that eventually he would turn against the prime minister, "whom he finds too popular and at the same time stubborn." Planning for that day, Kashani had lined up the support of various influential groups in Tehran. His son captained a "well-paid gang of professional hoodlums"—a useful element in the unmannered politics of the Tehran bazaar—and he himself had strengthened his tacit alliance with Razmara's assassins. In addition, the Tudehists appeared willing for the time being to throw their support to Kashani, evidently believing that whatever undermined Mosadeq would benefit them.

With Kashani's "weapons of organization and intimidation" growing more potent daily, Henderson expected the ayatollah to make a bid for primacy soon. Neither the parliament nor the shah, both cowed, would oppose him. Should Kashani succeed, the result would be "another defeat for Western policies in Iran," and it would "further reduce the chances of the country remaining outside the Soviet orbit." Although certainly not a communist himself, Kashani would open the way to a communist takeover. He had no experience administering a government. At best he would allow Iran's fiscal chaos to continue. More likely he would exacerbate it. He would probably pressure the shah to abdicate, and the latter "would be relieved at the opportunity." Kashani might well abolish the senate. He would undermine the army by installing his agents in key positions. As an advocate of neutralism and a sponsor of pan-Islamic solidarity, he would support efforts to eject the west entirely from the Middle East. Despite his proclaimed opposition to communism, he would continue to tolerate the Tudeh. He already was on record as saying he did not believe the Tudeh was really communist. In sum, a Kashani regime "might well create a situation that would give the Tudeh a unique opportunity to take over Iran."[17]

A Tudeh takeover could be prevented, Henderson thought, but prevention would require decisive action by the shah and the military, and this seemed ever more unlikely. Henderson could identify no officers who possessed the organizational ability, intelligence, and strength required for a successful blow. The essential difficulty, however, lay with the shah. The army would not move without a nod from the monarch, and he as usual could not make up his mind. Although he still spoke as though time were on his side, increasingly it seemed that he simply lacked the nerve to oppose Mosadeq and Kashani. The United States might hint at support for a general who would put the prime minister and the ayatollah in their places, but such hints would do no good without the crown's backing. Henderson predicted gloomily that the shah would "take fright at the very idea of a coup."[18]

17

What Gentlemen Will Do

The American elections of 1952 produced considerably less change in the direction of American foreign policy than the winning Republicans had promised their victory would bring. The Republicans labeled the containment policy of the Truman administration defeatist, but in office they embraced and extended it. Dwight Eisenhower and John Foster Dulles recognized that rolling back the iron curtain entailed unacceptable risks of superpower conflict. Moreover, they understood that simply holding the line against further communist advances—that is, containment—strained American ingenuity and resources to their limits. Eisenhower and Dulles were not so naïve as to think that communist leaders like Mao in China and Ho Chi Minh in Vietnam were simply puppets of the Kremlin. But they did believe, quite correctly, that communist regimes generally had more in common with each other than with democratic states, and they recognized in communism, especially in the third world, a profound challenge to a largely pro-western global status quo. Defending this status quo, as much as opposing communism per se, was what containment was all about.

Inconveniently, defending the status quo gained the United States few friends among those peoples for whom things-as-they-were promised little more than continued poverty and disfranchisement. Nor did it sit well with the keepers of America's anti-imperialist heritage. But if a realpolitik policy of holding what the west had lacked political appeal, an ideology of defending the ramparts of the free world against godless communism possessed plenty. Consequently challenges to the status quo came to be equated, partly by design and partly by self-persuasion, with communism. Sometimes the challengers were alleged to be communists in disguise. This approach gave rise to the notorious "duck test": if it walks like a duck, talks like a duck, and so forth. Jacobo Arbenz Guzman, who made the mistake of expropriating a quarter-million acres of land from the American-owned United Fruit Company, accepting a shipload of weapons from Czechoslovakia, and otherwise demonstrating an unwillingness to accept American dictation regarding the affairs of Guatemala, passed the test with flying feathers. In the process, Arbenz prompted Eisenhower to order the Central

Intelligence Agency to assist in his overthrow, which the agency did with delight.

A second approach applied where the challengers indisputably were not communists. Here the argument was that the incumbents, by ignorance or perversity, were playing into the hands of communists who waited just offstage. Not surprisingly, since any challenge to the status quo almost anywhere in the third world included elements espousing a Marxist philosophy, the possible uses of this approach were endless, and indeed it would be used by American leaders over the years to justify opposition to change in a wide variety of countries throughout Latin America, Asia, and Africa. Iran fell into this category.

II

Shortly after Eisenhower's November 1952 victory, the State Department called Henderson home for consultation. Before he left Tehran the ambassador paid his respects to Mosadeq and the shah. Mosadeq had nothing new to say beyond adding specifics to his tale of Iranian woe. Within the next few weeks his government would have to lay off some 20,000 workers. These "unfortunate people" would swell the ranks of "the malcontents in the country who were particularly vulnerable to Communist propaganda."[1]

Henderson had not spoken to the shah since the tumultuous events of 30 Tir, and he thought the monarch had aged visibly. Even so, the ambassador found him in better spirits than expected. "If his morale has been completely shattered as has been reported, his manner and statements did not so indicate." On the contrary, the shah discovered cause for optimism in the developments of the past several weeks. Now, he said, the issues were growing clear to the majority of the Iranian people. The Tudeh had come out in the open, and many individuals who had viewed the communists with indifference were being put off by their violence. As a consequence—predictably—he should bide his time and let the communists do themselves further damage.[2]

Henderson did not agree, as he explained to Acheson. The shah was merely making excuses for his timidity. For more than a year the ambassador had been predicting the imminent collapse of Iran's economy. Such a collapse had not occurred, and he was forced to admit that economically the country showed more resilience than he had guessed. He went so far, in fact, as to estimate that on purely economic grounds Iran could probably survive for another ten or twelve months with neither oil revenues nor foreign aid. But economics, he maintained, told just part of the story, and the smaller part at that. Henderson agreed with Mosadeq that government retrenchment would feed the fires of political discontent, and while it was not impossible that a government that laid off workers and raised taxes might keep dissent under control, it was unlikely. Such a regime would have to be "a skillful, strong and ruthless dictatorship" in order to deal successfully "with increased anti-westernism, with the dissatisfaction of taxpayers and ousted bureaucrats and other disgruntled elements, and with growing despondency among the masses." Aside from the fact that no one except the communists appeared capable of providing such leadership—certainly not the shah—the emergence of a regime of this sort "could hardly work out to our advantage in this area and could quickly undo all the constructive work we have put into Iran during recent years."[3]

Such were the opinions Henderson took to Washington in the autumn of 1952, and they helped persuade Acheson to make a final effort to solve the Iran problem. In talks with Anthony Eden the secretary of state declared that the American government remained committed to Anglo-American cooperation in Iran so long as such cooperation promised motion toward a settlement. But the situation was becoming so hazardous that Washington might be forced to move on its own to prevent a takeover by the left.

At a certain level, the stalemate between London and Tehran in the oil negotiations now hinged on the amount of compensation due the Anglo-Iranian company for its nationalized assets. At another and deeper level, the contest had become a test of strength: Mosadeq and Iranian nationalism, on one hand, versus Eden and Churchill and British imperialism on the other. In the latter struggle, neither side was any more inclined to compromise than it had been from the first.

This was not entirely apparent at the end of 1952, and Acheson sought to put together a package both sides might accept. With some difficulty he got Eden to go along with a scheme for submitting the compensation question to international arbitration and lifting the legal blockade of Iranian oil exports. To Mosadeq he held out the incentive of large-scale American assistance.

Henderson assumed the task of presenting the package to Mosadeq. This the ambassador did in marathon talks beginning on Christmas day and lasting until the middle of January 1953. Henderson later said he had "never worked harder" than in these negotiations. "I sat by the side of Mosadeq's bed one day for eight consecutive hours going over the proposal point by point and explaining to him the significance and advantages to Iran of each paragraph. Mosadeq seemed interested and even grateful and promised to give me an answer within the next few days after discussions with his advisers. When I left I was tired but somewhat encouraged." Unfortunately it soon became evident the prime minister was just being polite. Mosadeq had no intention of compromising with the British. Although Henderson might have predicted such an outcome, he found the whole affair extremely frustrating. "Mosadeq," he commented, "was like a rubber band which one could stretch but would go back to its original position when it was let go."[4]

Interpreting the failure of the talks, Henderson remarked that the prime minister correctly believed that his creaking coalition, held together only by a common xenophobia, probably could not survive a compromise on the oil issue. Beset by Kashani on one side and the Tudeh on the other, Mosadeq could still count on rallying the public only by inflaming hostility against the west. The oil card was his last.[5]

Washington might have improved its part of the deal, but Acheson was leaving office and the British were balking. Eden thought toleration of Mosadeq had gone far enough. The foreign secretary wrote afterward that he had determined by this time that Britain and the United States "should be better occupied looking for alternatives to Mussadiq rather than trying to buy him off."[6]

Although Henderson continued to differ strenuously with London on a number of points regarding Iran, he concurred on this. Some weeks earlier he had collaborated with George Middleton on a report indicating that a change from Mosadeq would not necessarily be a change for the worse. "We both agree too," the British diplomat wrote, "that the longer Mussadiq stays in power the greater will become the hold of the Tudeh on the country." Through the end of 1952 Henderson retained some hope that a Mosadeq-led Iran might avoid disaster. But the prime minister's refusal, or inability, to accept the Acheson compromise package in January 1953 convinced him Mosadeq had to go.[7]

During Eisenhower's initial months in office, problems other than Iran occupied the president and Secretary of State Dulles. Trying to liquidate the war in Korea came first on the list of priorities, not least because Eisenhower's pledge to go to Korea, and presumably find the answer to the deadlock that had been eluding the Democrats, had been a centerpiece of the 1952 campaign. Figuring out how to respond to Stalin's death in March 1953 also took time. Not until early spring, when Dulles embarked on a tour of the Middle East, did the new administration have a chance to turn major attention to Iran.

Meanwhile Henderson laid the groundwork for action. In February he sent Dulles a report of a conversation with Ala in which the court minister described a new campaign by Mosadeq to undermine the shah's authority. The prime minister had recently cut off funding of charitable organizations that dispensed aid in the name of the crown to the destitute of the country. Mosadeq defended the step on grounds of economy, but Ala considered it a transparent effort to alienate the people from the monarchy. Moreover, Mosadeq was insisting that control of the Meshed shrine, which for centuries had been under the special protection of Iran's shahs, be given to Kashani's associates in the interior ministry. In other respects, Ala continued, Mosadeq was growing increasingly cantankerous, and the shah was finding it nearly impossible to work with him. As a consequence the court again was looking for a replacement. Various names had surfaced, including Hassan Ali Mansur and Saleh. General Fazlollah Zahedi was intriguing for the job, but the shah considered him "an adventurer lacking judgment and balance." Ala asked Henderson for suggestions.

The ambassador replied that ousting Mosadeq would not be easy. It would take careful preparation. Henderson recommended lining up a new cabinet consisting of men of "complete integrity" and representing the most important political groups in the country. Once the cabinet was agreed upon, then the shah could strike at Mosadeq. The shah should publicly announce that the prime minister was being relieved. Henderson did not specify a replacement at this point, but he stressed that above all the shah must give the new premier "public and firm support."

Ala immediately objected that he could not advocate that the shah assume such an exposed position. Not only would the monarch personally be safer if he remained above the fray, his prestige would be better preserved. Ala said that every time the shah stooped to tangle with Mosadeq, his popularity diminished.

Henderson retorted that the issue at hand was not the popularity of the shah but the survival of the country. In any event, the ambassador added, public opinion in Iran was fickle and popularity ephemeral. If the shah had some "solid accomplishments" to his name he would not have to worry about the passions of the moment. Ala granted that Henderson might be right. Still, he said, he hesitated to suggest that the shah take unnecessary risks. The monarch would be much safer if he limited himself to charitable works and social-welfare programs. Henderson answered that such a course would lose Iran its freedom and the shah his crown.[8]

Despite the shah's, or Ala's—it was hard to know when the minister spoke for himself and when for his chief—low opinion of Zahedi, Henderson gradually concluded that the general was America's best bet for a successor to Mosadeq. A new outbreak of violence at the end of February prompted another offer by Mosadeq to resign and caused Henderson to evaluate Zahedi's prospects. The general, Henderson conceded to Dulles, was "not ideal," but neither was anyone else, and Zahedi seemed to possess "more chance of piloting Iran through the turbulent days following Mosadeq's resignation than any other candidate now on the horizon." Zahedi was about the only officer of any stature who had stood up to Mosadeq, openly denouncing the prime minister, branding his regime a "despotism," and declaring that Iran was "being dragged to the precipice by a feeble mind." Furthermore, according to Ardeshir Zahedi, the general's son, the elder Zahedi had negotiated the support of Kashani, who apparently believed that the prime minister was gathering too much power to himself.[9]

But this pregnant moment passed without issue. Neither Zahedi nor the shah made the decisive move necessary to displace the prime minister, and Mosadeq regained his footing. Ala appeared relieved. The court minister still shuddered at the thought of an assertive monarch.

Henderson was disgusted at the opportunity lost. In this latest confrontation with the prime minister, the shah had disgraced himself. "Mosadeq will continue to humiliate him almost systematically," Henderson predicted.[10]

What else the prime minister would do was hard to tell. "Mosadeq is so much a creature of his own emotion, prejudices and suspicions that attempts to analyze the motives of his various actions in light of ordinary rules of logic or on the basis of reason might well lead one astray." Nonetheless Henderson identified certain constants in Mosadeq's behavior. One, which accounted for much of the prime minister's extreme antipathy toward the shah, was a "secret contempt" on the part of a man of an old aristocratic family toward "the weakling son of an upstart tyrannical imposter." Another was the more practical desire to eliminate the shah as a last rallying point for Mosadeq's enemies. A third was the premier's oft-remarked habit of deflecting criticism by trying to pin responsibility on others.

> His career has been based on negative activities and slogans. As Prime Minister he has not been able to accomplish anything of a constructive character. When frustrated he searches for some new opponent to blame and destroy. He has thrown out the British; emasculated the Majlis, eliminated the senate; forced all well-known politicians out of public life; deposed all prominent civilian and military officials; sent various members of the Royal Family into exile, etc. Now he places the blame for his failures on the court and takes measures against the Shah.

The Iranian people, Henderson noted, had demonstrated a remarkable willingness to accept Mosadeq's arguments. But as the economy continued to decline and dissatisfaction continued to grow, the prime minister would need to find new devils. The United States was the obvious candidate.[11]

Indeed, Henderson suspected that he himself might be nominated. After all, not only did the ambassador represent the United States in Iran, he was actively conniving with the opposition. Although Mosadeq did not know the details of Henderson's conversations with Ala, the prime minister was certainly aware that the ambassador had been meeting with the shah's principal agent throughout the February crisis. (Years later Mosadeq accused Henderson of being part of an active conspiracy against him at this time.) Henderson half-expected to be declared persona non grata and kicked out of the country. He told Dulles he did not intend to cut off his contacts with the shah and Zahedi, but he remarked that "any moves which the Embassy might make in this glass house must be extremely discreet."[12]

Although the awkwardness of Henderson's position prompted discussion in Washington of sending a replacement, Dulles decided to keep the ambassador on. Henderson's connections to the court would not be easy to reproduce. The secretary of state commended Henderson for his adept handling of "a most difficult and confused situation." (Mosadeq, explaining why he had not asked for Henderson's withdrawal, said the American government would merely have replaced Henderson with "someone like him or even worse.") Dulles went on to request further information. The secretary worried that Mosadeq's recent triumph had essentially eliminated the shah from the Iranian picture. Would it be accurate, Dulles asked, to say that the shah had disappeared as a political factor?[13]

Henderson replied that the shah was down but not quite out. "Despite the aura of passivity which envelops him, the struggle for his survival and contacts between him and the opposition are still continuing." Henderson did not foresee any clearcut victory for the shah in the near term. Yet a compromise leaving him some influence remained a possibility. "Many army officers are very disturbed by the continuing inaction on the part of the Shah. Nevertheless, all of them have not given up the idea of making some move on his behalf, even though without his advance consent or knowledge."

Henderson believed that a move by the military, without the shah's knowledge, might be the only way to a solution. The shah, he said, "would undoubtedly be frightened at the thought of a military coup being attempted in his name and if given the opportunity would probably try to stop it." But even the military would require a nudge. Iranians, at least the educated ones, preferred talk to action, and they had little knack for organization. Henderson doubted that loyalists would act on their own.[14]

In April, the confrontation between Mosadeq and the shah took a new turn. The prime minister understood as well as Henderson that the army represented the primary threat to his power, and he sought to bring it under his control. The constitution left vague the issue of precisely who commanded the armed forces—the shah or the prime minister—and to clarify the matter Mosadeq arranged the appointment of a special majlis committee. When the group predictably reported in favor of the prime minister, Mosadeq made acceptance of the committee's report a test of confidence in his government. He mobilized his forces indoors and out, and carried the day. Although Mosadeq officially continued to distance himself from the Tudeh, Henderson noted that the communists had appeared "in full force" for a government-sponsored demonstration. "It is safe to assume that working arrangements were made," he concluded.[15]

A short while later Henderson learned that Ala was being pushed aside as the shah's principal adviser. Not infrequently Henderson had found the court minister's solicitude for the shah's political safety exasperating, but at least Ala's heart appeared to be in the right place, and recently he had begun advocating a mildly stiffer line—which had led to successful pressure from Mosadeq for his ouster. Occasionally Henderson thought Ala suffered from the same delusion of foreign omnipotence the prime minister labored under. Earlier in April, Ala had told the ambassador that the British were responsible for the current strife in Iran. London hoped to trigger a civil war, he claimed, so it could divide the country with the Russians. But Ala shared Henderson's concern for the growing strength of the left, and by the time of his resignation he had come around to the ambassador's view that Zahedi might be the man for the moment. In his parting interview with Henderson as minister of court, Ala recommended the general as "the only person at the present time capable of taking over from Mosadeq."[16]

Because Kashani had been identified as a potential supporter of Zahedi against Mosadeq, Henderson responded with particular interest when the ayatollah invited the ambassador for a visit. The conversation proved disappointing. Kashani said nothing about Zahedi, although he mentioned that he himself had been doing his best to persuade the shah to resist Mosadeq's attempts to take over the army. He spent most of the two-hour meeting denouncing Britain and criticizing the Americans for letting themselves be duped by the British. Kashani told Henderson that Washington did not fully appreciate the deviousness of the British. Despite their smooth talk of Anglo-American cooperation against the communists, their policies in Iran were "more anti-American than anti-Communist." Henderson thought Kashani was trying to provoke a response. The ambassador held his tongue. "Kashani is so unpredictable and untrustworthy," he commented afterward, "that in my opinion it would not be wise to discuss Iran's internal affairs with him except in the most guarded language. In fact it is somewhat dangerous for me to talk with him at all since I can never be sure how he will later describe our conversation."[17]

III

By the beginning of May the Eisenhower administration had cleared the decks for action in the Middle East. On Henderson's recommendation the secretary did not stop in Tehran during his tour of the region. Henderson guessed that Mosadeq would find a way of turning a high-level American visit to his advantage no matter what Dulles did. A formal and diplomatically correct approach would convict the Americans of arrogance, while a friendly greeting would imply endorsement of Mosadeq's regime. So Dulles stayed away. Henderson met the secretary in Karachi, where he reiterated in person what he had been cabling to Washington during the previous months.[18]

On this same tour Dulles stopped in Cairo, where he raised with Nasser the possibility of an anti-Soviet regional alliance. The Middle East Defense Organization, as its Anglo-American sponsors hopefully called it, would be centered on the British base at Suez. MEDO's prospects had dimmed somewhat following the 1952 revolt that brought Nasser and the Egyptian Free Officers to power, for they appeared as committed to removing Britain from Egypt as Mosadeq was to eliminating the British from Iran. But Nasser had not explicitly vetoed the plan, and Dulles thought it worth pitching.

Nasser quickly made plain that he would have nothing to do with MEDO. The colonel told Dulles he could not survive politically, perhaps not even personally, if he put opposition to communism ahead of getting the British out of Egypt. "How can I go to my people and tell them I am disregarding a killer with a pistol sixty miles from me at the Suez Canal to worry about someone who is holding a knife 5,000 miles away?" Dulles tried to change Nasser's mind, but to no avail. Writing to the State Department afterward, the secretary commented on the colonel and his associates in terms Henderson on occasion might have applied to Mosadeq in Iran: "Their emotions are so great that they would rather go down as martyrs than concede. . . . It is almost impossible to overemphasize the intensity of this feeling. It may be pathological, but it is a fact."[19]

Cairo's opposition forced Dulles to drop the idea of an Egypt-centered defense organization. As the secretary explained to Eisenhower upon his return to Washington, MEDO was dead. Even if by some miracle the Egyptians and British reached a peaceful settlement regarding Suez, popular antipathy toward Britain would prevent any meaningful cooperation.

But Dulles was not giving up on the collective approach to Middle Eastern security. His discussions with government officials in Turkey and Pakistan had convinced him that those countries would collaborate with the west. Syria and Iraq also seemed cooperative. The missing link—what Dulles called "the obvious weak spot in what could become a strong defensive arrangement of the northern tier of states"—was Iran. The northern-tier approach had much to recommend it, including the possibility of stopping the Russians before they got to the oil fields and warm waters of the Persian Gulf, and the advantages of fighting a defensive war in the mountains of the north rather than on the plains of the south. Besides, after Nasser's MEDO veto, the United States had few alternatives. The northern-tier scheme must be made to work. Dulles thought it would, if the United States could "save Iran."[20]

In Henderson's view, Nasser's veto and Dulles's conceptual shift to the northern tier could not have come at a more propitious moment. Although Mosadeq had triumphed in the April struggle for control of the army, Henderson detected in the prime minister's victory the possibility for a counteroffensive. By directly attacking the shah, Mosadeq had alienated important groups to whom the monarch remained the symbol of national unity. And while conventional wisdom accounted the replacement of Ala by Abol-Qasem Amini an advantage for the prime minister, Henderson was not so sure. Like Mosadeq, Amini came from an old and powerful family; Henderson saw the new court minister as being as much a potential rival to Mosadeq as an ally. To put the matter another way, Amini seemed to be working neither for Mosadeq nor for the shah, but for Amini. Under the right circumstances he might throw his weight against the prime minister. As for the political opposition, although Mosadeq's adversaries could not muster sufficient strength to challenge the premier in street fighting or open parliamentary combat, they had managed through what Henderson called "guerrilla tactics" to block a quorum in the majlis and prevent Mosadeq from further consolidating his power.

Henderson also noted a "sharp shift" in Mosadeq's political base. During the preceding half year most of the original National Front had defected, either explicitly or by inaction. Mosadeq now relied for support primarily on Iran's security forces, which, as he himself admitted to Henderson, he did not trust; on the bureaucracy, whose loyalty might last no longer than its paychecks; on the government-controlled radio, which still enabled him to reach the people; and on a remnant coalition of merchants and aristocrats, who for a variety of reasons found Mosadeq's coattails comforting. Henderson granted that the prime minister retained considerable appeal as the nationalist leader who was freeing Iran from the foreign yoke. But the ambassador contended that even in this area Mosadeq's prestige was wearing thin, and he predicted that holes would emerge as the premier continued to ignore the huge economic problems facing the country. Finally, people otherwise well disposed toward Mosadeq were beginning to resent his resorting to street violence and, more recently, to martial law to intimidate the opposition.

And yet, although Henderson was convinced that Mosadeq could be beaten, the prime minister remained for the moment "the outstanding political figure" in Iran. "His opponents thus far have not shown the courage and spirit and unity necessary seriously to threaten him." Amini posed the greatest challenge to Mosadeq's power, but the minister of court and his allies would have to demonstrate "exceptional skill" to bring the prime minister down. As for Zahedi, who had just been arrested by Mosadeq's agents, the general had appealed to the majlis and was making some progress there. But Zahedi had not obtained the blessing of the shah for a move against Mosadeq, and Henderson doubted he would mount a coup without such approval.[21]

To get a better handle on the situation, Henderson invited Amini to the embassy for tea. The ambassador asked the court minister for his assessment of how things stood. Choosing his words with care, Amini described his efforts to bridge the gap between Mosadeq and the shah. He could not claim great success, he said, because the prime minister was "stubborn and suspicious," while the shah was "suspicious and evasive." Even so, he thought the two had come to an agreement on the necessity for cooperation against the communists. Until recently Mosadeq had not hesitated to back the Tudeh in exchange for the party's support against the shah. Of late, however, the prime minister indicated a willingness to crack down on the communists. Amini told Henderson to ignore a report printed a few days before in an unfriendly newspaper to the effect that Iran's law courts had ruled that membership in the Tudeh did not constitute an imprisonable offense. Amini said that under martial law the ruling did not apply and that the government fully intended to keep Tudeh members behind bars. The published report was simply a piece of propaganda designed to prejudice the American embassy.

Amini raised the issue of American aid, asking whether Iran might expect any help from Washington in the near future. Henderson gave his stock response that Congress would vote no aid while Mosadeq refused a reasonable settlement of the oil dispute. Iran was a potentially wealthy country. Only the prime minister stood in the way of unlocking the country's resources.

To this Amini replied, "It might eventually be necessary for Iran to have a new prime minister." Henderson, not sure of Amini's game, declined to speak further, beyond saying that Iran's internal affairs were for Iranians to decide.[22]

A week later Henderson and Amini had another talk. The shah, it seemed, was getting cold feet and intended to leave the country for a vacation. Amini sought Henderson's help in persuading him to stay home. A sudden departure, the court minister declared, would destroy the balance that he—Amini—had contrived between the court and Mosadeq. Amini said that without the shah's knowledge he had talked the crown prince of Saudi Arabia into rescheduling a visit for the express purpose of keeping the monarch in town. But such stratagems could not work forever. Henderson refused to get involved, remarking that the shah was hardly more likely to listen to an American than to his own court minister.[23]

During this period Henderson made careful contact with Zahedi. Through an emissary Henderson considered reliable, the general outlined the policies he would pursue if he assumed a position of responsibility in the government. In particular Zahedi proposed "to take a strong stand toward the communists and to restore order in the country," after which he would turn to the economic and social reforms Mosadeq had ignored for so long. At that point he would require "substantial" American aid. As to how he might gain power, he indicated quite plainly his desire for American help. Zahedi believed that it would be necessary for the United States to intervene in Iranian affairs. It would be "impossible for Iranians to remove the present Government by their own efforts."[24]

Henderson's reports to Washington on these conversations had just the effect the ambassador desired. Combined with Dulles's shift away from Egypt toward the northern tier, they helped create a consensus in Washington for action against Mosadeq. Undersecretary of State Walter Bedell Smith, Eisenhower's wartime chief of staff and more recently director of the CIA, considered Henderson's commentary worthy of the president's personal attention. In a cover letter to Eisenhower, Smith said of the ambassador's remarks, "I believe they are a very accurate expression of the situation and national state of mind in Iran, so accurate in fact that I suggest you read them."

In one analysis Smith forwarded to Eisenhower, Henderson explained the distinctive interplay of psychology and politics in Iranian affairs. "There are extremely important psychological differences between the public mind of Iran and that of the United States," Henderson wrote. "Iranian distrust of foreigners is so intense that it is not difficult to stimulate resentment against any foreigners engaged in activities in Iran even though these activities are clearly beneficial to Iran." For this reason, whatever policy the United States adopted toward Iran would have to be quiet and unobtrusive. Furthermore, Americans would be blamed for every sort of evil that befell the country. "Relatively few politically conscious Iranians really believe that it is possible for a power like the United States to refrain from interference in Iran." Each group not receiving American assistance, covert or otherwise, assumed that the Americans were backing its opponents.

Since the United States would be damned for intervention, whether Americans intervened or not, Henderson thought Washington should get something for its trouble. Besides, he contended, responsible elements in Iran desired a more forceful American policy. "Most Iranian politicians friendly to the West would welcome secret American intervention." But the administration must act quickly. "The frustrations of practically all sections of the Iranian public, including those supporting as well as those opposing Dr. Mosadeq, as they note the deteriorating conditions of the country, fan the embers of xenophobia. Only those sympathetic to the Soviet Union and to international communism have reason to be pleased at what is taking place in Iran."[25]

Although Eisenhower and Dulles had other sources of intelligence regarding Iran—notably the CIA—Henderson's arguments, coming from one with no bureaucratic interest in covert operations, weighed heavily in the administration's decision to support Mosadeq's overthrow. In any event, Dulles wanted to hear more, and the secretary called Henderson home for consultation.

Before the ambassador left Tehran, Mosadeq handed him a letter for personal delivery to the president. By this time the prime minister had come to the conclusion that Iran's distress, not to mention his own, demanded a direct appeal for American aid. Blaming the low living standard of the Iranian people on "centuries-old imperialistic policies," he asserted that conditions would remain depressed without "extensive programs of development and rehabilitation." He conceded that the oil impasse constituted a barrier to Iranian prosperity. He hoped the American government would assist in the removal of this impediment, which resulted from the "illogical claims of an imperialistic company." Yet even if the oil did not flow, Iran could survive and prosper. "This country has natural resources other than oil. The exploitation of these resources would solve the present difficulties of the country." Exploitation, however, would be impossible without American economic aid. There was no time for delay. "If prompt and effective aid is not given this country now, any steps that might be taken tomorrow to compensate for the negligence of today might well be too late."[26]

Despite Mosadeq's plea for quick action, Eisenhower waited more than a month to reply. When he finally did, the president simply reaffirmed that compromise was the key to a settlement between Iran and Britain. In unmistakable terms Eisenhower asserted that Iran's failure to compromise continued to make American assistance politically impossible. "There is a strong feeling in the United States, even among American citizens most sympathetic to Iran and friendly to the Iranian people, that it would not be fair to the American taxpayers for the United States Government to extend any considerable amount of economic aid to Iran so long as Iran could have access to funds derived from the sale of its oil."[27]

Explaining his tardiness in replying, Eisenhower commented that his administration had wished to discuss Iranian affairs face-to-face with Ambassador Henderson. On June 25 Washington's foreign-policy heavyweights did just that. The president, who deliberately held aloof from discussions like the one at this meeting, stayed away, but nearly everyone else who counted was there: Secretary Dulles; his brother Allen Dulles, director of the CIA; Undersecretary Smith; Deputy Undersecretary of State Robert Murphy, the man who had coordinated covert activities for the 1942 allied landings in North Africa and who was principally responsible for the notorious but effective Darlan deal; Kermit Roosevelt, grandson of the Rough Rider, cousin of FDR, formerly of the wartime Office of Strategic Services and now the CIA's top hand for the Middle East; Robert Bowie, head of the State Department's policy planning staff and later deputy director of the CIA; Henry Byroade, assistant secretary of state for the Middle East; and Charles Wilson, secretary of defense. And Henderson, of course. Roosevelt, who later provided the fullest description of the gathering, remarked on Henderson's role: "Loy was, without question, a key person in the meeting. He had rushed to Washington from his post in Teheran for the express purpose of offering his recommendations. JFD [John Foster Dulles] might decide, but Loy would be most influential in that decision."[28]

The purpose of the session was to pass judgment on a CIA proposal for toppling Mosadeq. The plan Roosevelt described had been in the works for some time, originating with British intelligence in the wake of the nationalization of the Anglo-Iranian company's holdings. The British believed they had the expertise and the agents in Iran to make the scheme work, but they lacked the essential element to guarantee its success: money to prop up Mosadeq's successor. While British officials had raised the possibility of a joint undertaking in talks with the Truman administration, Truman's habitual distrust of the CIA, which he feared would become an American "Gestapo," and an American desire to exhaust other options first, caused the project to be postponed until Eisenhower's accession.[29]

The Republican administration had no qualms about covert warfare, and in the first half of 1953 the CIA moved forward with planning for a strike against Mosadeq. Henderson generally kept aloof from the CIA, partly because he had to maintain credibility with the governments to which he was accredited, and partly from the suspicion most career diplomats felt toward a competing bureaucracy, especially one whose charter implied that the diplomats were not up to the demands of the cold war. But Henderson knew Kermit Roosevelt, who headed the anti-Mosadeq planning, fairly well. For a short period early in World War II, Roosevelt had served in the State Department. Their paths crossed in Baghdad a few years later, and again in New Delhi. Although Henderson understood that both British intelligence and the CIA had contingency plans for dealing with Mosadeq, he avoided inquiring about the details. The less he learned, the less he had to hide. Henderson could massage the truth when duty required, but he preferred not to have to beat it to death.

Henderson discovered few more details at the June meeting in Foster Dulles's office, because Roosevelt had few to convey. In sketching Operation Ajax, Roosevelt warned his listeners that much of what he and the agency aimed to do would depend on the situation at the moment of action. In broad outline, they would encourage the shah to dismiss Mosadeq and appoint Zahedi in his place, while orchestrating support for the shah and opposition to Mosadeq among the military and the populace. The specifics would have to be worked out on the spot. To Allen Dulles's query about the "flap potential" of the operation—what would happen if the plan misfired—Roosevelt conceded that the operation entailed risks. "If we have badly misjudged the situation, the consequences are hard to gauge. The least one can say is that they would be very bad—perhaps terrifyingly so. Iran would fall to the Russians, and the effect on the rest of the Middle East would be disastrous." But he added, in line with what Henderson had been predicting, that the United States might well witness the same consequences by doing nothing.

When Roosevelt finished, Foster Dulles polled the group. The responses varied from hearty support to tepid acquiescence. Allen Dulles and Smith, not surprisingly, spoke most strongly in favor. Bowie and Byroade went along, with less enthusiasm. Henderson was uneasy about the whole situation. He could remember the time when Henry Stimson had ordered the closing of the State Department's "black chamber," the secret division for breaking the codes of foreign countries. "Gentlemen," Stimson had declared, "do not read each other's mail." Henderson regretted the passing of those days. But he had to admit they were gone. He said to Foster Dulles, "Mr. Secretary, I don't like this kind of business at all. You know that." Desperate circumstances, however, required desperate measures, and if the United States could provide the bit of stiffening the shah required to do his duty to Iran and the free world, it was worth the potential damage to America's good name. "We have no choice."[30]

IV

Foster Dulles and Eisenhower agreed, and Roosevelt, with the assistance of British and Iranian agents, put the plan into action. To distance himself from the scheme, as well as to preserve a modicum of deniability with Mosadeq in case the operation fouled up, Henderson insisted on staying out of Iran while the secret warriors went about their work. Gordon Mattison, the chargé d'affaires in Tehran in Henderson's absence, and Roy Melbourne, the chief political officer, were informed of the reason for the ambassador's vacation, but the rest of the embassy staff—aside from those working for the CIA, presumably—remained in the dark.[31]

Henderson never could leave his job at the office; present circumstances made his holiday in the Austrian Alps doubly fretful. A variety of mischances delayed the CIA operation past the initially scheduled date, and when Henderson, outside the loop of cable traffic, saw nothing in the newspapers about a change of government in Iran, he grew more restless than ever. On August 14 he flew to Beirut, then a calm tourist destination. He did not expect to be soothed himself, but at least he would be closer to the action. On the night of August 15, just before going to bed, he heard radio reports that the shah had called for the resignation of Mosadeq. During the thirty-six hours that followed, further reports indicated that Mosadeq had foiled a royalist coup and the shah had left the country. Believing, as he had feared it would, that the CIA plan had miscarried, he wondered if he might have done something to avert what appeared to be shaping up as a foreign policy debacle. "I did not sleep well last night," he wrote on August 16. "I worried. Perhaps I should have gone direct to Tehran instead of stopping over in Beirut." Intending to salvage what he could, he called the naval attaché at the American embassy and commandeered a plane.[32]

On arrival in Tehran he contacted Roosevelt to find out what was wrong. "We've run into some small complications," the latter replied, understating the case considerably. What had happened was that the shah had privately issued two decrees, one dismissing Mosadeq and the other naming Zahedi as his successor. A messenger, Colonel Nematollah Nasiri, delivered the decrees to Mosadeq, but the prime minister arrested the messenger and ignored the messages. Zahedi went into hiding, and the shah—typically, as Henderson viewed the situation—fled. Mosadeq dismissed the majlis and assumed plenary powers. Anti-shah and anti-American mobs roamed the streets of Tehran and other Iranian cities, attacking American property and threatening American lives.

But optimism was Roosevelt's stock in trade, and he assured Henderson that the situation could yet be saved. "I think we have things under control. Two or three days should see things developing our way."[33]

Although Roosevelt had entered the country secretly, Henderson had not—indeed, Mosadeq's son greeted him at the airport—and with Americans in danger the ambassador could hardly avoid calling on the prime minister. Henderson had been through some strange sessions with Mosadeq, but this was the most remarkable. Mosadeq's attire, if nothing else, would have indicated that something was up. He was fully dressed, "as though for a ceremonial occasion," Henderson reported to Dulles, rather than in the pajamas and robe he customarily wore during interviews. "He was as usual courteous, but I could detect in his attitude a certain amount of smoldering resentment." (In his private appointment log, Henderson described Mosadeq's attitude as more than smoldering. "We had some frank hot exchanges," the ambassador wrote.) Whether Mosadeq possessed specific knowledge of American efforts on behalf of the shah or merely made a good guess, Henderson could not tell. To the ambassador's expression of concern at the recent chain of events, he responded with what to Henderson seemed "a sarcastic smile."[34]

Henderson decided to take the offensive. He told Mosadeq that the wave of assaults on Americans was intolerable. Every hour, he said, the embassy received new reports of attacks in Tehran and elsewhere. Mosadeq replied smoothly that under the circumstances a certain amount of violence was inevitable. Iran was in the midst of a revolution, and the Iranian people believed that the Americans opposed them—hence the attacks. The prime minister reminded Henderson that during the American revolution the rebels had not always treated kindly those who adhered to the British crown. Henderson responded by reading aloud eyewitness accounts demonstrating the complicity of Iran's security forces in the attacks. He asked whether Mosadeq wanted the Americans to leave Iran. If the government could provide no better protection, he as ambassador would have to recommend their evacuation. Mosadeq said he wanted the Americans to stay. He would inquire into measures to protect them.

Henderson asked Mosadeq for his explanation of the recent developments. Mosadeq upbraided the Eisenhower administration for releasing his request for American aid to "the pro-British Iranian press." Henderson replied that he had understood that the initial leak came from Tehran, and that the White House had made the correspondence public to prevent a distorted version from gaining currency. Mosadeq vigorously denied Henderson's claim. Within his government, he said, knowledge of the letters was held extremely closely. He had kept them in his personal files, rather than among his official papers. Henderson hinted that perhaps his personal papers were not kept in such a way as to prevent "clever agents" from gaining access to them. In addition, the ambassador pointed out, electronic eavesdropping devices might have allowed "parties hostile both to Iran and the United States" to learn what was in the correspondence. Mosadeq continued to insist that American officials had published the letters to weaken his government. Henderson persisted in his denials.

This was as close as Mosadeq came during this interview to blaming the United States directly for the shah's effort to fire him. He was less reticent about alleging British complicity in the present turmoil. He defended his action in dissolving the majlis by declaring that British agents had "bought outright" nearly half the votes in the assembly. With ten votes more, he said, the pro-British faction would have gained a majority. When he learned of negotiations to complete the purchase, he decided the majlis was "unworthy of the Iranian people" and must be sent home.

To Henderson's query regarding the detention of Colonel Nasiri, Mosadeq answered that the officer had approached his house on the evening of August 15 with the aim of arresting him. But the prime minister had acted first, clapping the colonel in jail, along with several others involved in the plot. When Henderson asked whether he believed reports that the shah had issued decrees dismissing him as prime minister and appointing Zahedi in his place, Mosadeq replied that he had seen no such documents and would have ignored them if he had. The shah, he argued, wielded only empty, formal powers. On his own authority the shah could not demand a change in government. Henderson pursued this issue. Was the prime minister saying he had no official knowledge that the shah had sought to remove him, and that he would consider such an action invalid? "Precisely," Mosadeq responded.

Before leaving, Henderson commented that since his return to Tehran, various American officials had received intimations that the Iranian government believed that the American embassy was sheltering Iranian "political refugees." The ambassador said he wanted to refute these allegations "point blank." His policy was, first, to endeavor to stop anyone from entering the embassy; second, if someone nevertheless succeeded in gaining entrance, to attempt to persuade such a person to leave voluntarily; and third, to notify the Iranian government regarding anyone who refused to leave. Mosadeq replied that the Americans were free to keep anyone who came knocking. The ambassador could send him the bill for lodging.

Upon returning to his office, Henderson summarized the conversation for Dulles, focusing especially on what Mosadeq knew or guessed about American efforts to topple him.

> I am inclined to believe that he is suspicious that the United States Government or at least United States officials are either implicated in the effort to oust him or were sympathetically aware of such an effort in advance. His remarks to me were interspersed with a number of little jibes which, although semi-jocular in character, were, nevertheless, barbed. These jibes in general hinted that the United States was conniving with the British in an effort to remove him as Prime Minister. For

instance, he remarked at one point that the national movement was determined to remain in power in Iran and that it would continue to hold on to the last man, even though all its members would be run over by British and American tanks. When I raised my eyebrows at this remark, he laughed heartily.[35]

Tanks appeared, but they were Iranian. As rioting continued in Tehran, the Tudeh, assisted by CIA plants, played an ever more visible role, denouncing the shah for treason, demanding the abolition of the monarchy, and affirming the party's ties with the Soviet Union. This was too much for important segments of the army, and when Zahedi, still in hiding, held a press conference, at which he declared himself prime minister and, to support his claim, distributed copies of the shah's decree of appointment, the tide began to turn. Troops loyal to Zahedi mounted tanks and personnel carriers and drove to the majlis square. They were joined by an odd assortment of weight-lifters, wrestlers, and bazaar hangers-on who, with the encouragement of $100,000 distributed by Roosevelt's agents, rallied to the cause of the shah, threatening and in some cases inflicting immediate physical harm upon those who spoke ill of the monarch.

One branch of the mob, accompanied by an armored column, turned toward Mosadeq's house. When the bodyguards there refused to surrender, the troops opened fire with machine guns and anti-tank weapons. Mosadeq escaped just before the defenses gave way, taking refuge in the residence of a friend. By this time Zahedi had come out in the open, and now he set up command quarters in the Tehran officers' club. He ordered the dismissal and arrest of dozens of officials who had cooperated with Mosadeq, and he announced plans for the immediate return of the shah.[36]

From the American embassy, Henderson did not have the best view of all these events, and the circumstances did not lend themselves to his playing roving reporter. Yet what he could see of developments convinced him of the essentially popular nature of the uprising, despite the CIA assist. Describing the counterattack by the pro-shah, pro-Zahedi forces, Henderson wrote that the demonstration had begun "in a small way in the bazaar area," but that "the initial small flame found an amazingly large amount of combustible material and was soon a roaring blaze which during the course of the day swept through the entire city." When the crowd surged past the embassy, Henderson had an opportunity to assess its composition. Most of the leaders appeared to be civilians, and the rank and file were "not of the hoodlum type customarily predominant in recent demonstrations in Tehran." They appeared to come from "all classes of people, including workers, clerks, shopkeepers, students, et cetera." As to the temper on the streets, Henderson commented that the participants "seemed to be imbued with a strange mixture of resolution and gaiety." He added, however, that "the holiday mood which seemed to prevail did not prevent the execution of grim missions, which on at least two occasions resulted in loss of life."

Attempting to explain the sudden shift in Iranian fortunes, Henderson pointed out that not only the backers of Mosadeq but also supporters of the shah were "amazed" at the speed and comparative ease of the latter's victory. Henderson identified several contributing factors. First, Mosadeq and the left, by attacking the crown, had assaulted the sensibilities of the Iranian majority. "Iranian people of all classes were disgusted at the bad taste exhibited by the anti-Shah elements supporting Mosadeq. For instance, they were outraged when gangs of hooligans bearing red flags and chanting Communist songs began tearing down statues of the Shah and his father, and breaking into houses and shops for the purpose of destroying the Shah's pictures."

Second, the Mosadeq-Tudeh alliance had worried and alienated many members of the pro-western elite. "They were alarmed at seeing thousands of Tudeh demonstrators, whom they regard as agents of the Soviet Union, marching openly arm-in-arm through the streets denouncing the Shah and Western countries, particularly the United States. The Tudeh clearly overplayed its hand by causing the Iranian people to believe that the latter had to choose between Mosadeq and the Soviet Union on one hand and the Shah and the Western world on the other."

Third, the Iranian people had become "thoroughly tired of the stresses and strains of the last two years." Having seen their country lurch from crisis to crisis, "they yearned for a period of quietness which would give them a chance to improve their economic and social status." Under Mosadeq they could see no hope for such a respite.

Fourth, a falling-out between Mosadeq's partisans and the Tudeh had taken place in the midst of the rioting, preventing effective cooperation between those two anti-shah groups. Henderson thought that the Tudehists' strategy had been to hold Mosadeq's feet to the fire, rescuing him only in his extremity and after extorting sizable concessions. But they had waited too long and lost control of the situation.

Finally, and decisively, the majority of the armed forces and "great numbers" of the Iranian populace were inherently loyal to the shah. In a country wracked by turmoil and rapid change, the monarchy served as the fundamental symbol of stability and national unity. The officer corps and many civilian leaders, fearing "Iran's strong northern neighbor," looked to the United States for protection, and they believed that only under the shah would such protection become available.[37]

As to the American role in the turnabout, Henderson reported that "the impression is becoming rather widespread that in some way or other this Embassy or at least the United States Government has contributed with funds and technical assistance to overthrow Mosadeq and establish the Zahedi Government." This impression was accurate, of course, and at the moment it was working in America's favor, since Zahedi and the shah were riding a wave of mass approval. But Henderson advised against letting such a belief spread. He said he was "doing the utmost discreetly to remove this impression." For one thing, it would not serve America's interests for Washington to become known publicly as the manipulator of Iranian affairs. The United States had a reputation to uphold. For another, any such perception would prove an albatross for Zahedi. The new prime minister would face problems enough without being branded an American puppet. Finally, Zahedi might prove an albatross for the United States. Henderson, taking a rather cold-eyed view of a regime he had helped install, remarked, "Zahedi's Government, like all governments in Iran, eventually will become unpopular, and at that time the United States might be blamed for its existence."

Under present conditions Henderson judged silence the better part of diplomacy. He recommended that Washington refrain from commenting on charges of American intervention. Denials would simply raise the visibility of the issue. If the Eisenhower administration felt compelled to make a statement of some sort, it ought to confine itself to phrases along the lines of "the victory of the Shah was the result of the will of the Iranian people."[38]

Two days later Henderson visited the shah. As he recounted the meeting afterward, the monarch was "a changed man"—more confident and optimistic than Henderson had ever seen him. The ambassador relayed a goodwill message from Eisenhower, which he elaborated with an introduction of his own devising: "I congratulate you for the great moral courage which you displayed at a crucial time in your country's history. I am convinced that by your action you contributed much to the preservation of the independence and the future prosperity of Iran." In his cabled commentary, Henderson remarked, "The Shah wept as I read this."[39]

For several days Henderson thought it wise to stay away from Zahedi's office, but he communicated with the new prime minister through his son Ardeshir Zahedi. The ambassador urged the prime minister to assure Mosadeq's safety. The last thing the new government needed was to make a martyr of the former premier. Henderson also suggested that Zahedi adopt a business-as-usual approach to the whole affair. After all, the shah had simply exercised his prerogative in changing prime ministers. Nothing more—no revolution, no coup—had occurred.[40]

Zahedi took Henderson's advice, and by week's end the situation had calmed sufficiently that Henderson thought he could call on the prime minister in person. As the ambassador expected, Zahedi immediately asked about American aid. He asserted that his recent success in rescuing Iran from "the very brink of the Communist abyss" would mean nothing unless the Iranian people could see that the new government had something to offer them. His government looked to the United States to furnish help.

Henderson expressed sympathy but could make no commitment. He said that the prime minister would increase his chances of receiving American assistance by settling the oil dispute with Britain—if not immediately, then at least in due course. With Dulles's northern-tier concept in mind, the ambassador added that improved relations with Turkey and Pakistan would also facilitate matters. Zahedi indicated that he understood.[41]

Henderson came away from this meeting with a favorable opinion of the prime minister. Writing to Dulles, he described Zahedi as "clearly an activist who, if given wise guidance and not frustrated by the Shah, could do much to extricate Iran from its present political and economic morass." While Henderson had refrained from encouraging Zahedi to expect large amounts of American aid soon, he believed that the time had come for the United States to put up or stop talking. "Precisely the same issues are at stake regarding Iran as were at stake regarding Greece and Turkey in 1947." Had the American government not acted quickly to aid those countries then, the shape of the eastern Mediterranean would now be quite different. If the United States did not act with similar dispatch in the present case, it might lose Iran. Through the grapevine Henderson had learned that the administration was projecting somewhat less than $40 million in aid for Iran. This amount, he contended, would not suffice. "In view of the gravity of the issues at stake, we hope that the entire question will be considered at the highest levels." He strongly recommended a doubling of projected assistance, with delivery at once.[42]

The budget balancers in Washington would not go this far, although Dulles did succeed in squeezing a few million more out of the treasury. When Henderson told Zahedi to look for $45 million, the prime minister appeared, in Henderson's description, "genuinely disappointed." Zahedi's finance minister put the matter more forcefully, declaring that Iran would need "at least $300,000,000 to cope with the economic chaos left by the former government of Dr. Mussadeq."[43]

But some loose ends required attention first. One involved the fate of Mosadeq, who was arrested the day after Zahedi's coup. Henderson suggested to the shah that the government move "quickly and firmly" to bring the former prime minister to trial, since vacillation or delay would perpetuate "the myth that Mosadeq was the grand old man essential to Iran." The shah agreed. The plan nearly backfired, though, and Mosadeq, speaking in his own defense, again commanded center stage. He once more stirred the masses, with his closing appeal: "Since it is evident, from the way that this tribunal is being run, that I will end my days in the corner of some prison, and since this may be the last time that I am able to address myself to my beloved nation, I beseech every man and woman to continue on the glorious path which they have begun, and not to fear anything." Fortunately for the shah and Zahedi, the magic had vanished, and the trial court handed down a sentence of three years in solitary confinement.[44]

The other pressing problem was oil. With the enticement of greater American aid, the shah and Zahedi resumed discussions of the oil dispute with London. Progress toward a settlement came slowly. Even had the shah not determined to move cautiously in order to avoid rekindling anti-British feeling, a swift agreement would have been difficult. The amount of compensation due the Anglo-Iranian company raised one set of complications. Another resulted from the fact that there existed no immediate demand for Iran's oil. The world oil economy had adjusted to the absence of Iran's output, and Anglo-Iranian's competitors, including Aramco and other American firms, had appropriated that company's market share. Complex and lengthy negotiations eventually resulted in an agreement all parties could live with. London accepted the principle of nationalization, while a consortium of Anglo-Iranian and six other companies—four American, one Dutch, and one French—contracted to operate Iran's largest oil fields and the Abadan refinery.[45]

V

From Henderson's perspective the anti-Mosadeq coup was chiefly a means to the end of reestablishing order in Iran and protecting the country against Soviet encroachment. Though optimistic, Henderson in the first months after the coup reserved judgment on whether the shah and Zahedi would ultimately prove equal to the task. But by December most of his worries had vanished. In the last week of 1953 he reported with satisfaction that Zahedi and the shah were starting to move decisively. The government had "gathered courage" and "taken the plunge into the welter of its most serious problems." While American aid was beginning to ease the financial strain, on the political front the shah had dissolved the majlis and called for fresh elections, to root out those of Mosadeq's supporters who remained in the assembly.[46]

In a meeting with the shah, Henderson congratulated him on his initiative, especially with regard to the majlis. The shah replied that his decision to prorogue the parliament reflected a desire to reinvigorate the "royal prestige." During Mosadeq's sway, the prime minister had forced the crown to dismiss the parliament. Accomplishing the same end on his own account demonstrated that he was back in charge. Henderson complimented the shah on the government's recent successes in a variety of areas. As the ambassador paraphrased his remarks for Dulles, "It seemed to me that hardly a day had passed by without some significant, constructive decision having been made. I thought that as a result of the new attitude of decisiveness and the display of willingness to take action on the part of the Government it was now moving forward with renewed confidence and was creating public confidence in its ability to do things."[47]

The shah agreed, but added that public confidence needed further strengthening. Washington might help in this regard, he said, by allowing Iran to build up its military forces. After all, it was America's idea that Iran should contribute to regional defense. Without a respectable army this would be impossible. At present the country's military barely sufficed for internal security. Such an army could neither deter Soviet aggression nor give Iran "the self-assurance and confidence necessary for the maintenance of Iranian independence." If Washington assisted in revamping Iran's military establishment, his country could protect itself and help defend its neighbors as well.[48]

The shah's pitch for military aid was premature. The final settlement of the oil dispute, which would set both American arms and more American dollars flowing, remained several months in the future. In the meantime, though, the northern-tier alliance system began to take shape. In February 1954 the United States announced a weapons agreement with Pakistan. At about the same time Turkey and Pakistan, with American encouragement, took steps toward a mutual-security pact.

For the sake of diplomatic argument, the shah chose to interpret these developments, combined with the American refusal to grant his arms requests, as indications that Washington was losing interest in Iran. He complained to Henderson that so long as Mosadeq had been making trouble, the United States paid attention to Iran. But now that the situation was quiet, Iran was being treated as a "stepchild." He argued that when international communism had threatened Greece and Turkey in 1947, Washington rushed into the breach with hundreds of millions of dollars. When Europe had come under threat, the Americans sent billions via the Marshall plan. When Pakistan had required arms, the Eisenhower administration lost no time in responding. But Iran was being pushed behind the door, at precisely the moment when his country mattered more than ever. Pointing out that two thousand kilometers separated Turkey from Pakistan, the shah argued that Iran was the "keystone of the arch" of the northern tier. The loss of Iran to international communism would result in the collapse of the arch, endangering the entire region. In light of America's apparent indifference to Iran's fate, he was beginning to conclude that Washington had already written off his country. The purpose of the Turkey-Pakistan agreement, it seemed, was to protect those two nations while conceding Iran to the Russians.

To these comments, and to the shah's further remark that Iran was not asking for much—merely funds and equipment for an army of 150,000, so his people would not feel "entirely defenseless"—Henderson responded that he was sorry to hear the shah doubt the devotion of the United States to Iran's independence and security. For the better part of a decade Washington had consistently taken Iran's side against the communists. It would continue to do so. But it could not fulfill all the requests of all its friends at once. To provide the aid the shah wanted required consideration of "numerous international and domestic political, military and economic factors." Congress and the American public supported emergency assistance to countries facing crises, as the 45-million-dollar package to Iran amply demonstrated. Long-term commitments, however, were another issue. As a matter of principle, Americans did not like foreign governments to become chronically dependent on American assistance.[49]

The shah made no appreciable headway until the summer of 1954, when the petroleum deadlock broke. At that point, American aid, both economic and military, began gushing along with the oil. Within the next two-and-a-half years Iran received over $300 million in American assistance, divided about evenly between the economic and military sectors.[50]

Meanwhile the shah continued to consolidate his power. The majlis elections of early 1954 returned a pliable body of legislators, and the assembly, after its heyday under Mosadeq, entered a long period of eclipse. When those few dissidents who remained at large took to the streets in July, the shah cracked down hard and fast, scattering the demonstrators with little difficulty. Henderson commented, "It would appear that the no-nonsense attitude of the security forces is so well established that the government has only to show it has force at its command and spread word that demonstrators will be shot to maintain city-wide order and force troublemakers to remove their feeble efforts to outlying areas."

By August the crackdown had spread to the security units themselves. Henderson described a series of "sensational arrests" of police and military officers alleged to be spies for the Tudeh and probably the Kremlin. In September the government continued its "dramatic moves" against the communists, including the destruction of a previously secret Tudeh publishing operation. In October the regime laid plans for still more severe sanctions against the left, including the death penalty for individuals convicted of agitating for the overthrow of the government.[51]

VI

CONCLUSION

18

The Ends of Empire

In October 1954, Henderson received a cable asking him to travel to Washington. President Eisenhower, the message said, wished to confer upon him a citation for distinguished service. Henderson was ordinarily the most obedient of public servants in following State Department directives, but in this case he objected. Since the advent of the Republicans eighteen months before, trimmers in the budget bureau and Congress had cut appropriations for the State Department. The department had responded by laying off many junior foreign-service officers, seemingly regardless of merit and promise. In 1954, Henderson turned sixty-two, and with retirement approaching he began thinking of the future of the service. On grounds of common sense alone, he judged the decision to waste the seed corn, as it were, of the service ill-advised. Moreover he considered the junior officers, many of whom he had personally trained overseas and in Washington, his heirs—a sentiment strengthened by the fact that he had no children. Naturally he resented unfair treatment of them. In any event, he replied to Dulles's request that he come to the United States to be honored, by remarking somewhat sarcastically that in these times of strained finances the department would do well to save the thousands of dollars a ceremonial visit home for him would cost, and devote the savings to keeping the foreign service well staffed. Unamused, Dulles changed his request to an order.[1]

Arriving in December, Henderson met Herbert Hoover, Jr., at the airport. Henderson had become acquainted with Hoover during the last phase of the oil negotiations in Tehran, when Hoover handled the business end of the talks on the American side while Henderson managed the politics of the deal. Hoover, currently undersecretary of state, informed Henderson that there was more to this visit than an award ceremony. Dulles needed a person to oversee a reorganization of the State Department and the foreign service, and now that the situation in Iran was under control the secretary had decided on Henderson.

Henderson had last returned to Washington for reassignment in 1945. Then he had landed in the middle of the Palestine dispute and barely survived with career intact. Nearly a decade later he found himself drawn into another thankless task. Since the late 1940s, and especially since the communist victory in China, the State Department and foreign service had come under severe and sustained attack. Conservatives and demagogues on the make had discovered in the diplomatic corps convenient scapegoats for America's inability to reshape the world entirely according to its desires. Although departmental morale had suffered serious injury under Acheson, the fact that Acheson himself took many of the hits tended to shelter the career people from the worst of the criticism. When the Republicans arrived, however, the damage escalated to devastating proportions. John Foster Dulles dodged the blows, letting them crash down on those in the lower echelons. Dulles approved the appointment of Scott McLeod, a McCarthy henchman, as the department's security officer. When repeated investigations failed to produce evidence that China hands John Carter Vincent and John Paton Davies were risks to national security, Dulles bowed to protests from the China lobby and fired them for bad judgment.

The latter action had an especially chilling effect on the foreign service, for it suggested a political test for continued employment, and certainly for advancement. Henderson's old friend Charles Bohlen, who early found himself the object of McLeod's harassment, came away with a distinct distrust and distaste for Dulles. Bohlen later declared that the new secretary and his Republican associates entered the State Department "like a wagon train going into hostile Indian territory, and every night they'd group their wagons around the fire." When Dulles warned department personnel at his first meeting that he expected their "positive loyalty," the undefined phrase evoked shudders. U. Alexis Johnson, a Far Eastern veteran who weathered the purges, considered it part of a campaign to buy off the Republican right. Bohlen agreed, explaining that Dulles "was a man who really had one obsession: to remain Secretary of State." This obsession caused him to take great pains to appease the McCarthyites. "Mr. Dulles," Bohlen asserted, "throughout his entire life had really been pointing toward the Secretaryship of State, and he had it, and he was concerned lest the right wing of the Republican Party might mount a campaign against him which might cause him to leave the office."[2]

Dulles compounded his difficulties in the department by what Henderson delicately described as a "lack of personal graciousness." As Henderson explained,

> Mr. Dulles never went out of his way to make himself popular with the various officers. When he would come into a staff meeting in the morning, he wouldn't say, "Good morning, gentlemen." He would grunt. He would sit at the table and almost inarticulately he would indicate that the meeting had begun. His remarks were essentially short, sometimes gruff, and almost invariably to the point. When we would adjourn for Christmas, he wouldn't say, "Well, gentlemen, I hope you have a Merry Christmas." He would say, "Ugh"—and that would be the end of the meeting.

After working with Dulles on a daily basis for four years, Henderson came to recognize this reserve as indicative of painful shyness. The secretary's hard exterior, Henderson suggested, was "merely a protective shield covering a kind heart." When Henderson received special presidential dispensation to remain in the department past his sixty-fifth birthday, Dulles sent him a warm letter of congratulations. But few of those who worked under the secretary detected his generous streak. To the rest he seemed aloof and insensitive.[3]

Additional damage to department morale resulted from the Republican budget cuts. In part, the reductions reflected the perennial and not unfounded belief of the party of business that any government bureaucracy contains slackers and timeservers. Equally, however, the enthusiasm for streamlining, which by the beginning of 1954 had pared one-fifth from the department's operating funds, resulted from a suspicion that after two decades of Democratic control the State Department was hopelessly infested by New Dealers. Under government rules the easiest way to throw the rascals out was to abolish their jobs. Whatever the causes of the cuts, the consequence was a further erosion of morale. One career officer remarked at the time, "If I had a son, I would do everything in my power to suppress any desire he might have to enter the Foreign Service." George Kennan, whose identification with containment convinced Dulles he had to go—despite, or rather because of, Dulles's inability to offer anything better—commented that the housecleaning weakened the foreign service "beyond hope of recovery."[4]

Dulles, even while contributing to the morale problem, recognized that something must be done to solve it. This was where Henderson fitted in. The secretary intended to name Henderson deputy undersecretary for administration. Henderson possessed three particular qualifications for the job. First, although he protested that he had never held a departmental administrative post, Henderson's files contained three decades' worth of commendations for aptitude and efficiency in administering the affairs of legations and embassies. Dulles himself called Henderson "particularly able." Second, no individual commanded greater respect in the career service than Henderson. In his long climb to the top of the profession, he had taken care not to push others aside or to aggrandize himself. He had repeatedly stood up for the integrity of the service and its mission. Fully aware of the political costs, he had insisted on calling spades spades, particularly regarding the Soviet Union during the war and Palestine afterward. If Dulles intended to restore his credibility with the foreign service, Henderson was just the man for the job.[5]

Finally, and perhaps most importantly, Henderson's conservative credentials were in impeccable order. When McCarthy was still a Democrat and decades before the Wisconsin politician discovered communism, Henderson was warning against Moscow's global designs. Henderson's distrust of New Dealers was a matter of record. Henderson had been among the first to suspect Alger Hiss of disloyalty: in 1945 he quietly ordered the members of NEA to keep confidential materials and information away from the man who would become a symbol of bureaucratic treachery. Although his handling of the Hiss matter was probably not known to the McCarthyites, Henderson's reputation as a hardline anticommunist certainly was. In such esteem, in fact, did the right-wingers hold Henderson that when McCarthy blasted the department and foreign service as a group, on more than one occasion he excepted Henderson by name. Nevertheless Dulles, chancing nothing, directed the FBI to do a security check on Henderson. Not surprisingly Henderson came up clean.[6]

While Henderson could expect not to run afoul of the communist-hunters, he recognized that administering the State Department would entail no end of headaches. At first he tried to duck the assignment. He reminded Dulles that his specialty was political reporting and diplomacy. Administration, he said, was not his field. If the secretary wished to install another person as ambassador in Tehran, of course he would not object. But he doubted that he would best serve the department in an administrative capacity. Dulles told him to think the matter over.

Henderson did, consulting friends and former associates, including Kennan, who had nothing nice to say about Dulles. But ultimately the question was settled by presidential intervention. Eisenhower invited Henderson to the White House for a talk. After congratulating him for his good work in Iran, Eisenhower raised the issue of Henderson's promotion. He said he heard Henderson had doubts about the job of deputy undersecretary. Henderson explained the reasons for his reluctance. Eisenhower replied that he desired to rebuild the spirit of the diplomatic corps. He hoped Henderson would play a major part in the process. By way of conclusion he added, "I understand that you are a Foreign Service Officer and that it is the tradition of Foreign Service Officers to take any position that the President asks them to take." Henderson accepted the job.[7]

II

Henderson's reassignment to Washington coincided with the period of most rapid growth and greatest elaboration of the American empire. In overt political terms, the Republican administration extended the American alliance network until it girdled the earth. In Europe, Eisenhower and Dulles made permanent the division of the continent by engineering the integration of West Germany into NATO. They intruded America into the Middle East by their alliance with Pakistan and their sponsorship of the Baghdad pact, and later by the enunciation of the Eisenhower doctrine and the landing of American troops in Lebanon. From the Pacific they worked west, adding bilateral defense treaties with South Korea and Taiwan to a similar arrangement with Japan, and complementing the ANZUS (Australia-New Zealand-United States) agreement with the broader and more ambitious Southeast Asia Treaty Organization. By the middle of Eisenhower's second term, the American system comprised a chain of allies and clients that began in North America, hopscotched the Atlantic via Iceland and Britain, ran down through Europe to the Mediterranean, spanned the northern tier of the Middle East to Iran and Pakistan, arced across Southeast Asia to the Pacific, and climbed the offshore island chain from New Zealand through Australia, the Philippines, Taiwan, Japan, and back to Canada and the United States. All this, of course, was in addition to the traditional American domination of the western hemisphere. The world had never seen such an arrangement. No country had ever defined its interests so sweepingly as to prepare to fight nearly from pole to pole all around the globe.

By covert methods as well, the Eisenhower administration expanded American influence abroad. Iran, where Washington's secret warriors helped add a crucial client to the American system, was one case among several. In Guatemala, the CIA underwrote the rightist rebellion that replaced the refractory Arbenz with a complaisant and reliably pro-American regime, and demonstrated the difficulty—impossibility, in this case—of defecting from the American empire. In Vietnam, covert operatives assisted in the entrenchment of Ngo Dinh Diem in Saigon and the harassment of Diem's rivals in North Vietnam. In Indonesia, American agents funded and armed a revolt against the neutralist government of Sukarno. In the Philippines and Syria, the CIA helped rig elections. In the Congo, American hit-men plotted the assassination of Lumumba. In Cuba, individuals backed by Washington added preparations for a military assault to designs on Fidel Castro's life.

Quite obviously, this breathtaking extension of the American writ had numerous causes and multiple authors. Events—the Viet Minh victory at Dienbienphu, the Communist Chinese shelling of Quemoy and Matsu, the abortive Anglo-French-Israeli invasion of Egypt, to name three of the most spectacular—conspired to convince American leaders of the need to push America's defense perimeter ever-farther forward. Various voices, from the White House down through the State Department, the Pentagon, and the CIA, in Congress and the press, called for greater vigilance in the defense of the "free world."

But the fundamental premises on which the grand strategy of the American empire rested were the ones Henderson had been expounding for twenty years. Henderson had argued that the Soviet Union posed an abiding threat to the security and values of the United States, and that only active American opposition to communist pressure could prevent the contraction and eventual collapse of the area of democracy. During much of those twenty years Henderson's preaching of his one big truth had fallen on ears disinclined to listen. Twice it had earned him exile. But by the mid-1950s it was solid American orthodoxy.

III

Henderson's earlier postings had afforded him plenty of experience in the corridor skirmishing that accompanied policy-making in the State Department. He carried scalps as well as scars from his days in the Eastern Europe and Near East divisions. His position as deputy undersecretary for administration, however, entailed politics of another kind, namely, the struggle over the budget. This aspect of his work, which among other things required cultivating congressional committees, took some getting used to. After most of a year on the job he could still tell a friend, "It is with perpetual surprise that I find myself engaged in the task of administering the Department and the Foreign Service."[8]

His principal assignment was to effect what was called "Wristonization." Responding to Eisenhower's expressed desire for economizing, a presidential panel headed by Henry Wriston of Brown University recommended the amalgamation of the civil-service employees of the State Department with the officers of the foreign service. The job was among the most thankless Henderson had ever been called upon to perform. Melding the two career ladders into one required unenviable and sometimes unfair apples-oranges comparisons—not unlike those involved in the implementation of the Rogers Act, which had opened a diplomatic career to Henderson during the 1920s. Henderson did his best, fielding the complaints of the individuals affected and of interested legislators, and righting the correctable wrongs.

Even while occupied chiefly with administration, Henderson kept his hand in on the policy side. Dulles, recognizing Henderson's expertise on matters relating to the Soviet Union and the Middle East, and sharing Henderson's general outlook on the world, solicited Henderson's opinions during the daily staff meetings the secretary convened with the department's top officials. Dulles also delegated to Henderson primary responsibility for liaison with the Baghdad pact organization, to which Henderson was accredited as American observer.

Henderson was a natural choice for the Baghdad job. He got along well with the British, but not too well. Iraq he knew from personal experience as American minister. The Turks remembered him as an architect of the Truman doctrine. The Pakistanis liked him for opposing Nehru's neutralism. His reputation as an anti-Zionist enhanced his standing among all the Muslims in the organization. The shah of Iran held him in especial esteem as a major factor in the shah's still being shah. Indeed, upon Henderson's departure from Tehran the shah had issued a royal *farman* hailing Henderson as a great friend of Iran. Iran would miss a man so "thoroughly acquainted" with the special needs of the country, one fully cognizant of "the real cause of communist incursions" against its people, the message declared. The ambassador's actions during the dark days between 1951 and 1953 assured him an honored place in the pages of Iran's history. Iran wished him well in his new position in Washington. "Now his activities must reach the four corners of the world, but all the same it is hoped that he will not forget Iran." (To ensure that he not forget, the shah kept in close contact with Henderson, beyond activities related to the Baghdad pact. In 1957, for example, Henderson received an invitation to the wedding of the shah's daughter to Zahedi's son. Henderson would have been happy to attend, but with the situation in the Middle East in one of its recurrent stages of unrest, he was obliged to beg off.)[9]

The Baghdad alliance failed to live up to American expectations, partly because of differences among the members and partly because the United States, after heavily promoting the organization, could not bring itself to join. Israel objected to anything that strengthened Iraq, as did Saudi Arabia—Israel for military and territorial reasons, Saudi Arabia for dynastic and political ones. Britain was happy to be the large western frog in this relatively small Middle Eastern pond. The Eisenhower administration chose not to strain relations any more than necessary with either Tel Aviv and the Israel lobby or Jiddah and the royal family of oil, or to encroach on what little remained of Britain's Middle Eastern sphere. Upon Henderson's recommendation, though, the adminis-tration accepted an American seat on various committees connected to the alliance.[10]

On several occasions during the last half of the 1950s, Henderson attended meetings of the Baghad group. His presence served to demonstrate that despite America's formal diffidence the United States was deeply interested in the secu-rity of the group's members. He closed a June 1957 session with a vigorous affirmation of the right of small countries to cooperate against aggression. The "menace of international Communism" was a compelling fact of life, he asserted. As long as it remained so the United States would extend its "close cooperation" to the nations of the Baghdad pact.

In certain respects America's outsider status militated to the benefit of the alliance. Among the Turks, Pakistanis, Iranians, Iraqis, and British, divergent views frequently arose. Henderson often served as mediator. After one espe-cially divisive meeting of April 1956, an American naval officer remarked that "Mr. Henderson's smooth diplomatic efforts were particularly effective" in keeping the session on track.[11]

In addition to his duties with the Baghdad organization, Henderson took an important part in the great Middle East crisis of the 1950s: the Suez affair of 1956. Despite Nasser's rejection of MEDO, the Eisenhower administration had persisted in the hope of coaxing the Egyptian leader in the direction of the west. Even after Nasser announced an arms-for-cotton deal with the Soviet bloc in September 1955, Washington sought to arrange a modus vivendi. At the begin-ning of the next year Eisenhower sent one of his most trusted advisers, Robert Anderson, to Cairo to offer Nasser a variety of inducements to come to terms with Israel and thereby enhance the stability of the region. But Nasser spurned the American overtures, leading Eisenhower and Dulles to conclude that he was a lost cause. By the summer of 1956 the administration had shifted into an oppositionist mode. The clearest signal of the new policy came in the middle of July, when Dulles told the Egyptian ambassador that the United States was no longer offering to subsidize the construction of a new dam at Aswan on the Nile.

Although it had not been so intended—not primarily, at least—Nasser took the withdrawal of the offer of American subsidies as a personal challenge. "This is not a withdrawal," he told Nehru, fortuitously present when Nasser received the news. "It is an attack on the regime and an invitation to the people of Egypt to bring it down."[12]

A week later Nasser launched his counterattack by seizing control of the Suez canal and nationalizing the company that ran it. As had Mosadeq's nationalization of the Anglo-Iranian company, Nasser's strike against the canal company, also largely British-owned, set the military planners in London to work designing an operation to undo the dastardly deed.

Eisenhower and Dulles managed to postpone the escalation, pending a conference in London of the principal countries using the canal. After considerable wrangling the conference proposed a scheme for international operation and control of the canal. The conference nominated a commission to travel to Cairo to persuade Nasser to accept the plan. As the lead American on the commission, Dulles nominated Henderson.

Henderson knew the situation surrounding the Suez dispute well enough to realize that the chances of Nasser's accepting the eighteen-power proposal were nearly nil. He told Dulles as much when the secretary summoned him to London for briefing. But Dulles wanted to go ahead, so Henderson did. Before leaving England he joined Anthony Eden for lunch. Eden, now prime minister, declared that Nasser would not get away with his theft. "We are determined to secure our just rights in Suez," he said. "If necessary we will use force, because I would rather have the British Empire fall in one crash than have it nibbled away, as it seems is happening now."[13]

Joining Henderson in Cairo were the foreign ministers of Iran, Sweden, and Ethiopia and the prime minister of Australia, Robert Gordon Menzies. Menzies, who headed the delegation, was florid and bluff, a landmark among Commonwealth leaders for three decades. As likely to antagonize as to please, he proved less than inspired as a choice for chairman. Even before the London conference began, Menzies had announced his opinions regarding Nasser and the canal controversy. On British television he declared, "To leave our vital interests to the whim of one man would be suicidal. We cannot accept either the legality or the morality of what Nasser has done." But Menzies liked and trusted Henderson, and he recognized Henderson's usefulness. Henderson, Menzies wrote afterward, "was a calm, level-headed career diplomat with wide experience in the Middle East. He had good and cautious judgement, and was an excellent corrective to my own occasional impetuosity."[14]

From the first, Henderson set out to provide this corrective. On the mission's initial evening in Cairo he requested a private interview with Nasser. The Egyptian president, noting that Menzies sounded even more like a nineteenth-century imperialist than Eden, complained that the western powers were trying to back him into a corner. They were presenting a package on what amounted to a take-it-or-leave-it basis. "This is nonsense," he said. "I don't work that way. I want to reach an agreement. Instead you send this Australian mule to threaten me."

Henderson replied that Nasser had misinterpreted Menzies's intentions. Menzies was a "blunt man," Henderson acknowledged, but he did not mean to threaten or dictate terms. "He is a man you can talk to. Don't be misled by his bluntness. If he felt he was simply being used as a messenger boy he would go straight back to Australia. His mission can amount to very nearly the same thing as negotiations, and he wants to make a success of it. He is not here to act as a spokesman for the British empire."

Somewhat mollified, Nasser consented to go ahead with the talks, but on his own terms. He received the delegation in Cairo's Kubba Palace, instead of the foreign office, driving home the point that any concessions by Egypt would be at his pleasure only. When Menzies requested two meetings per day, one in the morning and one in the evening, Nasser asserted that more pressing matters left him free only in the evening. "It looks as if I may have a war on my hands," Nasser declared. "In the morning I must be preparing for it."[15]

Menzies did little to diminish Nasser's war worries. The Australian prime minister suggested that London and Paris were seriously contemplating the use of force; the only preventive was Egypt's acceptance of international control. Nasser responded that international control would imply an international control agency, which would demand international protection, requiring foreign troops and the reoccupation of Egypt. His people would never stand for this. "You think that an international administration would end the trouble, but I think that an international administration would be the beginning of trouble." Menzies rejoined, in what could only seem a menacing tone, "Mr. President, your refusal of an international administration will be the beginning of trouble." At this point Nasser nearly terminated the talks. He gathered his papers, saying, "You are threatening me. Very well, I am finished. There will be no more discussions. It is all over."

Henderson intervened. "Mr. President," he explained to Nasser, "I don't think Mr. Menzies meant what he said as a threat." The delegation did not wish to impose a solution upon Egypt, nor did it threaten reprisals if Egypt did not agree. Pointing to his own presence as an American on the committee, Henderson continued, "I want to explain that the United States is not a colonial power. Our policy has been against colonization since independence. We could never agree to join any colonialist arrangement, and if the American government had thought that the purpose of this committee was to impose a solution on Egypt we would not have taken part in it. Our only wish is to reach a solution that is compatible with Egypt's full exercise of its sovereignty."[16]

Henderson's intervention saved the moment, and Nasser consented to continue the discussions. But Henderson could not rescue the mission, which, as he had foreseen, was doomed from the beginning. Dulles's international control scheme was only partly serious. Although the secretary certainly would have been happy to have his plan implemented, he knew that Nasser could not accept it. Dulles was playing for time, hoping postponement would diminish the threat of war, as it had in the Iranian case. In this instance, though, postponement simply sharpened the positions of the contesting powers. "The salient fact," Menzies remarked later, "was that Nasser had been left in possession of the field, that there was no room left for negotiation, except at a disadvantage, and that Britain and France were left with a grim choice between surrender, or force."[17]

After further unproductive talks, the members of the committee retired to prepare their report. The report placed the burden of failure on Nasser. "We encountered with regret an immovable resistance to any control or management other than the Government of Egypt itself," the report declared. "In spite of our best and most patient efforts, we constantly came up against such phrases as 'collective colonialism,' 'domination,' and 'seizure' and what seemed to be an unwillingness to meet reason with reason. In the result, therefore, the central proposals of the eighteen powers were completely rejected."

Henderson concurred in the report, although on his own he might not have chosen to place Nasser's rejection in the realm of the irrational. Whatever the limitations of Henderson's world-view, he rarely resorted to the crutch of calling his opponents irrational. If anything, he conceded a surfeit of rationality to those he considered most threatening. Communists might be misguided, but if one granted their premises, their conclusions followed in logical lockstep. This was what gave them the sense of mission that made them so dangerous. In Nasser, Henderson, like most American officials engaged in policy-making toward the Middle East, perceived a shrewd nationalist who intended to play the great powers against each other, to Nasser's and Egypt's benefit. Yet shrewdness did not rule out volatility. Henderson did not doubt that Nasser would make the most of his reputation for emotionalism. "My impression," he cabled Dulles, "is that Nasser could be entirely irresponsible if he should conclude he has little to gain by exercising restraint." He added that it was "impossible to predict what this rather unpredictable person might do." He thought Nasser was waiting for the west to make the next move.[18]

The west—specifically Britain and France, with Israel's help—did move next, attacking Egypt at the end of October. The operation aborted, however, when the most important western power—the United States—refused to go along. The Eisenhower administration turned the diplomatic and financial screws on London and Paris and brought the operation to a shuddering halt. The immediate result was humiliation for Britain and France and glory for Nasser. America won some propaganda points by refusing to countenance aggression, even when quiet committed by its closest allies.

But the most significant long-term consequence of the Suez crisis was Washington's final immersion in the affairs of the Middle East. Suez begat the Eisenhower doctrine of 1957, which represented an attempt to replace vanishing British influence in the region with an American anticommunist protectorate. The Eisenhower doctrine gave rise to the American intervention in Lebanon in 1958, when the president responded to a radical revolt in that country by landing 14,000 American troops. Between them, the Eisenhower doctrine and its Lebanese application demonstrated the still-expanding character of the American definition of national security. The empire continued to grow.

IV

In addition to his responsibilities relating to the Middle East, mundane matters of administration occupied most of Henderson's time from 1957 to 1959. Budgets had to be prepared and justified, personnel patted and poked, embassies and consulates staffed and supervised. This last category of activity became a full-time job as the decade wound down. Although Henderson turned sixty-five, normal retirement age, in the summer of 1957, he remained in the department at the particular request of the White House. In Eisenhower's last years in office the president grew increasingly aware of the importance of the third world, and as the number of newly independent countries burgeoned he wanted American relations with the post-colonial governments to get off to the best start possible. He gave Henderson the task of arranging the details of establishing those relations.

The final full year of Eisenhower's administration witnessed a flood of new members into the United Nations. From Africa alone, eighteen new countries joined. Each country, no matter how small, unstable, or relatively insignificant, required attention. In the autumn of 1960 Henderson headed a thirteen-person party on a six-week sweep across Africa, visiting twenty-one countries in an effort to extend America's best wishes and to ascertain the lay of the political land.

Following a rough flight over the Atlantic the Americans touched down in Morocco. "Quiet and uneventful," Henderson noted in a trip diary, adding that the most pressing need at the moment was a new file clerk for the embassy in Dakar. From Morocco the party traveled to Mali, thence to Guinea. In Guinea they encountered what struck Henderson as "considerable inefficiency and confusion" among officials charged with checking credentials. The plane carrying the Americans had just taken off from the airport at Conakry when the pilot received orders to return. The person examining the visitors' passports had not found what his superiors were looking for, so he hauled the Americans and their papers back for closer scrutiny. Henderson never learned what the problem was, but after a few hours he and his associates were on their way again.

The group survived engine trouble over Chad and "fatalistic" driving by chauffeurs in Malagasay. Henderson had a "frank" exchange with the president of Liberia, who thought the Americans ought to be more generous to the country they had helped found a century and a half before. In Gabon the head of state missed a dinner engagement; the Americans later learned he had been arrested for trying to poison the prime minister.[19]

Not unexpectedly, four days spent in the Congo produced the greatest excitement of the journey. Arriving in the third week of November, the group flew into the middle of that country's civil war. Even before independence on June 30, fighting had broken out among several factions determined to grab a share of power from the decamping Belgians. Patrice Lumumba and Joseph Kasavubu had formed an uneasy—very uneasy—alliance as prime minister and president. Their differences soon led to a murderous falling out. Joseph Mobutu, a rising figure in the army, challenged the control of the civilians. Antoine Gizenga, a radical with close ties to Moscow, denounced the regime in the capital Leopoldville and established his own provisional government in Stanleyville. Moise Tshombe, a favorite among many westerners for his charm and support of Belgian copper interests, led a rebel movement in the southeastern province of Katanga.

The Congo conflict assumed international proportions when Brussels, fearing for the safety of the nearly 100,000 Belgian nationals residing in its erstwhile model colony, sent in paratroopers, not thirty days after granting independence. Lumumba decried the invasion and called on the UN to vindicate Congolese independence. Hedging his bets, the prime minister also appealed to the Soviet Union.

The appeal to the Kremlin particularly caught the attention of the United States. When Russian and Czech equipment and advisers shortly arrived on the scene, the Congo became a regular battlefield in the cold war. Publicly, Washington backed the UN, which by now had inserted its own forces into the country. Secretly, the CIA conspired against Lumumba, hatching a variety of plots to assassinate the charismatic leftist. As matters developed, Lumumba's Congolese enemies caught up with him first. After living under what amounted to house arrest for several weeks in the autumn, Lumumba escaped just about the time of Henderson's visit, only to be captured by Mobutu's men, who had him killed.[20]

Henderson had no reason to know of the CIA's assassination efforts against Lumumba, and he almost certainly would have disapproved. Toppling governments, as in Iran, was one thing, especially since the Iran operation could be justified on grounds that the shah had acted within his constitutional rights in dismissing Mosadeq. But assassinating people was a different matter. Henderson would have been quite content to see Lumumba neutralized—politically.

Without question, however, Henderson found the chaos in the Congo deeply disturbing, all the more since it touched directly the group he headed. As the Americans drove to the airport prior to their departure, they encountered a car flying an American flag, in a ditch by the roadside, burning. A body lay in the middle of the pavement. Henderson first thought the corpse was one of his group. "I was horrified," he scribbled in his log book, "afraid some member of our party killed or injured." But the deceased turned out to be a bicyclist careless of the traffic. The American car had swerved, unsuccessfully, to avoid him and had landed in the ditch. A crowd of bystanders immediately surrounded the car, which was carrying the American military attaché and his wife. The driver, an American warrant officer named St. Laurence, left the car to reason with or distract the onlookers. They set upon him at once, beating him for several minutes and stabbing him besides. Another young man in the car, a foreign-service officer named Frank Carlucci, went to St. Laurence's aid, and was also attacked with fists and knives. Finally the embassy's administrative officer, an imposing former law-enforcement official from Alaska, became the "hero of the occasion," in Henderson's words. "He got out of his car and singlehanded put the mob of ten or fifteen armed with knives out of commission."[21]

The incident, while tragic to the dead bicyclist and nearly so to St. Laurence and Carlucci, provided Henderson an unexpected opportunity to see more of the country, and especially to gain greater insight into its politics. The American ambassador, Clare Timberlake, had been planning a visit to Elizabethville to interview Tshombe. Arguing that Timberlake should not leave two wounded Americans in Leopoldville, Henderson offered to substitute.

When he touched down at the Elizabethville airport, Henderson was met by an honor guard and by Tshombe in diplomatic attire. Military salutes followed, then a motorcycle escort to rebel headquarters. Tshombe introduced Henderson to his provisional cabinet, and for an hour they discussed the political situation in the Congo and its relation to American policy. "They were critical but not vindictive," Henderson commented later. They offered "friendly criticism in a way to make us strengthen our policy."

Henderson found Tshombe as charming as most westerners did. Recounting the meeting, Henderson said,

Tshombe wanted a fair rapport. He didn't want to have an independent country. He was perfectly willing to talk and come to an understanding with any leader of

Leopoldville who showed that he had a sense of real responsibility and also showed that he had some power. He said, "Mobutu is the man I have found so far who seems to me to have both those qualities." He said to me, "I have been somewhat disappointed in Kasavubu, because Kasavubu just made a speech to the United Nations in which he criticized the Belgians as invaders," and he said, "Really, we need the Belgians as advisers and Kasavubu knows that as well as I do, and when he stands up and says that to the UN he is not sincere, and I like to deal with sincere people.[22]"

Henderson also liked to deal with sincere people, which was one reason why he liked Tshombe, who certainly seemed sincere. Another reason was that Tshombe could be counted on to oppose communism. Yet Mobutu was equally satisfactory on the latter score, and, as events proved, he was more clever. Mobutu eventually managed to eliminate all his rivals, including Tshombe, and establish a long-lived and pro-western, if hugely corrupt, regime.

Much before he did, Henderson had left. From the Congo the American group traveled to East Africa, where they discovered much disarray but nothing so exciting as the Congolese civil war. At the end of November they wound up their tour and headed home.

On arrival in Washington, Henderson held a background press conference. He commented on his impressions of Africa. Although he described Tshombe in favorable terms, many of the officials he had met he judged weak or dishonest or both. Their governments professed nonalignment but in practice tilted toward Moscow and Beijing. Speaking not for publication, he cited the examples of Mali, which exhibited "very strong leanings to the left," and Guinea, which was "overrun with Soviet Communists, Czechs and Chinese."

Henderson found this leftist tendency distressing, since evidence increasingly indicated that the outcome of the cold war might hinge on events in Africa and other decolonized regions. "I don't know of anything that has happened in recent years that has been more dramatic in international affairs than the opening of these new countries," he declared. "Practically a new world has opened to us from a diplomatic point of view." The communists were staking their claim to this new world. The United States could do no less.[23]

V

The United States did no less. By the time Henderson took his postponed retirement in January 1961, American leaders had thoroughly embraced the anticommunist activism he had been advocating for a quarter-century. When John Kennedy spoke of paying any price and bearing any burden to assure the survival of liberty against its foes, Henderson might have written the speech. When Kennedy sent thousands of American troops into Vietnam, beginning the escalation that would produce the largest and bloodiest imperial war of the century,

Henderson might have been advising him. In a way Henderson was, since Kennedy's—and Johnson's and Nixon's—approach to Vietnam followed directly from the Truman doctrine and other initiatives in whose formulation Henderson had played a principal part.

Because of this close connection between Henderson and the central theme of the foreign policy of the United States during the cold war, an evaluation of the man and his career amounts in large part to an evaluation of the policy. The touchstone of the policy, and the cornerstone of the American anticommunist empire, was the belief that the Soviet Union and its allies were bent on global conquest. As Henderson had conceded in his 1936 assessment of the Kremlin's conduct of international affairs, events might sometimes force the Soviets to set their ultimate objective momentarily aside. But however devious the route they were required to traverse, Moscow and its agents never abandoned the goal of world revolution. Collaboration with the communists, for instance during World War II, could only be temporary. Communist ideology could accommodate a truce with the democratic nations; it could not countenance permanent peace.

From this premise followed the conclusion that informed American cold-war policy: that the United States must be as vigilant in resisting communist expansion as the Soviets were in promoting it. Whether resistance involved money, as in the Greek-Turkish aid package, the Marshall plan, and wheat for India; military alliances, as in NATO, SEATO, and the smaller treaty arrangements of the 1950s; direct combat, as in Korea and Vietnam; or covert operations, as in Iran, Guatemala, the Congo, and elsewhere, the United States could never let down its guard. To falter would yield short-term defeat and long-run disaster—the former by loss of territory to totalitarianism, the latter by the major war that irresolution inevitably would produce, as Munich had produced World War II.

Was the premise correct? Did the conclusion logically follow?

The answer to the first question is yes and no. When Henderson argued that Stalin intended world conquest, he was right, in a certain sense. Like Marx and Lenin, Stalin looked to and worked toward the day when a communist state would govern the earth. But Henderson's observation was not particularly helpful. The important issue in international affairs is almost never what governments want, but what they are willing to chance to get what they want. Wishes are not tanks. Henderson himself admitted that Stalin was not prepared to stake the security of the Soviet Union on an unlikely world revolution.

Henderson did have a point, though, in contending that as the Soviet system was constituted under Stalin, the west could never have relations with it of the sort Americans considered normal. So long as Moscow retained its revolutionary ideology, however attenuated that ideology might become in the course of month-to-month decision-making, the Russians would always probe the weak spots in the political and military defenses of the capitalist west. For this reason, vigilance definitely was in order.

Yet vigilance did not necessarily imply the ambitious anticommunist measures Henderson urged. To beware the Kremlin's probes was one thing; to commit the United States to defend the status quo of half the world was something else. Would Iran have slipped into the Soviet sphere in 1946 without America's diplomatic support, or in 1953 without the help of the CIA? Would communists have captured the government of Greece in 1947? Would a North Korean triumph in 1950 have critically altered the balance of power in Asia? Would a Lumumba-led Congo in 1960 have seriously endangered the security of the American alliance system?

The answer to these questions is maybe yes, maybe no. Whether what didn't happen would have happened under different circumstances is impossible to tell. Henderson was not willing to take the risk, and neither were other American policy-makers. Fortunately for them, their country was so situated at the end of World War II as to allow them to indulge their preference for avoiding risk. Controlling nearly 50 percent of humanity's industrial output, Americans could conceive the notion of eliminating, or at least considerably diminishing, the ambiguity of life on the same planet as Stalin and the communists. Did communism threaten Athens and Ankara? Write a check for $400 million. Would economic dislocation lead to bolshevism-by-the-ballot in Western Europe? Send $12 billion. Would a North Korean victory endanger Japan? Dispatch MacArthur and crack the whip over the UN. Would the Russians be tempted to jump Germany? Dig foxholes along the Elbe and fill them with American troops. Might Mosadeq lose his grip and let the Tudeh take over? Bribe the mob. Was Lumumba getting too friendly with the Soviets? Put out a contract on him.

Henderson's motives, and those of others who advocated strong action to block communist expansion, were honorable. Henderson had been over into the Stalinist future and knew it didn't work. He rightly judged that the earth would be a better place the less territory the communists controlled.

The deficiency of Henderson's approach was not its desire to keep as much of the world as possible out of the communist grip. The deficiency was its mistaken belief that the United States possessed the capacity to respond decisively to every sign of slippage in Moscow's direction. Even if Americans overestimated Stalin's role in the Greek insurgency, there was a strong probability that a communist regime in Athens would look to the Soviet Union for support. Yet was the probability sufficient, and would the resulting threat to American interests have been sufficiently dire, to warrant a major commitment of American resources and prestige? Perhaps. But perhaps not. And definitely not, if the response to Greece's troubles should set a precedent for similar commitments in a score of other places—as it did.

The problem was not any individual commitment—not the commitment to Greece, to Turkey, to Western Europe, to Korea, to Iran, to Lebanon, to Vietnam—but the whole constellation of commitments taken together. This was too much for any country to bear, even one as wealthy and well placed as the United States.

The oversubscription of American resources became undeniable only after Henderson retired, but the misfit between promises and capacity showed itself occasionally from the beginning. It appeared first as a willingness to cut ethical corners. In 1942 Henderson had written to Samuel Harper, "It is possible that before this war is over we shall be compelled like various other nations to think only of what we can do to win rather than to bother too much about principles." The temptation to expedience did not disappear as the world war gave way to the cold war. Henderson questioned the Truman administration's decision to acquiesce in French pressure on Syria and Lebanon, wondering what it would do to American ideals "to tolerate one of our Allies engaging in a policy which partakes of aggression because we do not wish to give offense to that Ally." He recognized the undemocratic character of the Tsaldaris government in Greece. He expressed genuine reluctance at turning the CIA loose in Iran.

In such cases and others, Henderson understood that policy decisions entailed costs. He realized that the costs included damage to America's values and good name. But he consistently judged the costs of containing communism to be less than the dangers of failing to do so. Once, in describing Nehru's efforts to persuade Asian leaders to stay out of the cold war, Henderson guessed that the Indian prime minister might win over persons "without profound convictions based on their own experiences." Henderson's convictions were based on his own experience of Stalinism, and these convictions caused him to conclude that however questionable the means of an anticommunist policy, the end almost always justified them.[24]

Henderson's narrow focus on the communist threat was instrumental in producing an essentially negative foreign policy. Henderson knew what he was *against*. It was less clear what he was *for*. He claimed to be acting on behalf of democracy and individual liberty, but many of the regimes America backed in the late 1940s and 1950s—from Greece and Iran to South Korea and South Vietnam, not to mention Latin America's customary clutch of colonels—belied the claim. So repugnant and dangerous did Henderson and other American officials consider the communist challenge that they were willing to confer membership in the "free world" on almost any government calling itself anticommunist. Once the Rhees, Pahlevis, Diems, and Batistas caught on to America's game, the neighborhood went downhill fast.

The narrowness of Henderson's—and America's—perspective produced another problem that grew increasingly evident with time. By concentrating to such a degree on the Soviet Union and its allies, American officials missed the significance of certain events and phenomena that would transform the land-scape of international politics. To considerable degree, the system that emerged from the dust and smoke of 1945 did in fact exhibit the bipolar, zero-sum prop-erties asserted by Henderson and other hardline anticommunists. But over time the emergence of new—a "third"—world changed the rules of diplomacy. To many individuals in countries like India and Iran, the superpower struggle was just one facet of life in the modern age, and by no means the most important. Henderson spent three years in India and three in Iran. He gained more than a little grasp of the politics and foreign policy of each. Yet, thinking the relation-ship of India and Iran to the superpower conflict the only issue that really mat-tered, he failed to appreciate the importance of much that would shape the attitudes of Indians and Iranians for years to come—the idealistic component of Indian nonalignment, for example, and the cultural antiwesternism of Islamic fundamentalists in Iran.

Henderson's strength, and the source of his impact on American policy, was his single-mindedness. In Washington under Robert Kelley, later in Riga, and most importantly in Moscow, he learned one big lesson. The subsequent course of his career—especially his exile from European affairs in 1943 and his removal from NEA in 1948—drove the lesson deeper. Even after he retired, through years of lecturing and an extended stint as director of the Center for Diplomacy and Foreign Policy at American University in Washington, he held to the views he had formed long before. At a reunion of Moscow old-timers in 1975 he stated, without the least apology, that so far as the Soviet Union was concerned, "My mind hasn't been changed much during the last fifty years." What he called "the ultimate foreign policy objective" of the Soviets remained as before. "That objective has been and continues to be the creation of a communist-controlled world with headquarters hopefully in Moscow."[25]

Yet if Henderson's single-mindedness, which allowed him to persist in advo-cating an anticommunist policy in the face of concerted opposition and even-tually to claim victory over the hands-across-the-caviar types, was his strength, it was also his weakness. It was the weakness as well of American cold-war policy. Obsessed with the struggle against communism, the United States black-ened its reputation by engaging in activities incompatible with America's pro-fessed ideals. Intent on securing support against Moscow, Washington allied itself with a cast of unsavory characters. Dividing the globe conceptually into east and west, American officials misunderstood issues that informed the out-look of the third world. Interpreting international events chiefly in the frame-work of America's rivalry with the Soviet Union, Henderson and other Amer-ican policy-makers imposed an awkward globalism on circumstances better perceived in local or regional terms.

The limitations of Henderson's perspective were becoming evident by the time he quit the State Department; the American empire was ratting at the edges. In 1961 the Kennedy administration was getting ready to carry the policy of anticommunist activism to its logical conclusion in Southeast Asia. By all reasonable measures, the conflict in Vietnam was fundamentally a struggle between indigenous forces for control of that country. But first Truman, then Eisenhower, then Kennedy, and finally Johnson and Nixon re-cast the war there as a test of America's fortitude in opposing communist aggression. Despite the fact that a coherent world-communist movement no longer existed, American leaders contended that global peace would begin to disintegrate in the jungles of Indochina unless the United States intervened—just as Henderson had claimed that the disintegration would begin in the hills of Greece or in the Tehran bazaar.

Henderson lived to the age of ninety-three, dying in 1986. He witnessed the conclusion of the Vietnam war, and he recognized its importance for American foreign policy. "History may well regard the outcome of our war in Vietnam as the watershed of United States influence on world politics," he said in 1975. America's defeat in Vietnam was not predestined; under different but not inconceivable circumstances South Vietnam might have become another South Korea. Henderson, characteristically, thought the war winnable, attributing the defeat to a "lack of determination and decisiveness."[26]

But in another sense Vietnam was unavoidable. If the defeat the United States suffered had not happened in Vietnam, it would have occurred elsewhere along the imperial frontier. From the time Truman took the advice of Henderson and his associates and defined American security primarily as a matter of opposing communism, the United States assumed a task that could end only in defeat, somewhere, sometime.

Henderson sought the spread of self-determination and democracy. Unfortunately, reality refused to cooperate, and he settled for less. Instead of an empire of freedom, he helped create an empire of anticommunism. The former might have yielded order and peace; the latter produced manipulation and ceaseless struggle. For all the worthiness of Henderson's intentions, the United States could not maintain the struggle forever. During forty years of effort, Henderson contributed greatly to the rise of the American empire. At the same time, he helped make inevitable its eventual end.

Notes

On Sources

By far the most important source on Henderson is Henderson himself: the many thousands of letters, diary entries, memoranda, cables, and reports he produced during his long life. These documents reside in two collections primarily: the Henderson papers at the Library of Congress in Washington, and the records of the Department of State at the National Archives in Washington and at the National Records Center in Suitland, Maryland. The Henderson papers (cited below as HP) include the typescript of a memoir Henderson began but never completed, covering his career to the beginning of World War II. (An abridged version of this memoir has been published as *A Question of Trust* (Stanford, Cal., 1986).) The typescript is the source of general background information for the early chapters of the present book; citations of specific passages are abbreviated HM and take the form volume/part/page. References to State Department records (record groups 59 and 84) always give the file number; where the extension does not simply repeat the date, this also is given.

Henderson participated in oral history projects conducted by the Truman Library in Independence, Missouri; by Columbia University in New York; and by Princeton University in Princeton, New Jersey. Transcripts of interviews with Henderson by members of these projects are cited in the notes. Because Henderson died before research for the present book began, the author did not interview him directly.

Henderson appears in various manuscript collections beyond his own and the State Department's. Those consulted include the records of the American Red Cross and of the American Joint Chiefs of Staff, located at the National Archives; the Laurence Steinhardt, Joseph Davies, William Standley, and Charles Bohlen papers at the Library of Congress; the John Foster Dulles and George Kennan papers at Princeton University; the Robert Kelley and George McGhee papers at Georgetown University in Washington; the William Bullitt, Arthur Bliss Lane, Chester Bowles, and Dean Acheson papers at Yale University in New Haven, Connecticut; the Samuel Harper papers at the University of Chicago; the Franklin Roosevelt and John Wiley papers at the Roosevelt Library in Hyde Park, New York; the Harry Truman, Dean Acheson, and George McGhee papers at the Truman Library; the Dwight Eisenhower and John Foster Dulles papers at the Eisenhower Library in Abilene, Kansas; and the British Foreign Office and Cabinet records at the Public Record Office in Kew, England.

Additional sources include the memories of Henderson's associates, as related in conversation and correspondence with the author. Were he alive today, Henderson would be pushing one hundred; hence few of his contemporaries remain. But Elbridge Durbrow, Edward Elson, Parker Hart, George Kennan, Henry Mattox, Armin Meyer, Bart Stephens, William Taft III, T. Eliot Weil, and Fraser Wilkins have provided valuable insight into Henderson the man, as well as Henderson the professional. Interviews with other individuals, conducted by the oral history projects mentioned above, are cited in the notes.

Published works are described in customary fashion. Titles in the State Department's *Foreign Relations of the United States* series are abbreviated, giving year, volume and pages—e.g., FR 45:7, 368–69.

Much of Henderson's correspondence was conveyed by cable. To render this material easier to read, understood but deleted words (primarily articles) and punctuation have been reinserted.

1. Beyond Arkansas

1. HM, 1/3/2–3.
2. LWH report, 3/14/19, HP.
3. Taylor to LWH, 3/29/19, HP.
4. Directive by Tidbury, 4/18/19, HP; LWH diary, 4/20/19, HP.
5. LWH diary, 4/21/19, HP.
6. LWH diary, 4/21/19, HP.
7. LWH diary, 4/21/19, HP.
8. LWH diary, 4/22/19, HP.
9. LWH diary, 4/22/19–4/26/19, HP.
10. LWH diary, 4/24/19, HP.
11. LWH report, 4/30/19, HM, 1/3/46–49; LWH report, 5/1/19, file 948.08, American Red Cross (ARC) records.
12. LWH diary, 5/5/19 and 5/6/19, HP.
13. Ryan to LWH, undated (Aug. 1919), HP; LWH notes, 10/3/19, 948.08, ARC records.
14. Ryan to Olds, 10/12/19 and 10/26/19, 948.08, ARC records; HM, 1/4/7.
15. HM, 1/4/13.
16. HM, 1/4/46–51; LWH to Ryan, 12/2/19, 12/21/19, and 12/26/19, 948.08, ARC records.
17. HM, 1/4/57–60.
18. Ryan to Olds, 1/4/20, 948.08, ARC records; Landtoff report, undated (Apr. 1920), HP.

19. Ryan to Olds, undated (Jan. 1919), 948.53, ARC records.
20. HM, 1/4/80–86.
21. LWH to McCaffrey, 8/20/20, HP; LWH to Mason, 11/26/20, HP; LWH to Kelly, 11/5/20, HP.
22. LWH to Kelly, 11/5/20, HP; LWH to Heup, 3/22/21, HP; LWH to Houghton, 7/9/21, HP.
23. HM, 1/4/93–94; *Red Cross Bulletin* (Riga), 11/27/20, HP; Heup to LWH, 2/17/21, HP; LWH to Heup, 3/22/21, HP.
24. Salmonovitz to LWH, undated (1921), HP.

2. Initiation

1. HM, 2/5/1–6.
2. LWH to Civil Service, 11/5/21, HP; Carr to LWH, 11/17/21, HP.
3. Dearing to LWH, 2/28/22, HP; LWH to Carr, 3/25/22, HP; Carr to LWH, 3/30/22, 7/22/22, 7/27/22 and 8/2/22, HP; LWH to Carr, 8/5/22, HP; LWH to Guaranty Trust Co., 8/7/22, HP.
4. State Department to LWH, 8/29/22, HP.
5. HM, 2/6/17–20.
6. HM, 2/6/24.
7. Young to LWH, 11/18/24, HP.
8. LWH to Propas, 2/5/77, HP.
9. George F. Kennan, *Memoirs*, vol. 1 (Boston, 1967), 33.
10. Charles E. Bohlen, *Witness to History* (New York, 1973), 28; Daniel Yergin, *Shattered Peace* (Boston, 1977), 20; Kennan, *Memoirs*, 84.
11. Samuel N. Harper, *The Russia I Believe In* (Chicago, 1945), 201; Yergin, 20.
12. Ibid., 99–100.
13. Lee H. Burke, "Homes of the Department of State and its Predecessors," *Department of State Newsletter*, June 1976, 30–34.
14. HM, 2/7/8–18.
15. HM, 2/7/53 ff.

3. Recognizing Reality

1. American Red Cross (Mulhall) to State Department, 4/24/24, 123 H 383/16, State Department records.
2. Coleman to Kellogg, 9/19/27, 123 H 383/35; Kennan, *Memoirs*, 29–30.
3. Grew to Coleman, 4/23/27, 123 H 383/30a.
4. Study of Latvian political parties, 3/7/28, HP.
5. Coleman to Kellogg, 4/30/28, 123 H 383/43; LWH to Kellogg, 11/6/28, 123 H 383/44; Sussdorff to Kellogg, 11/6/28, 123 H 383/44.
6. Sussdorff to Kellogg, 11/8/28 and 11/6/28, 123 H 383/48 and 123 H 383/44.
7. HM, 3/8/77 ff.
8. LWH to Kellogg, 2/27/28, 123 H 383/38.
9. Harding inaugural address, *New York Times*, 3/5/21.
10. Hoover State of the Union address, 12/3/29, *Public Papers of the Presidents: Hoover 1929* (Washington, 1974), 406–407.

11. Joan Hoff Wilson, *American Business and Foreign Policy, 1920–1933* (Lexington, Ky., 1971), 93–98.
12. Melvyn P. Leffler, *Elusive Quest* (Chapel Hill, N.C., 1979), 198.
13. Bohlen, *Witness to History*, 11.
14. List of LWH dispatches and cables, undated, 123 H 383/64.
15. LWH report, 1/3/30, 861.44 Stalin/1; HM, 3/8/101 ff.
16. Sussdorff to Stimson, 3/6/30, 123 H 383/64; LWH purport file, 2/4/30, 123 H 383.
17. LWH report, 5/3/30, HP; Stimson to Sussdorff, 7/14/30, HP.
18. LWH to Stimson, 5/9/30, 123 H 383/65; Kelley to Hengstler, 7/22/30, 123 H 383/79; HM, 3/8/128–29.
19. LWH to Kelley, 7/28/30, 123 H 383/69.
20. *New York Post* article in *Congressional Record* 94 (1948): 559–560; HM, 3/8/133–39.
21. Robert Paul Browder, *The Origins of Soviet-American Diplomacy* (Princeton, N.J., 1953), 77–79.
22. Kelley memo, 7/27/33, FR 33–39, 6.
23. Robert Dallek, *Franklin D. Roosevelt and American Foreign Policy* (New York, 1979), 79; John Morton Blum, *From the Morgenthau Diaries* (Boston, 1959), 56–57.
24. Roosevelt to Kalinin, 10/10/33, 711.61/287a.
25. Kalinin to Roosevelt, 10/17/33, 711.61/287 1/2.
26. HM, 4/9/74 ff.
27. Welsh to Roosevelt, 10/23/33, FW 711.61/299; Thomas R. Maddux, *Years of Estrangement* (Tallahassee, 1980), 16–17; Dallek, 79–80.
28. HM, 4/9/81–83.
29. Litvinov to Roosevelt, 11/16/33, and Roosevelt to Litvinov, 11/16/33, 711.61/343 1/8–7/8 and 343a 2/8–6/8; Elbridge Durbrow oral history (Truman Library), 13.
30. LWH purport file, 2/7/34, 123 H 383; HM, 4/9/126–32.

4. Red Tape

1. HM, 5/10/6–10.
2. HM, 5/10/83–84.
3. Charles W. Thayer, *Bears in the Caviar* (Philadelphia, 1950), 30.
4. Bohlen, *Witness to History*, 19.
5. Bullitt to Phillips, 1/4/34, FR 33–39, 55.
6. Bohlen, 19–20; memorandum by Messersmith, 12/30/37, FR 33–39, 453.
7. Bullitt to Hull, 3/28/34, FR 33–39, 71.
8. Memorandum by Roosevelt and Litvinov, 11/15/33, FR 33–39, 26; Beatrice Farnsworth, *William C. Bullitt and the Soviet Union* (Bloomington, Ind., 1967), 121–27; Bullitt to Hull, 10/10/34, FR 33–39, 157.
9. Bullitt to House, 7/2/34, Bullitt papers; Bullitt to Kelley, 6/20/34, Kelley papers; Bullitt to House, 9/2/34, Bullitt papers.
10. Bohlen, 24; Thayer, 115, 156–164.
11. Kelley to Hull, 5/7/35, 711.61/521.
12. Bullitt to Hull, 7/13/35 and 7/19/35, FR 33–39, 223; Adam Ulam, *Expansion and Coexistence* (New York, 1968), 229–30.
13. Bullitt to Hull, 8/21/35, 861.00 Congress of Communist International VII/57.
14. Bullitt to Kelley, 9/22/34, Kelley papers.
15. HM, 5/10/175–80.
16. LWH to Messersmith, 12/4/37, HP; LWH to Lane, 6/25/37, Lane papers.

17. LWH to Messersmith, 12/4/37 and 2/2/38, HP.
18. HM, 6/10/149–51.
19. Memo by Kennan, 11/24/37, *FR 33–39*, 398.
20. LWH to Hull, 2/18/38, *FR 33–39*, 633.

5. Moscow Trials

1. Robert Conquest, *The Great Terror* (New York, 1968), 100; LWH summary of Zinoviev-Kamenev trial, 12/31/36, HP.
2. LWH summary, 12/31/36, HP; LWH to Hull, 9/1/36, HP.
3. LWH to Hull, 9/1/36, HP; LWH memo, 1/14/39, HP.
4. LWH summary, 12/31/36, HP.
5. LWH to Hull, 8/27/36, *FR 33–39*, 300.
6. Kennan memo, 2/13/37, *FR 33–39*, 362.
7. Bohlen, *Witness to History*, 48–51.
8. Kennan, *Memoirs*, 82; Bohlen, 44; Martin Weil, *A Pretty Good Club* (New York, 1978), 92.
9. LWH to Wishnatski, 3/18/72, HP; LWH to Lane, 2/6/37, Lane papers.
10. Davies to Hull, 3/2/38 and 2/13/38, *FR 33–39*, 527, 532.
11. Davies to James Roosevelt, 2/19/37, Davies papers.
12. Bullitt to LWH, 11/1/39, HP.
13. Bohlen, 44; LWH to Kelley, 4/29/37, Lane papers.
14. LWH to Kelley, 4/29/37, Lane papers.
15. LWH to Hull, 6/13/37, *FR 33–39*, 383.

6. Minuet for Dictators

1. LWH to Hull, 11/16/36, Kelley papers.
2. Moore to Davies, 3/25/37, 711.61/611; LWH to Kelley, 4/27/37, Lane papers.
3. LWH to Kelley, 4/27/37, Lane papers.
4. LWH oral history (Truman Library), 17 ; Weil, *Pretty Good Club*, 90–93.
5. HM, 4/9/38; Graham H. Stuart, *The Department of State* (New York, 1949), 329–31.
6. HM, 7/10/328.
7. LWH to Lane, 5/19/37, Lane papers.
8. Ibid.
9. LWH to Kelley, 4/29/37, Lane papers.
10. Ibid.
11. LWH to Hull, 12/22/37, *FR 33–39*, 401.
12. Messersmith to LWH, 1/5/38, HP; LWH to Messersmith, 2/5/38, HP.
13. LWH memo, 7/2/38, in Moscow to State Department, 7/13/38, 711.61/657.
14. Harper to LWH, undated (1936), Harper papers.
15. HM, 8/16/704–6.

7. Between the Devil and the Deep Blue Sea

1. Robert A. Divine, *The Illusion of Neutrality* (Chicago, 1962), 211–13.
2. HM, 7/10/324.

3. Kirk to Hull, 10/31/38 and 11/25/38, FR 33–39, 591 ff.

4. Kirk to Hull, 3/30/39, FR 33–39, 747.

5. Kirk to Hull, 4/6/39, FR 33–39, 750.

6. LWH memo, 6/2/39, HP; Ulam, Expansion and Coexistence, 270.

7. LWH memo, 6/2/39, HP.

8. LWH to Wiley, 7/10/39, Wiley papers.

9. LWH memo, 7/22/39, FR 33–39, 773.

10. Harper to LWH, 8/22/39, Harper papers; LWH to Harper, 8/24/39, ibid.; LWH to Wiley, 8/17/39, Wiley papers.

11. Bohlen, Witness to History, 69–87; Ulam, 275; LWH to Steinhardt, undated [Oct. 1939], HP.

12. LWH to Steinhardt, [Oct. 1939] and 12/15/39, HP.

13. LWH to Harper, 12/9/39, HP; LWH to Steinhardt, 12/15/39, HP.

14. Berle in John Lewis Gaddis, Russia, the Soviet Union, and the United States (New York, 1978), 139–40.

15. Maddux, Years of Estrangement, 119.

16. LWH to Steinhardt, 12/15/39, HP.

17. LWH to Harper, 2/16/40 and 5/23/40, Harper papers.

18. Moffat memo, 4/1/40, 711.61/726; LWH memos, 4/9/40, 711.61/732; 8/1/40, 711.61/743 1/4; 8/15/40, 711.61/743 1/2; 4/4/40 and 5/22/40, FR 40:3, 268 ff.

19. LWH memos, 4/16/40 and 6/7/40, FR 40:3, 289, 311.

20. Welles memo, 7/27/40, FR 40:3, 327.

21. LWH memo, 3/2/40, HP.

22. LWH memo, 7/26/40, 711.61/776.

23. Steinhardt to LWH, 10/20/40, in Atherton to Welles, 11/26/40, FR 40:3, 406.

24. LWH to Steinhardt, 12/13/40, Steinhardt papers.

25. Welles to Oumansky, 1/21/41, FR 41:1, 696; Gaddis, 143–44; Hull to Steinhardt, 3/1/41, FR 41:1, 712.

26. LWH to Steinhardt, 3/31/41, Steinhardt papers.

27. LWH memo, 4/18/41, 711.61/814 1/2.

28. Memo by division of European affairs, 6/21/41, FR 41:1, 766.

29. LWH memo, 7/2/41, 711.61/824.

30. LWH to Harper, 6/27/41, Harper papers.

31. LWH to Harper, 7/21/41, Harper papers; LWH to Steinhardt, 8/18/41, Steinhardt papers.

32. LWH to Harper, 7/21/41, Harper papers.

33. Berle to Welles, 7/30/41, FR 41:1, 798.

8. Comrades in Arms

1. LWH to Harper, 12/10/41, Harper papers.

2. LWH to Harper, 12/26/41 and 1/22/42, Harper papers.

3. LWH to Welles, 4/9/42, FR 42:3, 435.

4. LWH to Harper, 3/2/42, Harper papers.

5. LWH to Harper, 3/2/46, Harper papers; New Masses, 2/24/42.

6. LWH to Harper, 3/16/42, Harper papers.

7. LWH to Harper, 3/23/42, Harper papers.

8. LWH to Harper, 12/16/41, Harper papers; William H. Standley, Admiral Ambassador to Russia (Chicago, 1955), 498; Standley to Hull, 6/22/42, FR 42:3, 598.

9. W. Averell Harriman, *Special Envoy to Churchill and Stalin* (New York, 1975), 147–50.
10. LWH to Harriman, 1/22/75, HP; Winston S. Churchill, *The Hinge of Fate* (Boston, 1950), 475.
11. Harriman to Roosevelt, 8/13–14/42, FR 42:3, 618.
12. Stalin to Harriman, 8/13/42, FR 42:3, 621.
13. Roosevelt to Stalin in Standley to Molotov, 8/19/42, FR 42:3, 626.
14. Ulam, *Expansion and Coexistence*, 334–38.
15. Standley, 97.
16. Standley, 265–94; Standley memo, undated, FR 42:3, 637.
17. Standley, 295–96.
18. Hurley to Roosevelt in LWH to Hull, 11/15/42, FR 42:3, 655.
19. Roosevelt to Stalin in LWH to Molotov, 11/21/42, FR 42:3, 662.
20. Stalin to Roosevelt in Molotov to LWH, 11/27/42, FR 42:3, 663; Roosevelt to Stalin in LWH to Molotov, 12/5/42, FR 42:3, 665.
21. Roosevelt to Stalin in LWH to Molotov, 12/10/42, FR 42:3, 675; Stalin to Roosevelt, 12/13/42, FR 42:3, 675.
22. LWH memo, 3/22/43, FR 43:3, 352; Polish ambassador to LWH, 3/22/43, FR 43:3, 354.
23. For the tiff at Davies' house, see "Atmosphere About Henderson Aggressively Anti-Soviet," reprinted from *New York Post* in *Congressional Record* 94 (1948), 559.
24. LWH to Atherton, 6/11/43, FR 43:3, 543.
25. LWH oral history (Truman Library), 22–27.

9. "Head net, mosquito bar, sun helmet . . ."

1. Equipment requisition filed for LWH in Havens to Case, 8/4/43, 123 H 383/341.
2. Hubert Herring, "The Department of State," *Harper's* 174 (February, 1937), 225–38, quote from 231–32; Evan M. Wilson, *Decision on Palestine* (Stanford, Cal., 1979), 4.
3. Phebe Marr, *The Modern History of Iraq* (Boulder, Colo., 1985), 29–86; Leonard Mosley, *Power Play* (New York, 1973), 126–31.
4. LWH to Hull, 1/12/44, 890G.00/685; LWH to Hull, 4/13/44, 890G.00/695.
5. LWH to Hull, 4/13/44, 890G.00/695.
6. Marr, 47–48.
7. Wm. Roger Louis, *The British Empire in the Middle East* (Oxford, 1984), 309–11, 315.
8. LWH memos, 11/20/43 and 11/23/43, 123 H 383/375.
9. Marr, 77–78; Louis, 308; LWH to Hull, 12/1/43, 123 H 383/375.
10. LWH to Hull, 1/1/44, 890G.00/679; LWH to Hull, 2/28/44, 890G.00/694.
11. LWH to Hull, 12/1/43, 123 H 383/375.
12. LWH to Hull, 2/29/44, 890G.001/22.
13. LWH to Hull, 4/21/44, 890G.001/23.
14. LWH to Hull, 12/24/43, 890G.00/677; 6/5/44, 890G.00/702; and 7/15/44, 890G.00.
15. LWH to Hull, 11/4/44, 890G.00.
16. LWH to Hull, 11/25/44, 890G.00.
17. Thompson to Eden, 7/31/44, FO 371/E4881, British Foreign Office records.
18. Nissim Rejwan, *The Jews of Iraq* (Boulder, Colo., 1985), 217–27.
19. LWH to Hull, 2/7/45, 890G.00; Rejwan, 227–230.

20. Marr, 90.
21. LWH to Hull, 11/30/44, 890G.00.
22. LWH to Hull, 1/2/45 and 1/11/45, 890G.00.
23. LWH memo, 2/8/45, 890G.00.
24. LWH to Hull, 2/8/45, 890G.00.

10. The Great Game

1. LWH oral history (Truman Library), 31–32; LWH to Hull, 3/27/44, 123 H 383/386; LWH to Hull, 4/21/44, 123 H 383/388; Foreign Office to British embassy in Baghdad, 3/18/44, FO 371/E1657, British Foreign Office records; Baghdad to Foreign Office, 3/21/44, FO 371/E1836, ibid.; LWH to Cornwallis, 3/25/44, FO 371/E2427, ibid.
2. LWH oral history (Truman Library), 32.
3. Wilson, Decision on Palestine, 7.
4. James G. McDonald, My Mission to Israel (New York, 1951), 12–13.
5. Humphrey Trevelyan, Diplomatic Channels (Boston, 1973), 78–79; Louis, British Empire in the Middle East, 38.
6. Elizabeth Monroe, Britain's Moment in the Middle East (Baltimore, 1981 ed.), 32–33; Louis, 147.
7. Satterthwaite to Department of State (DOS), 4/30/45, 890D.01.
8. LWH to Phillips, 5/21/45, 890D.01; LWH memo, 5/3/45, 890D.01/5–345.
9. British embassy to DOS, 5/21/45, 890D.01; memo of conference, 5/23/45, FW 890D.01; LWH memo, 5/3/45, 890D.01.
10. LWH to Grew, 5/23/45, 890D.01.
11. Phillips to Grew, 5/3/45, 890D.01; memo of conversation, 5/28/45, 890D.01; DOS to Paris, 5/26/45, 890D.01.
12. Paris to DOS, 5/29/45, 890D.01.
13. LWH to Truman, 5/31/45, 890D.01.
14. DOS to Paris, 5/31/45, 890D.01.
15. Louis, 169–70; Wright to LWH, 5/31/45, 890D.01.
16. Louis, 148; Paris to DOS, 6/3/45, 890D.01.
17. Memo of conversation, 6/20/45, 890D.01.
18. Memos of conversations, 8/17/45, 890D.20 Missions; 8/22/45, FR 45:8, 1159; and 8/28/45, FR 45:8, 1161.
19. LWH to Dunn, 8/17/45, FR 45:8, 1201; memo of conversation, 7/11/45, FR 45:8, 1199.
20. Memo of conversation, with attachment, 10/5/45, 890D.20 Missions.
21. LWH to Matthews, undated (11/13/45), 890D.20 Missions/12–2145.
22. Matthews to LWH, 11/17/45, 890D.20 Missions/12–2145.
23. Memo of conversation, 12/3/45, 890D.01; DOS to British embassy, 12/13/45, 890D.01.
24. LWH to de Santis, 1/24/76, HP.
25. Memos of conversations, 6/1/45, 800.24591, and 6/18/45, FR 45:8, 380; LWH to Grew, 6/1/45, FW 800.24591.
26. Bruce Robellet Kuniholm, The Origins of the Cold War in the Middle East (Princeton, N.J., 1980), 272–73.
27. Durbrow oral history (Truman Library), 77.
28. LWH memo, 8/7/45, 800.24591; LWH to Byrnes, 8/23/45, FR 45:8, 393.

29. LWH to Byrnes, 8/23/45, *FR* 45:8, 393.
30. LWH to Acheson et al., 9/17/45, *FR* 45:8, 410.
31. Tehran to DOS, 9/25/45, 891.00; LWH to Murray, 9/27/45, *FR* 45:8, 423.
32. LWH to Dunn, 11/19/45, *FR* 45:8, 430.
33. Byrnes to Harriman, 11/23/45, *FR* 45:8, 448.
34. Tehran to DOS, 10/11/45, 11/17/45 and 11/20/45, 891.00; Kuniholm, 280–282.
35. LWH to Byrnes, 12/11/45, *FR* 45:8, 488.
36. Kuniholm, 284; Moscow to DOS, 12/23/45, *FR* 45:8, 510.
37. LWH memo, 12/28/45, *FR* 46:7, 1.
38. Ibid.
39. John Lewis Gaddis, *The United States and the Origins of the Cold War* (New York, 1972), 299–304; Walter Millis, ed., *The Forrestal Diaries* (New York, 1951), 134.
40. Yergin, *Shattered Peace*, 176; Walter LaFeber, *America, Russia, and the Cold War* (New York, 1985 ed.), 39.
41. Tabriz to DOS, 3/7/46, *FR* 46:7, 344; Kuniholm, 318–319.
42. Wright memo, 8/16/65 (written at the behest of the State Department's historian), summarized in *FR* 46:7, 346.
43. Ibid.
44. DOS to Moscow, 3/8/46, *FR* 46:7, 348.
45. Harry S. Truman, *Memoirs,* vol. 2 (Garden City, N.Y., 1956), 94–95; Kuniholm, 320–21; Ulam, *Expansion and Coexistence,* 425–27; William Taubman, *Stalin's American Policy* (New York, 1982), 145–50.
46. LWH to Acheson, 10/8/46, *FR* 46:7, 523.
47. Ibid.
48. LWH to Acheson, 10/18/46, with attachment: "Implementation of United States Policy Toward Iran," *FR* 46:7, 533.

11. The Battle Joined

1. C. M. Woodhouse, *The Struggle for Greece* (New York, 1979), 161–163.
2. Kuniholm, *Origins of the Cold War in the Near East,* 109–125; Woodhouse, 149–73.
3. LWH memo, 11/10/45, 868.00.
4. Ibid.
5. See Byrnes to Truman, 11/10/45, 868.00; DOS to Athens, 11/28/45, 868.50; LWH to Acheson, 12/19/45, 868.50.
6. Athens to DOS, 12/14/45, 868.51, and 12/15/45, 868.50.
7. DOS to London, 1/10/46. 868.51.
8. Athens to DOS, 1/1/46, 868.51, and 1/11/46, 868.00.
9. Memo of conversation, 9/5/46, 868.00.
10. Clayton to Byrnes, 9/12/46, *FR* 46:7, 209.
11. LWH memo with attachment, 10/21/46, 868.00.
12. Memo of conversation, 10/29/46, *FR* 46:7, 255; Louis, *British Empire in the Middle East,* 97.
13. NEA memo in Acheson to Byrnes, 8/15/46, *FR* 46:7, 840; LWH oral history (Truman Library), 234–235.
14. Kuniholm, 359–63; Dean Acheson, *Present at the Creation* (New York, 1969), 195–96; LWH oral history (Truman Library), 235–36; Acheson to Orekhov, 8/19/46, *FR* 46:7, 847.
15. Smith to DOS, 1/8/47, *FR* 47:5, 2; Wilson to DOS, 1/17/47, *FR* 47:5, 7.

16. Paul Porter to Clayton, 2/17/47, FR 47:5, 17.
17. Joseph M. Jones, *The Fifteen Weeks* (New York, 1955), 3–4.
18. Acheson, 217; British aides-memoire, 2/21/47, FR 47:5, 32.
19. LWH to Marshall, 2/20/47, 868.00; LWH oral history (Truman Library), 78.
20. Jones, 132–34; Kennan, *Memoirs*, 330–32.
21. "Crisis and Imminent Possibility of Collapse in Greece," FR 47:5, 29.
22. LWH memos, 2/24/47, FR 47:5, 42.
23. Minutes of meeting, 2/24/47, 868.00.
24. LWH to Acheson, with attachments, 2/25/47, 868.00.
25. Reams to Marshall, 2/24/47, 868.00; minutes of State-War-Navy meeting, 2/26/47, FR 47:5, 56.
26. FR 47:5, 58 n. 2; LWH to Acheson, 2/27/47, FR 47:5, 64.
27. Aide-memoire by LWH, 3/1/47, FR 47:5, 72; memo of conversation, 3/4/47, FR 47:5, 79.
28. Memo by Howard with attachment, 3/4/47, FR 47:5, 84.
29. Acheson to Truman, undated (3/7/47), FR 47:5, 98; *Department of State Bulletin*, 3/23/47, 534.
30. Kennan, 333–35; Jones, 142–45; Kennan to Acheson, 3/6/47, 868.00; Arthur H. Vandenberg, Jr., ed. *The Private Papers of Senator Vandenberg* (Boston, 1952), 340; Acheson, 219–21; LWH oral history (Truman Library), 85–88.
31. LWH oral history (Truman Library), 91–93.
32. Marshall memo, 7/10/47, FR 47:5, 218.
33. Athens to DOS, 9/2/47, FR 47:5, 323.
34. LWH oral history (Truman Library), 92–94.
35. Athens to DOS, 9/2/47, FR 47:5, 323.
36. Ibid.
37. Lawrence S. Wittner, *American Intervention in Greece* (New York, 1982), 112–13; Constantine Tsaldaris oral history, Truman Library, 9.
38. LWH oral history (Truman Library), 95–96.
39. Ibid.
40. LWH memo, 12/22/47, FR 47:5, 458.

12. In the Palestine Labyrinth

1. LWH oral history (Truman Library), 102; Peter Grose, *Israel in the Mind of America* (New York, 1983), 139–54.
2. LWH oral history (Truman Libary), 43–44.
3. Memo of conversation, 6/20/45, 867N.01.
4. Memo of conversation, 6/27/45, 867N.01; also memo of conversation, 6/28/45, 867N.01.
5. Zvi Ganin, *Truman, American Jewry, and Israel* (New York, 1979), 32.
6. Truman, *Memoirs*, 2:161–65.
7. LWH memo, 8/17/45, 867N.01.
8. LWH to Byrnes with attachment, 8/31/45, 867N.01.
9. Kenneth Ray Bain, *The March to Zion* (College Station, Tex., 1979), 80.
10. Ibid.
11. LWH to Acheson, 10/1/45, lot 56 D 359; Baghdad to DOS, 9/26/45, 867N.01; Beirut to DOS, 8/31/45, 867N.01.

12. LWH memo, 10/3/45, 867N.01.
13. Weizmann to Byrnes with attachment, 10/3/45. 867N.01.
14. LWH to Byrnes, 8/21/45, 867N.01; LWH to Acheson, 9/26/45, 867N.01; LWH to Byrnes, 10/9/45, lot 56 D 359.
15. Wadsworth to LWH, Eddy to LWH, Tuck to LWH, all 10/26/45, *FR* 45:8, 790 ff.
16. Acheson to Byrnes, 10/10/45, 867N.01; LWH to Byrnes, 10/10/45, lot 56 D 359. See Saud to Roosevelt, 3/10/45, and Roosevelt to Saud, 4/5/45, 867N.01/1–1045.
17. Bain, 84–85; John Snetsinger, *Truman, the Jewish Vote, and the Creation of Israel* (Stanford, Cal., 1974), 16.
18. *Department of State Bulletin*, 10/21/45, 623; Allen to LWH, 9/26/45, 867N.01. See also Merriam to LWH, 9/26/45, 876N.01.
19. Louis, *British Empire in the Middle East*, 428.
20. LWH to Byrnes with attachments, 8/24/45, 867N.01.
21. Wilson, *Decision on Palestine*, 3; Walter Isaacson and Evan Thomas, *The Wise Men* (New York, 1986), 449.
22. Bain, 101–5; Louis, 396–413; Wilson, 68–79.
23. Jerusalem to DOS, 5/2/46, 867N.01; Cairo to DOS, 5/3/46, 867N.01.
24. LWH memo, 5/10/46, 867N.01.
25. Wilson, 93–95; Louis, 436.
26. *Public Papers of the Presidents: Truman 1946*, 442–44.
27. LWH to Acheson, 10/21/46, with attachments: exhibit A, 10/15/46; exhibit B, 10/21/46; exhibit C, 10/19/21, lot 56 D 359.
28. Memo of conversation, 11/6/46, 867N.01.
29. LWH to Lehrs in DOS to Basel, 12/6/46, 867N.01; Lehrs to LWH in Basel to DOS, 12/30/46, 867N.01.
30. LWH to Acheson, [12/27/46], 867N.01; NEA memo in LWH to Acheson, 12/27/46, 877N.01.
31. LWH to Acheson, 2/10/47, 867N.01/2–747.
32. LWH to Acheson, 2/17/47, 867N.01.
33. Memo of conversation, 4/23/47, 867N.01. Also LWH memo, 4/15/47, 867N.01.
34. Epstein to LWH, with attachment, 5/28/47, 867N.01.
35. LWH to Acheson, 5/29/47, 867N.51/5–2847.
36. Memo of conversation, 5/29/47, 867N.01.
37. Michael J. Hogan, *The Marshall Plan* (Cambridge, Eng., 1987), 40–45.
38. Memo of conversation, 6/19/47, 867N.01.
39. LWH to Acheson, with attachments, 7/7/47, 867N.01.
40. Minutes of meeting, 9/15/47, *FR* 47:5, 1147; LWH oral history (Truman Library), 125–127.
41. Eleanor Roosevelt to Acheson, 6/5/46, 867N.01; Acheson reply (drafted by LWH), 6/14/46, 867N.01/6–546.
42. LWH to Marshall, 9/22/47, 501.BB Palestine; LWH to Lovett, 9/18/47, 501.BB Palestine.
43. LWH oral history (Truman Library), 131–134.
44. Memo of conversation, 10/3/47, 501.BB Palestine.
45. LWH to Lovett, 10/3/47, 501.BB Palestine.
46. Memo of conversation, 10/13/47, *FR* 47:5, 1180.
47. LWH to Lovett, 10/22/47, *FR* 47:5, 1195; LWH memo, 11/4/47, 867N.01.
48. LWH to Marshall, 9/22/47, 501.BB Palestine.
49. LWH to Marshall, 11/10/47, *FR* 47:5, 1249; Wilson, 123–127.

50. LWH to Lovett, 11/24/47, 867N.01.
51. Tuck to LWH, 12/3/47, 867N.01; LWH memo, 12/11/47, 867N.01.
52. LWH memo, 12/8/47, 501.BB Palestine/12-947.
53. NEA memo, 3/24/48, 501.BB Palestine.
54. LWH to Ehrlich, 3/31/48, HP.
55. Hadlow to Mason, 5/2/48, FO 371/E5986; Beeley to Burrows, 5/3/48, FO 371/ E6677, British Foreign Office records.
56. Hadlow memo, 5/10/48, FO 371/E6420, British Foreign Office records.
57. Marshall memo, 5/12/48, FR 48:5, 972.
58. LWH to Lovett, 5/16/48, FR 48:5, 1001.
59. LWH to Horan, 3/31/48, HP.
60. Congressional Record 93:4728-29.
61. Bartley C. Crum, Behind the Silken Curtain (New York, 1947), 36-41; New York Times, 8/22/46.
62. New York Times, 5/4/48; Congressional Record 94:557-61.
63. LWH oral history (Truman Library), 157.
64. LWH to Mandel, 7/24/46, HP.
65. Allen H. Podet, "Anti-Zionism in a Key United States Diplomat," American Jewish Archives 30 (1978):155-87.
66. LWH to Horan, 3/31/48, HP; LWH to Grady, 6/4/48, lot 67 D44 7381.
67. LWH to Offie, 2/2/48, lot 67 D44 7381; LWH to Ehrlich, 3/31/48, HP.
68. Clifford memo, 6/17/48, LWH interview, 1/12/76, both FR 48:5, 1117 ff.
69. Wright oral history, 41.

13. The Most Charming Man He Ever Despised

1. Memo of conversation, 1/10/48, 501.BC.
2. LWH to Lovett, 1/9/48, 501.BC.
3. Memo by Jones, undated, attached to Jones to LWH, 5/10/61, HP.
4. C. S. Jha, From Bandung to Tashkent (Madras, 1983), 74.
5. LWH oral history (Truman Library), 170-74.
6. Millis, Forrestal Diaries, 387.
7. LWH to Marshall, 12/22/48, FR 48:5, 520.
8. LWH to Acheson, 1/24/49, FR 49:6, 1690.
9. LWH to Acheson, 7/29/49, FR 49:6, 1726.
10. C. L. Sulzberger, A Long Row of Candles (New York, 1969), 791; LWH to Acheson, 11/18/48, lot 67 D44 7381; LWH to Acheson, 7/29/49, FR 49:6, 1726.
11. LWH to Acheson, 8/15/49, FR 49:6, 1732; and 8/16/49, 845.00.
12. LWH memo for the record, 8/11/76, HP; LWH to Acheson in background memo on Nehru visit, 10/3/49, Truman papers; LWH memo, undated, McGhee papers (Georgetown).
13. George McGhee, Envoy to the Middle World (New York, 1983), 47; Acheson, Present at the Creation, 334-36; British high commissioner in India to Foreign Office, 12/9/49, FO 371/F18771, British Foreign Office records.
14. Attachment to Hillenkoetter to Truman, 12/20/49, Truman papers.
15. LWH to Acheson, 5/26/49, 893.01.
16. LWH to Acheson, 6/21/49, 893.01.
17. LWH to Acheson, 9/6/49, 893.01.

18. Donovan to Acheson, 10/13/49, 893.01; memo of conversation, 10/19/49, 893.01; Butterworth to Acheson, 12/5/49, 893.01; LWH to Acheson, 12/2/49 and 12/6/49, 893.01.
19. LWH to Acheson, 12/11/49, 893.01.
20. LWH to Acheson, 12/19/49 and 12/29/49, 893.01.
21. LWH address: "Objectives of U.S. Policies Toward Asia," *Department of State Bulletin*, 4/10/50, 562.
22. LWH to Acheson, 4/12/50, *FR* 50:5, 1461.

14. Their Finest Hour

1. LWH to Acheson, 6/27/50, 330.
2. Madras to DOS, 6/28/50, 795.00; LWH to Acheson, 6/28/50, 330.
3. Hare to LWH, 6/28/50, 795.00; LWH to Acheson, 6/28/50, 330.
4. LWH to Acheson, 6/29/50, 795.00; and 6/29/50, 791.00.
5. LWH to Acheson, 6/30/50, 791.00.
6. LWH to Acheson, 7/3/50, 795.00.
7. *New York Times*, 7/4/50; LWH to Acheson, 7/6/50, 791.00.
8. LWH to Acheson, 7/5/50, 791.00; and 7/8/50, 795.00.
9. *New York Times*, 7/7/50; Acheson to LWH, 7/5/50, 791.00; LWH to Mathews, 8/8/50, lot 67 D44 7381.
10. LWH to Acheson, 7/11/50, 330.
11. LWH to Acheson, 7/13/50, 791.00; and 7/13/50, *FR* 50:7, 376.
12. LWH to Acheson, 7/16/50, 795.00.
13. Acheson to LWH, 7/16/50, *FR* 50:7, 406; Acheson to LWH, 7/17/50, 795.00.
14. LWH to Acheson, 8/18/50, 330.
15. Webb to LWH, 9/16/50, *FR* 50:7, 733.
16. LWH to Acheson (two cables), 9/20/50, *FR* 50:7, 742 ff.
17. LWH to Acheson, 9/18/50, 795B.5; 9/23/50, *FR* 50:7, 763; and 10/1/50, 795.00.
18. LWH to Acheson, 9/27/50, 795B.5.
19. LWH to Acheson, 9/30/50, *FR* 50:7, 831; and 10/3/50, 795.00.
20. LWH to Acheson, 9/30/50, *FR* 50:7, 831; and 10/3/50, 795.00.
21. LWH to Acheson, 10/4/50, *FR* 50:7, 869.
22. LWH to Acheson, 10/5/50, *FR* 50:7, 880.
23. Webb to LWH, 10/4/50, 795.00.
24. Merchant memo, 10/6/50, lot 67 D44 7381; LWH to Poteat, 3/21/72, HP.
25. Webb to LWH, 10/4/50, *FR* 50:7, 875; LWH to Acheson, 11/16/50, *FR* 50:7, 1167.
26. LWH to Acheson, 10/6/50, *FR* 50:7, 886.
27. LWH to Acheson, 10/6/50, 10/7/50 and 10/10/50, *FR* 50:7, 889, 901, 921.
28. Robert J. Donovan, *Tumultuous Years* (New York, 1982), 307–10; Acheson to LWH, 12/4/50, 791.13.
29. LWH to Acheson, 12/5/50, *FR* 50:7, 1418.
30. Acheson to LWH, 12/13/50, 795.00; LWH to Acheson, 12/18/50, 795.00.
31. LWH to Acheson, 1/20/51 and 1/28/51, *FR* 51:6, 2087, 2092.
32. LWH to Acheson, 1/29/51, *FR* 51:6, 2093.
33. LWH to Acheson, with enclosure, 2/21/51, lot 67 D44 7381.
34. Report of South Asia conference, 2/26/51–3/3/51, McGhee papers (Truman Library).

35. Henderson remarks, South Asia conference, lot 67 D44 7381; notes for Ceylon conference, undated, HP.
36. Government of India to United States government, 8/23/51, *Department of State Bulletin*, 9/3/51, 385; Sarvepalli Gopal, *Nehru* (Cambridge, Mass., 1976–84), 2:137–38.
37. LWH to Acheson, 9/6/51, 611.91.

15. Nationalization and Its Discontents

1. Bombay *Current*, 9/5/51, clipping in HP.
2. Cabinet minutes, 9/27/51, CAB 128/20, British Cabinet records.
3. LWH to Acheson, 10/22/51, 350 Iran; LWH oral history (Truman Library), 200.
4. Amin Saikal, *The Rise and Fall of the Shah* (Princeton, N.J., 1980), 19–26.
5. LWH to Acheson, 10/22/51, 350 Iran.
6. LWH to Acheson, 10/22/51, 350 Iran; Tehran to London, 7/28/52, in Cabinet paper, 8/5/52, CAB 129/54, British Cabinet records.
7. LWH to Acheson, 10/22/51, 350 Iran.
8. Barry Rubin, *Paved with Good Intentions* (New York, 1980), 54.
9. LWH to Acheson, 11/12/51, 350 Iran.
10. LWH to Acheson, 12/4/51, 350 Iran.
11. Ibid.
12. LWH to Acheson, 12/12/51, 350 Iran.
13. Ibid.
14. Memos of conversation, 10/23/51 and 10/28/51, *FR* 52–54:10, 241–49; cabinet minutes, 1/17/52, CAB 128/24, British Cabinet records.
15. Brief by Ross, undated (Feb. 1952), FO 371/EP1022; Middleton to Furlonge, 11/19/51, FO 371/EP1024, British Foreign Office records.
16. Shepherd to Strang, 9/11/51, FO 371/EP1024, British Foreign Office records.
17. Joint estimate from U.K.-U.S. embassies in Tehran, 11/19/51, FO 371/EP1024; Middleton to Furlonge, 11/19/51, FO 371/EP1024; British Foreign Office records.
18. LWH to Acheson, 11/29/51 and 1/12/52, 350 Iran.
19. LWH to Berry, 1/12/52, 611.88.
20. Farhad Diba, *Mohammad Mossadegh* (London, 1986), 151; LWH to Acheson, 12/12/51, 350 Iran.
21. LWH to Acheson, 12/12/51, 350 Iran; LWH to Acheson, 12/4/51, 888.00-TA.
22. LWH to Acheson, 1/26/52, 350 Iran.
23. LWH to Acheson, 2/27/52, 350 Iran.
24. LWH to London, 3/19/52, 350 Iran.
25. LWH to Acheson, 3/22/52, 350 Iran.
26. LWH to Acheson, 3/26/52, 350 Iran.
27. LWH to Acheson 4/10/52, 350 Iran.

16. The Wily Premier, the Angry Ayatollah, and Hamlet on the Peacock Throne

1. LWH to Acheson, 4/14/52, 350 Iran.
2. LWH to Acheson, 4/23/52, 350 Iran; and 4/15/52, 611.88.
3. LWH to Acheson, 4/26/52, 350 Iran.

4. LWH to Acheson, 5/24/52, 350 Iran.
5. LWH to Acheson, 5/28/52, 350 Iran.
6. LWH to Acheson, 7/2/52, 350 Iran.
7. Diba, *Mossadegh*, 154–57; Rubin, *Paved with Good Intentions*, 73.
8. Sepehr Zabih, *The Mossadegh Era* (Chicago, 1982), 56–66; Richard W. Cottam, *Nationalism in Iran* (Pittsburgh, 1979 ed.), 216.
9. LWH to Acheson, 7/24/52, 350 Iran.
10. LWH to Acheson, 7/28/52, 350 Iran
11. LWH to Acheson, 7/30/52, 350 Iran.
12. Middleton to Foreign Office, 7/28/52, in cabinet paper, 8/5/52, CAB 129/54; cabinet papers, 6/17/52 and 7/29/52, CAB 128/25, British Cabinet records; summary by Sarell, 4/29/52, FO 371/EP10345, British Foreign Office records.
13. LWH to Acheson, 8/3/52 and 8/8/52, 350 Iran.
14. LWH to Acheson, 9/26/52, 350 Iran.
15. LWH to Acheson, 8/8/52, 350 Iran.
16. LWH to Acheson, 8/20/52, 611.88.
17. LWH to Acheson, 9/26/52, 350 Iran.
18. LWH to Acheson, 8/3/52, 350 Iran.

17. What Gentlemen Will Do

1. LWH to Acheson, 11/10/52, 350 Iran.
2. LWH to Acheson, 11/15/52, 350 Iran.
3. LWH to Acheson, 11/5/52, 888.00.
4. Acheson, *Present at the Creation*, 683–85; LWH oral history (Truman Library), 226–227; also LWH to Acheson, 12/27/52, 12/31/52, 1/2/53, 1/17/53, and 1/19/53, *FR* 52–54:10, 557 ff.
5. LWH to Acheson, 1/7/53 (two cables) and 1/8/53, 788.00.
6. Anthony Eden, *Full Circle* (Boston, 1960), 232–36.
7. Middleton to Foreign Office, 10/6/52, in Cabinet paper, 10/13/52, CAB 129/55, British Cabinet records.
8. LWH to Dulles, 2/19/53, 788.00.
9. LWH to Dulles, 2/23/53, 2/27/53 and 2/28/53, 788.00.
10. LWH to Dulles, 2/23/53 and 2/28/53, 788.00.
11. LWH to Dulles, 3/4/53 and 3/10/53, 788.00.
12. Mohammad Mussadiq, *Mussadiq's Memoirs* (London, 1988), 273–74; LWH to Richards, 3/15/53, 788.00; LWH to Dulles, 3/4/53 and 3/14/53, 788.00.
13. Byroade to Dulles with attachment, 3/6/53, 788.00; Dulles to LWH, 3/3/53, 788.00; Mussadiq, 278.
14. LWH to Dulles, 3/6/53 and 3/8/53, 788.00.
15. Zabih, *Mossadegh Era*, 100; LWH to Dulles, 4/21/53, 788.00.
16. LWH to Dulles, 4/15/53 and 4/25/53, 788.00.
17. LWH to Dulles, 4/4/53 and 4/25/53, 788.00.
18. LWH oral history (Princeton), 7–8.
19. Muhammad Heikal, *Nasser* (London, 1972), 51; Dulles to Smith, 5/13/53, *FR* 52–54:9, 25.
20. NSC minutes, 6/1/53, Eisenhower papers.
21. LWH to Dulles, 5/8/53, 788.00.

22. LWH memo, 5/18/53, attached to LWH to Dulles, 5/28/53, 788.00.
23. LWH memo, 5/26/53, attached to LWH to Dulles, 5/28/53, 788.00.
24. LWH to Dulles, 5/20/53, 788.00.
25. Smith to Eisenhower, 5/23/53, 611.88.
26. Roy M. Melbourne, "America and Iran in Perspective," *Foreign Service Journal* 57:4 (April 1980), 14; Mosadeq to Eisenhower, 5/28/53, *Department of State Bulletin* 7/20/53, 75.
27. Eisenhower to Mosadeq, 6/29/53, *Department of State Bulletin* 7/20/53, 74.
28. Kermit Roosevelt, *Countercoup* (New York, 1979), 3–8.
29. C. M. Woodhouse, *Something Ventured* (London, 1982), 111–23; James A. Bill, *The Eagle and the Lion* (New Haven, 1988), 85–86. Mark J. Gasiorowski, "The 1953 Coup d'Etat in Iran," *International Journal of Middle East Studies* 19:3 (August 1987): 261–86; Donald N. Wilber, Adventures in the Middle East: Excursions and Incursions (n.p., 1986), 187–89.
30. Roosevelt, 14–18; Henry L. Stimson with McGeorge Bundy, *On Active Service in Peace and War* (New York, 1948), 188.
31. Melbourne, 15.
32. LWH appointment book, 8/15–16/53, HP.
33. Roosevelt, 182–183.
34. LWH appointment book, 8/17–18/53, HP; LWH to Dulles, 8/18/53, 788.00.
35. LWH to Dulles, 8/18/53, 788.00.
36. Zabih, 120–23; Diba, *Mossadegh*, 181–87; Rubin, *Paved with Good Intentions*, 85; Gasiorowski, "The 1953 Coup d'Etat in Iran," 274.
37. LWH to Dulles, 8/20/53, 788.00.
38. LWH to Dulles, 8/21/53, 788.00.
39. LWH to Dulles, 8/23/53, *FR* 52–54:10, 762–5.
40. LWH oral history (Truman Library), 215.
41. LWH to Dulles, 8/27/53, 611.88.
42. LWH to Dulles, 8/28/53 and 8/30/53, 888.00TA.
43. LWH to Dulles, 8/30/53, 888.00TA; Rouhollah K. Ramazani, *Iran's Foreign Policy* (Charlottesville, Va., 1975), 262.
44. LWH to Dulles, 9/8/53, 788.00; Diba, 189.
45. Ramazani, 264–67; Mark Hamilton Lytle, *The Origins of the Iranian-American Alliance* (New York, 1987), 214; David S. Painter, *Oil and the American Century* (Baltimore, 1986), 192–98.
46. LWH to Dulles, 12/24/53, 788.00.
47. Ibid.; LWH memo, 12/22/53, attached to LWH to Dulles, 12/28/53, 788.00.
48. LWH memo, 12/22/53, attached to LWH to Dulles, 12/28/53, 788.00.
49. LWH to Dulles, 3/8/54, 788.00.
50. Rubin, 94–95.
51. LWH to Dulles, 7/22/54, 8/31/54, 9/28/54; Rountree to Dulles, 10/19/54; 788.00.

18. The Ends of Empire

1. LWH oral history (Columbia), 26.
2. Charles E. Bohlen oral history, Columbia University, 1–2; Bohlen oral history, Princeton University, 14, U. Alexis Johnson oral history, Princeton University, 5.
3. LWH oral history (Princeton), 38; Dulles to LWH, 6/27/57, Dulles papers (Princeton).

4. Barry Rubin, "Constant and Changing," *Foreign Service Journal*, November 1984, 29.
5. Dulles telephone conversation with Adams, 6/18/54, Dulles papers (Eisenhower Library).
6. LWH to Barros, 3/24/78, HP; LWH oral history (Columbia), 32; Jerolaman to Willis, 1/14/54, Eisenhower papers.
7. LWH to Kennan, 1/18/55, HP; Kennan to LWH, 1/24/55, HP; LWH oral history (Columbia), 26–27.
8. LWH to Jones, 9/27/55, HP.
9. Attachment to Warner to LWH, 1/21/55, HP; Zahedi to LWH, 9/25/57, HP; LWH to Zahedi, 10/9/57, HP.
10. Collins to Radford, 11/23/56, file 092.2 Baghdad Pact, Arthur Radford files, Joint Chiefs of Staff records; LWH oral history (Princeton), 17–18.
11. LWH speech, undated (June 1957), HP; Cassady to CNO, 4/27/56, 092.2 Baghdad Pact, Radford files, Joint Chiefs of Staff records.
12. Heikal, *Nasser*, 73–74, 228–29; Heikal, *Cutting the Lion's Tail* (New York, 1987), 117.
13. Herman Finer, *Dulles over Suez* (Chicago, 1964), 176, 192.
14. Donald Neff, *Warriors at Suez* (New York, 1981), 300; Robert Gordon Menzies, *Afternoon Light* (New York, 1968), 160.
15. Finer, 193; Heikal, *Lion's Tail*, 148.
16. Heikal, *Nasser*, 102–103; *Lion's Tail*, 151–52.
17. Menzies, 168–69.
18. Finer, 200; LWH to Dulles, 9/9/56, Eisenhower papers. See also *FR* 55–57: 16, 441–47.
19. African trip diary, 10/17/60–11/28/60, HP; background press briefing, 11/30/60, HP.
20. Madeleine G. Kalb, *The Congo Cables* (New York, 1982), 175–96.
21. African trip diary, HP.
22. African trip diary, HP; background press briefing, 11/30/60, HP.
23. Background press briefing, 11/30/60, HP.
24. LWH to Harper, 3/23/42, Harper papers; LWH to Grew, 5/23/45, 890D.01; LWH to Acheson, 2/21/51, 791.13.
25. Foy D. Kohler and Mose L. Harvey, eds., *The Soviet Union: Yesterday, Today, Tomorrow* (Miami, 1975), 28–29.
26. Kohler and Harvey, 78–79.

Index

Abdullah, Emir, 126
Acheson, Dean, x, 139, 144–46, 149, 152–59, 170–71, 174, 179, 188, 201, 205, 212, 217, 271, 296
Ajax, Operation, 282
Ala, Hussein, 143, 145, 241–42, 255–58, 272, 275–77
Albania, 150
Alexander, A. V., 128
Ali, Rashid, 116, 118, 123–25
Allen, George, 145–46, 172–73
American Red Cross, v–viii, 4–16
Amini, Abol-Qasem, 277–79
Anderson, Robert, 301
Arab League, 175
Arbenz Guzman, Jacobo, 269, 299
Attlee, Clement, 170, 197, 222
Australia, 299
Australia–New Zealand–United States pact (ANZUS), 298–99
Austria-Hungary, viii

Baghdad pact, 298
Bajpai, Girja, 200–201, 205–7, 212–21
Balfour declaration, 116
Barkley, Alben, 204
Beck, Joseph, 76
Beeley, Harold, 187
Ben Gurion, David, 167–68, 177
Beria, Lavrenti, 65
Berle, Adolph, 93, 100
Berlin blockade and airlift, 199
Bevin, Ernest, 173, 178, 185, 198–99, 204
Bohlen, Charles, 24, 35, 37, 45, 50, 52, 54, 64–67, 89, 92, 144, 296
Borah, William, 39
Bowie, Robert, 281–82
Briand, Aristide, 33
Britain, 3, 8, 80, 86–87, 89–91, 108, 116–19, 122–25, 128–36, 136–45, 147–48, 152–60,
167–70, 173, 176–79, 184–87, 196–99, 233–38, 249, 253–54, 258–59, 299
British Guiana, 3
Bukharin, Nikolai, 35, 64–65
Bulgaria, 150
Bullitt, William, 40–41, 44–45, 49–55, 60, 66–67, 86
Byrnes, James, 138–41, 144, 148, 152, 154, 169–77, 188
Byroade, Henry, 281–82

Canada, 299
Carlucci, Frank, 307
Carr, Wilbur, 18–19, 23, 45
Castro, Fidel, 299
Celler, Emanuel, 189
Central Intelligence Agency, viii, x, 205, 269–70, 279–86, 299, 306–7
Chad, 306
Chamberlain, Neville, 87
Chiang Kai-shek, 77, 200
China, 77, 88, 151, 200, 205–7, 217–22, 226–27, 299
Churchill, Winston, viii–ix, 104–6, 108–9, 132, 143–44, 147, 167, 196–97, 204, 242, 265, 271
Clay, Lucius, 199
Clayton, William, 150
Cleveland, Grover, 3
Clifford, Clark, 182, 184, 187, 192
Colby, Bainbridge, 39
Coleman, Frederic, 29, 35
Collins, Harold, 21
Communist International (Comintern), 27, 44, 54–55, 71, 74, 76
Congo, vii–viii, x, 299, 306
Coolidge, Calvin, 25–26, 33
Crain, James, 157
Crossman, Richard, 175
Crum, Bartley, 189

Cuba, vii, 3, 299
Czechoslovakia, 76, 89, 199

Daladier, Edouard, 87
Davies, John Paton, 296
Davies, Joseph, 60, 65–67, 78–79, 103, 110, 221
Dawes, Charles, 33
De Gaulle, Charles, 129–36
Diem, Ngo Dinh, 299
Douglas, Lewis, 199
Douglas, William O., 143
Dulles, Allen, 281–82
Dulles, John Foster, 180, 269, 272, 274–77,
	281–83, 295–98, 301–2
Durbrow, Elbridge, 43, 51, 66, 138

Eden, Anthony, 147, 242, 265, 271–72, 302
Egypt, x, 168–69, 276–77, 301–2
Einstein, Albert, 175
Eisenhower, Dwight, 269, 272, 280–83, 295,
	301–2
Eisenhower doctrine, 298, 305
Emami, Jamal, 239–41
Epstein, Eliahu, 187

Faisal, King, 119, 126
Farouk, King, 185
Fawzi, Mahmoud, 168–69
Faymonville, Philip, 82
Finland, 92–93
Fischer, Louis, 62
Florinsky, Dmitri, 55–56
Forrestal, James, 153, 159
France, 8, 76, 80, 86–87, 89–91, 95, 126, 128–
	36
Frankfurter, Felix, 40

Gade, John, v–vii, 12
Gamon, John, 22
Gandhi, Mohandas, 197
Germany, viii, 6–8, 15–16, 76–81, 86, 89–95,
	98–99, 101, 118
Ghazi, King, 119
Gizenga, Antoine, 306
Goldmann, Nahum, 167–68, 177
Gorky, Maxim, 65
Grady, Henry, 176, 197
Greece, ix, 147–64, 165
Green, William, 39
Grew, Joseph, 29, 127, 131–32
Gromyko, Andrei, 94–95, 145
Grummon, Stuart, 89
Guatemala, 269–70, 299
Guinea, 305–6, 308

Henderson, Elise (Mrs. Loy Henderson), 37–
	38, 84, 86, 110
Henderson, Loy: administrative duties in State
	Department, 305; administrative problems
	in Moscow, 52–53, 55–59; and Africa, 305–
	8; and Ala, 255–58; and American Red
	Cross, v–viii, 4–16; on American public's
	fickle nature, 92–93; on American society,
	17; on America's responsibilities as great
	power, 134; appearance (physical), 251; on
	Arab nationalism, 121–23; on Asian
	criticism of American policy, 208–10; and
	Baghdad pact, 300–301; and Bevin, 198–99;
	and Britain in Iran, 237–38; career
	frustrations, 29, 32, 36–37, 85; career
	planning, 17–19; on China, 207–8;
	complains about unfair criticism, 100;
	controversy in India, 208–10; criticized by
	Communists, 103; and Davies, 66; death of,
	313; distinguished-service citation, 295; and
	Dulles, 296–98; on duty, 14–15, 18–19;
	early life and family, 3–6, 14; on Finland,
	92; on foreign service's future, 295; and
	foreign-service reorganization, 296–98; on
	Germany, 94; on global role for United
	States, 211–12; on great powers in Middle
	East, 141–43, 146; and Greece and Turkey,
	147–64; and idea of American empire, 195–
	96; illness and injury, 5–6, 13–15, 31–33,
	101; and India, 196–230; on Indian-famine
	relief, 223–24; as intermediary between
	Roosevelt and Stalin, 108–9; on
	international relations, 90–91; and Iran,
	137–46, 233–92, 300–301; on Iranian
	psychology, 280; and Iraq, 116–25; and
	Islamic fundamentalism, 236–37, 266–67;
	and Kashani, 238–39; and Kashmir, 197,
	201–2, 209; on Korean war, 212–13, 217,
	220; on Levant, 129–30; (minor) love affair,
	16; and Middleton, 243; on "moral
	embargo," 94; and Mosadeq, 234–36, 243–
	50, 253–75, 278–86; and Moscow show
	trials, 60–63; and Nasser, 302–4; and
	Nehru, 202–4, 222–30; and Pahlevi, 235–
	36, 248, 252–55, 258, 260, 268, 270–71, 275,
	290–92; and Palestine and Israel, 166–92;
	and Poland, 109–10; reaction to criticism,
	103, 190–92; on Saleh, 256; on Soviet
	agriculture, 36; on Soviet aims, 74–76, 309,
	312; on Soviet-American collaboration, 136;
	on Soviet domestic conditions, 81–84; on
	Soviet foreign policy, 76–81, 96–103; on
	Soviet military fortunes, 99; and Soviet
	NKVD, 56–57; on Soviet purges, 67–73;

and Soviet recognition, 42–44; on Soviet threat, 165–66; on Soviet trade, 42, 94–96; on Soviet Union in Middle East, 134; on Stalin, 35–36, 69–72, 82–84; and Suez affair, 301–5; on Trotsky, 71–72; on Truman, 182, 185; and Tsaldaris, 161–63; on United Nations, 129–31, 135, 143, 164; on Vietnam war, 313; views summarized and evaluated, 299, 309–13; and Zionism and Zionists, 123–26, 128, 188–92
Henderson, Roy, 3–6, 14–15, 33
Hadlow, R. N., 187
Harding, Warren, 33
Harper, Samuel, 25, 86, 92, 94, 99–104
Harriman, W. Averell, viii–ix, 104–6, 140, 234
Harvey, Oliver, 129
Hathaway, Charles, 19–20
Havlik, Hubert, 157
Hawley-Smoot tariff, 34–35
Hickerson, John, 155
Hiss, Alger, 144, 297–98
Hitler, Adolf, viii, 76–78, 87
Ho Chi Minh, 269
Hoover, Herbert, vi, 22, 33–35, 39
Hoover, Herbert Jr., 295
Huddle, J. Klahr, 79
Hull, Cordell, 42, 111–12
Hurley, Patrick, 108

Iceland, 299
Illah, Abdul, 167
India, ix–x, 196–230
Indonesia, 299
Inverchapel, Lord, 156
Iran, ix–x, 136–46, 150, 233–92, 299
Iraq, 116–25, 132
Isabel, Colonel, 12–13
Israel, 187–88, 301
Italy, 80

Jamali, Fadhil, 185
Japan, viii, 37–38, 76–77, 80, 88, 91, 95, 96, 101, 108, 226–27, 298–99
Jha, C. S., 199
Johnson, Herschel, 183–85
Johnson, Lyndon, 309
Johnson, U. Alexis, 296
Jones, Joseph, 155

Kamenev, Lev, 61–63
Karakhan, Lev, 54–55, 57
Kasavubu, Joseph, 306–7
Kashani, Abol-Qasem, 237–39, 240–41, 246–47, 259–60, 266–68, 272, 276

Kashmir, 197, 201, 203
Kelley, Robert, 24–28, 32, 35, 36–45, 49–50, 54, 67, 78–79, 88–89, 128
Kellogg, Frank, 33
Kennan, George, 24, 37, 45, 50, 52, 58, 64–66, 141, 143, 155, 160, 165, 297–98
Kennedy, John, 308–9
Kerensky, Alexander, 7
Kirk, Alexander, 89–90
Kirov, Sergei, 60, 62
Kissinger, Henry, x–xi
Klein, Arthur, 189–90
Korean war, 211–12
Krestinsky, Nikolai, 65
Krock, Arthur, 215
Kuniholm, Bertel, 45, 51
Kurds, 119–20

Lacoste, François, 133–35
Lane, Arthur Bliss, 66
Laski, Harold, 204
Lebanon, 128–36, 298
Lehrs, John, vi, 177
Lenin, V. I., 8, 35–36
Liberia, 306
Litvinov, Maxim, 25, 41–44, 53–55, 69, 85–86, 110–11
Locarno treaty, 91
Lovett, Robert, 174, 183–84, 187, 197, 242
Lozovsky, Solomon, 107
Lumumba, Patrice, viii, 299, 306–7

MacArthur, Douglas, 222–23
MacGowan, David, 35–36
Macmillan, Harold, 147
MacVeagh, Lincoln, 148–49, 159, 162
Makki, Hussein, 247
Malagasy, 306
Mali, 305, 308
Malik, Charles, 133–34
Malitsky, Valentine, 58
Mansur, Hassan Ali, 272
Mao Zedong, 200, 269
Marshall, George, 154–59, 174, 179–80, 184, 187–89, 192
Marshall plan, 179, 251
Masaryk, Jan, 199
Matthews, H. Freeman, 135–36
Mattison, Gordon, 283
McCarthy, Joseph, 188, 208, 296–98
McDonald, James, 128
McGhee, George, 204, 227–28, 234
McKinley, William, 3–4
McLeod, Scott, 296

Melbourne, Roy, 283
Menon, K. P. S., 207
Menon, V. K. Krishna, 199
Menzies, Robert Gordon, 302–4
Merchant, Livingston, 221
Messersmith, George, 57, 84
Middle East Defense Organization (MEDO), 276–77, 301
Middleton, George, 243, 265, 272
Mikhailsky, Pavel, 55–56
Mobutu, Joseph, 306–8
Moffat, Jay Pierrepont, 89
Molotov, V. M., 97, 106, 140
Molotov-Ribbentrop pact, 92
Morgenthau, Henry, 40–41
Morocco, 305
Morrison, Herbert, 176
Mosadeq, Mohammed, x, 233–89
Mozayeni, Mansur, 240–41
Murphy, Robert, 281–82
Murray, Wallace, 128, 139–40, 145
Mustapha, Mulla, 120

Nasiri, Nematollah, 283, 285
Nasser, Gamal Abdel, x, 251, 276–77, 301–4
Nehru, Jawaharlal, ix–x, 198, 202–8, 213–17, 222, 225–27
New Zealand, vii, 299
Niles, David, 182, 184, 192
Nixon, Richard, 309
Noel-Baker, Philip, 197
Norstad, Lauris, 155
North Atlantic Treaty Organization (NATO), 211, 251
Nye, Archibald, 199

Oumansky, K. A., 95–99, 110

Packer, Earl, 44–45
Pahlevi, Mohammed Reza, 138, 235–36, 240–41, 283, 286–92
Pahlevi, Reza Shah, 138
Pakistan, 196–97, 201–3, 277, 289, 291, 299
Palestine, 116, 128, 166–92
Pandit, Vijaya Lakshmi, 202, 223
Panikkar, K. M., 206, 215–20
Patterson, Robert, 159
Patton, George, 51
Paul, King, 163
Philippines, 3, 299
Phillips, William, 130–31, 191
Piatakov, Grigori, 64
Pinchot, Gifford, 39
Podet, Allen, 190–91

Point Four program, 210
Poland, 11, 86, 90–93
Potsdam conference, 137
Puerto Rico, 3

Qavam, Ahmad, 145–46, 255–56, 259–63

Radek, Karl, 54, 64
Rajagopalachari, Chakravarti, 200–201
Rapallo treaty, 80
Razmara, Ali, 238, 266
Rhee, Syngman, 212
Richards, Arthur, 238
Rogers Act, 23, 300
Romania, 90
Roosevelt, Eleanor, 78, 82, 111, 180–81
Roosevelt, Franklin, ix, 39, 41–44, 65, 85, 88, 96, 98, 105–6, 108, 110, 116, 121, 147, 166–67, 172, 196–97
Roosevelt, Kermit, 281–84
Roosevelt, Theodore, 4, 34, 39
Rosenman, Samuel, 172
Rossow, Robert, 144
Royall, Kenneth, 153
Russo-Japanese war, 4
Ryan, Edward, v–vii, 11–16

Said, Nuri al-, 116–21, 126–27, 167, 182–83
Saleh, Allahyar, 255–58, 272
Saud, Ibn, 126, 166, 172, 189
Saudi Arabia, 132
Shaw, George Bernard, 204
Shepherd, Francis, 243
Sherman, Forrest, 155
Shertok, Moshe, 178–80, 186
Sichel, Herbert, 154
Silver, Abba Hillel, 177, 179–80
Skvirsky, Boris, 41
Smetona, Antanas, 9
Smith, Al, 39
Smith, Walter Bedell, 153, 279, 281–82
Sokolnikov, Grigori, 64
Sophoulis, Themistocles, 161–63
South Korea, 298
Southeast Asia Treaty Organization (SEATO), 299
Soviet Union (and Russia), vii–x, 7–8, 24–27, 30–31, 35–44, 89–92, 95, 118–20, 125–26, 131–33, 136–46, 150–54, 157, 164, 173–74, 182–85, 214–15, 236, 254, 306
Spanish-American war, 3
Stalin, Joseph, 35–36, 41, 53–55, 61–63, 67, 69–72, 99, 105–7, 137, 141, 143–45, 147
Standley, William, 105–7, 109–11

Steiger, Boris, 68–69
Steinhardt, Laurence, 92–94, 96–97
Stettinius, Edward, 130, 138
Stimson, Henry, 34, 282
Suez affair, 301–5
Sukarno, 299
Sulzberger, C. L., 202
Sussdorf, Louis, 31, 36
Sykes-Picot agreement, 129
Syria, 126, 128–36, 299

Taiwan, 298
Thayer, Charles, 51–52, 54, 60
Thornburn, Major, 9
Timberlake, Clare, 307
Tito, Josip Broz, 165, 200
Transjordan, 126, 128
Trevelyan, Humphrey, 128
Trotsky, Leon, 8, 35, 62, 63
Truman, Harry, ix, 131–32, 144, 148, 153, 156,
     159–61, 166–69, 172, 174, 176–77, 182,
     185–86, 192, 201, 204–5, 221–23, 242
Truman doctrine, ix, 155, 160, 165, 195, 309
Tsaldaris, Constantine, 161–63, 180
Tshombe, Moise, 306–8
Tudeh party, 235–37, 245–46, 250, 259–60,
     265, 268, 271–72, 286–87, 292
Turkey, ix, 144, 150–61, 277, 289, 291

Umari, Arshad al-, 122
United Nations, 152, 179–85, 188

Valera, Eamon de, 21
Vanderberg, Arthur, 161
Venezuela, 3
Versailles treaty, 15

Vietnam, 299, 308–9, 313
Vincent, John Carter, 296
Vishinsky, Andrei, 61–62, 64, 107
Voroshilov, Klimenti, 54, 62

Wallace, Henry, 143
Walsh, Edmund, 39, 42
Ward, Angus, 51
Webb, James, 217–18
Weizmann, Chaim, 166–67, 171, 177
Welles, Sumner, 78–79, 95–96, 102, 110–11
West Germany, 298
Wiley, John, 50, 92
Williams, Francis, 173
Willkie, Wendell, 107
Wilson, Charles, 281–82
Wilson, Edwin, 154
Wilson, Evan, 115, 128, 174
Wilson, Woodrow, 4–6, 15, 49, 116
Wise, Stephen, 167
Wright, Edwin, 144, 155
Wright, Michael, 198
Wriston, Henry, 300
Wristonization, 300

Yagoda, Genrikh, 64
Yezhov, Nikolai, 64–65
Young, Evan, 18, 23
Young, Owen, 34
Yugoslavia, 147, 150, 200

Zahedi, Ardeshir, 273, 288
Zahedi, Fazlollah, 272–79, 282–90, 301
Zhou Enlai, 77, 218–19
Zinoviev, Grigori, 35, 61–63
Zorin, Valerian, 214